T0338386

ORGANIC SYNTHESES

ORGANIC SYNTHESES

AN ANNUAL PUBLICATION OF SATISFACTORY
METHODS FOR THE PREPARATION OF
ORGANIC CHEMICALS
VOLUME 91
2014

KAY M. BRUMMOND
VOLUME EDITOR

ORGANIC SYNTHESES

Out of print.
†*Deceased.*

Out of print.
†*Deceased.*

Out of print.
†*Deceased.*

Out of print.
†*Deceased.*

NOTICE

Beginning with Volume 84, the Editors of *Organic Syntheses* initiated a new publication protocol, which is intended to shorten the time between submission of a procedure and its appearance as a publication. Immediately upon completion of the successful checking process, procedures are assigned volume and page numbers and are then posted on the Organic Syntheses website (www.orgsyn.org). The accumulated procedures from a single volume are assembled once a year and submitted for publication. The annual volume is published by John Wiley and Sons, Inc., and includes an index. The hard cover edition is available for purchase through the publisher. Incorporation of graphical abstracts into the Table of Contents began with Volume 77. Annual volumes 70-74, 75-, 80-84 and 85-89 have been incorporated into five-year versions of the collective volumes of *Organic Syntheses*. Collective Volumes IX, X, XI and XII are available for purchase in the traditional hard cover format from the publishers.

Beginning with Volume 88, a new type of article, referred to as Discussion Addenda, appeared. In these articles submitters are provided the opportunity to include updated discussion sections in which new understanding, further development, and additional application of the original method are described. Organic Syntheses intends for Discussion Addenda to become a regular feature of future volumes.

Organic Syntheses, Inc., joined the age of electronic publication in 2001 with the release of its free web site (www.orgsyn.org). The site is accessible through internet browsers using Macintosh and Windows operating systems, and the database can be searched by key words and sub-structure. John Wiley & Sons, Inc., and Accelrys, Inc., partnered with Organic Syntheses, Inc., to develop a database (www.mrw.interscience.wiley.com/osdb) that is available for license with internet solutions from John Wiley & Sons, Inc. and intranet solutions from Accelrys, Inc.

Both the commercial database and the free website contain all annual and collective volumes and indices of *Organic Syntheses*. Chemists can draw structural queries and combine structural or reaction transformation queries with full-text and bibliographic search terms, such as chemical name, reagents, molecular formula, apparatus, or even hazard warnings or phrases. The contents of individual or collective volumes can be browsed by lists of titles, submitters' names, and volume and page references, with or without structures.

The commercial database at www.mrw.interscience.wiley.com/osdb also enables the user to choose his/her preferred chemical drawing package, or to utilize several freely available plug-ins for entering queries. The user is also able to cut and paste existing structures and reactions directly into the structure search query or their preferred chemistry editor, streamlining workflow. Additionally, this database contains links to the full text of primary literature references via CrossRef, ChemPort, Medline, and ISI Web of Science. Links to local holdings for institutions using open url technology can also be enabled. The database user can limit his/her search to, or order the search results by, such factors as reaction type, percentage yield, temperature, and publication date, and can create a customized table of reactions for comparison. Connections to other Wiley references are currently made via text search, with cross-product structure and reaction searching to be added in the near future. Incorporations of new preparations will occur as new material becomes available.

INFORMATION FOR AUTHORS OF PROCEDURES

Organic Syntheses welcomes and encourages submissions of experimental procedures that lead to compounds of wide interest or that illustrate important new developments in methodology. Proposals for *Organic Syntheses* procedures will be considered by the Editorial Board upon receipt of an outline proposal as described below. A full procedure will then be invited for those proposals determined to be of sufficient interest. These full procedures will be evaluated by the Editorial Board, and if approved, assigned to a member of the Board for checking. In order for a procedure to be accepted for publication, each reaction must be successfully repeated in the laboratory of a member of the Editorial Board at least twice, with similar yields (generally $\pm 5\%$) and selectivity to that reported by the submitters.

Organic Syntheses Proposals

A cover sheet should be included providing full contact information for the principal author and including a scheme outlining the proposed reactions (an *Organic Syntheses* Proposal Cover Sheet can be downloaded at orgsyn.org). Attach an outline proposal describing the utility of the methodology and/or the usefulness of the product. Identify and reference the best current alternatives. For each step, indicate the proposed scale, yield, method of isolation and purification, and how the purity of the product is determined. Describe any unusual apparatus or techniques required, and any special hazards associated with the procedure. Identify the source of starting materials. Enclose copies of relevant publications (attach pdf files if an electronic submission is used).

Submit proposals by mail or as e-mail attachments to:

Professor Charles K. Zercher
Associate Editor, *Organic Syntheses*
Department of Chemistry
University of New Hampshire
23 Academic Way, Parsons Hall
Durham, NH 03824

For electronic submissions: *org.syn@unh.edu*

Submission of Procedures

Authors invited by the Editorial Board to submit full procedures should prepare their manuscripts in accord with the Instructions to Authors which are described below or may be downloaded at orgsyn.org. Submitters are also encouraged to consult this volume of *Organic Syntheses* for models with regard to style, format, and the level of experimental detail expected in *Organic Syntheses* procedures. Manuscripts should be submitted to the Associate Editor. Electronic submissions are encouraged; procedures will be accepted as e-mail attachments in the form of Microsoft Word files with all schemes and graphics also sent separately as ChemDraw files.

Procedures that do not conform to the Instructions to Authors with regard to experimental style and detail will be returned to authors for correction. Authors will be notified when their manuscript is approved for checking by the Editorial Board, and it is the goal of the Board to complete the checking of procedures within a period of no more than six months.

Additions, corrections, and improvements to the preparations previously published are welcomed; these should be directed to the Associate Editor. However, checking of such improvements will only be undertaken when new methodology is involved.

NOMENCLATURE

Both common and systematic names of compounds are used throughout this volume, depending on which the Volume Editor felt was more appropriate. The Chemical Abstracts indexing name for each title compound, if it differs from the title name, is given as a subtitle. Systematic Chemical Abstracts nomenclature, used in the Collective Indexes for the title compound and a selection of other compounds mentioned in the procedure, is provided in an appendix at the end of each preparation. Chemical Abstracts Registry numbers, which are useful in computer searching and identification, are also provided in these appendices.

ACKNOWLEDGMENT

Organic Syntheses wishes to acknowledge the contributions of Amgen, Inc. and Boehringer Ingelheim to the success of this enterprise through their support, in the form of time and expenses, of members of the Board of Editors.

INSTRUCTIONS FOR AUTHORS

All organic chemists have experienced frustration at one time or another when attempting to repeat reactions based on experimental procedures found in journal articles. To ensure reproducibility, *Organic Syntheses* requires experimental procedures written with considerably more detail as compared to the typical procedures found in other journals and in the "Supporting Information" sections of papers. In addition, each *Organic Syntheses* procedure is carefully "checked" for reproducibility in the laboratory of a member of the Board of Editors.

Even with these more detailed procedures, the experience of *Organic Syntheses* editors is that difficulties often arise in obtaining the results and yields reported by the submitters of procedures. To expedite the checking process and ensure success, we have prepared the following "Instructions for Authors" as well as a **Checklist for Authors** and **Characterization Checklist** to assist you in confirming that your procedure conforms to these requirements. Please include a completed Checklist together with your procedure at the time of submission. Procedures submitted to *Organic Syntheses* will be carefully reviewed upon receipt and procedures lacking any of the required information will be returned to the submitters for revision.

Scale and Optimization

The appropriate scale for procedures will vary widely depending on the nature of the chemistry and the compounds synthesized in the procedure. However, some general guidelines are possible. For procedures in which the principal goal is to illustrate a synthetic method or strategy, it is expected, in general, that the procedure should result in at least 5 g and no more than 50 g of the final product. In cases where the point of the procedure is to provide an efficient method for the preparation of a useful reagent or synthetic building block, the appropriate scale also should not exceed 50 g of final product. Exceptions to these guidelines may be granted in special circumstances. For example, procedures describing the preparation of reagents employed as catalysts will often be acceptable on a scale of less than 5 g.

In considering the scale for an *Organic Syntheses* procedure, authors should also take into account the cost of reagents and starting materials. In general, the Editors will not accept procedures for checking in which the cost of any one of the reactants exceeds $**500** for a single full-scale run. Authors are requested to identify the most expensive reagent or starting material on the procedure submission checklist and to estimate its cost per run of the procedure.

It is expected that all aspects of the procedure will have been optimized by the authors prior to submission, and it is required that each reaction will have been carried out at least twice on exactly the scale described in the procedure, and with the results reported in the manuscript.

It is appropriate to report the weight, yield, and purity of the product of each step in the procedure as a range. In any case where a reagent is employed in significant excess, a Note should be included explaining why an excess of that reagent is necessary. If possible, the Note should indicate the effect of using amounts of reagent less than that specified in the procedure.

The Checking Process

A unique feature of papers published in *Organic Syntheses* is that each procedure and all characterization data is carefully checked for reproducibility in the laboratory of a member of the Board of Editors. In the event that an editor finds it necessary to make any modifications in an experimental procedure, then the published article incorporates the modified procedure, with an explanation and mention of the original protocol often included in a Note. The yields reported in the published article are always those obtained by the checkers. In general, the characterization data in the published article also is that of the checkers, unless there are significant differences with the data obtained by the authors, in which case the author's data will also be reported in a Note.

Reaction Apparatus

Describe the size and type of flask (number of necks) and indicate how *every* neck is equipped.

"A 500-mL, three-necked, round-bottomed flask equipped with an 3-cm Teflon-coated magnetic stirbar, a 250–mL pressure-equalizing addition funnel fitted with an argon inlet, and a rubber septum is charged with … "

Indicate how the reaction apparatus is dried and whether the reaction is conducted under an inert atmosphere. Note that balloons are not acceptable as a means of maintaining an inert atmosphere. The description of the reaction apparatus can be incorporated in the text of the procedure or included in a Note.

"The apparatus is flame-dried and maintained under an atmosphere of argon during the course of the reaction."

In the case of procedures involving unusual glassware or especially complicated reaction setups, authors are encouraged to include a photograph or drawing of the apparatus in the text or in a Note (for examples, see *Org. Syn.*, Vol. 82, 99 and Coll. Vol. X, pp 2, 3, 136, 201, 208, and 669).

Use of Gloveboxes

When a glovebox is employed in a procedure, justification must be provided in a Note and the consequences of carrying out the operation without using a glovebox should be discussed.

Reagents and Starting Materials

All chemicals employed in the procedure must be commercially available or described in an earlier *Organic Syntheses* or *Inorganic Syntheses* procedure. For other compounds, a procedure should be included either as one or more steps in the text or, in the case of relatively straightforward preparations of reagents, as a Note. In the latter case, all requirements with regard to characterization, style, and detail also apply. Authors are encouraged to consult with the Associate Editor if they have any question as to whether to include such steps as part of the text or as a Note.

Authors are encouraged to consider the use of "substitute solvents" in place of more hazardous alternatives. For example, the use of *t*-butyl methyl ether (MTBE) should be considered as a substitute for diethyl ether, particularly in large scale work. Authors are referred to the articles "Sanofi's Solvent Selection Guide: A Step Toward More Sustainable Processes" (Prat, D.; Pardigon, O.; Flemming, H.-W.; Letestu, S.; Ducandas, V.; Isnard, P.; Guntrum, E.; Senac, T.; Ruisseau, S.; Cruciani, P. Hosek, P. *Org. Process Res. Dev.* **2013**, *17*, 1517–1525) and "Solvent Replacement for Green Processing" (Sherman, J.; Chin, B.; Huibers, P. D. T.; Garcia-Valis, R.; Hatton, T. A. *Environ. Health Perspect.* **1998**, *106* (Supplement I, 253–271) as well as the references cited therein for discussions of this subject. In addition, a link to a "solvent selection guide" can be accessed via the American Chemical Society Green Chemistry website at http://www.acs.org/content/acs/en/greenchemistry/research-innovation/tools-for-green-chemistry.html.

In one or more Notes, indicate the purity or grade of each reagent, solvent, etc. It is highly desirable to also indicate the source (company the chemical was purchased from), particularly in the case of chemicals where it is suspected that the composition (trace impurities, etc.) may vary from one supplier to another. In cases where reagents are purified, dried, "activated" (e.g., Zn dust), etc., a detailed description of the procedure used should be included in a Note. In other cases, indicate that the chemical was "used as received".

"Diisopropylamine (99.5%) was obtained from Aldrich Chemical Co., Inc. and distilled under argon from calcium hydride before use. THF (99+%) was obtained from Mallinckrodt, Inc. and distilled from sodium benzophenone ketyl. Diethyl ether (99.9%) was purchased from Aldrich Chemical Co., Inc. and purified by pressure filtration under argon through activated alumina. Methyl iodide (99%) was obtained from Aldrich Chemical Co., Inc. and used as received."

The amount of each reactant must be provided in parentheses in the order mL, g, mmol, and equivalents with careful consideration to the correct number of **significant figures**. Avoid indicating amounts of reactants with more significant figures than makes sense. For example, "437 mL of THF" implies that the amount of solvent must be measured with a level of precision that is unlikely to affect the outcome of the reaction. Likewise, "5.00 equiv" implies that an amount of excess reagent must be controlled to a precision of 0.01 equiv.

The concentration of solutions should be expressed in terms of molarity or normality, and not percent (e.g., 1 N HCl, 6 M NaOH, not "10% HCl").

Reaction Procedure

Describe every aspect of the procedure clearly and explicitly. Indicate the order of addition and time for addition of all reagents and how each is added (via syringe, addition funnel, etc.).

Indicate the temperature of the reaction mixture (preferably internal temperature). Describe the type of cooling (e.g., "dry ice-acetone bath") and heating (e.g., oil bath, heating mantle) methods employed. Be careful to describe clearly all cooling and warming cycles, including initial and final temperatures and the time interval involved.

Describe the appearance of the reaction mixture (color, homogeneous or not, etc.) and describe all significant changes in appearance during the course of the reaction (color changes, gas evolution, appearance of solids, exotherms, etc.).

Indicate how the reaction can be monitored to determine the extent of conversion of reactants to products. In the case of reactions monitored by TLC, provide details in a Note, including eluent, R_f values, and method of visualization. For reactions followed by GC, HPLC, or NMR analysis, provide details on analysis conditions and relevant diagnostic peaks.

"The progress of the reaction was followed by TLC analysis on silica gel with 20% EtOAc-hexane as eluent and visualization with p-anisaldehyde. The ketone starting material has $R_f = 0.40$ (green) and the alcohol product has $R_f = 0.25$ (blue)."

Reaction Workup

Details should be provided for reactions in which a "quenching" process is involved. Describe the composition and volume of quenching agent, and time and temperature for addition. In cases where reaction mixtures are added to a quenching solution, be sure to also describe the setup employed.

> "The resulting mixture was stirred at room temperature for 15 h, and then carefully poured over 10 min into a rapidly stirred, ice-cold aqueous solution of 1 N HCl in a 500-mL Erlenmeyer flask equipped with a magnetic stirbar."

For extractions, the number of washes and the volume of each should be indicated as well as the size of the separatory funnel.

For concentration of solutions after workup, indicate the method and pressure and temperature used.

> "The reaction mixture is diluted with 200 mL of water and transferred to a 500-mL separatory funnel, and the aqueous phase is separated and extracted with three 100-mL portions of ether. The combined organic layers are washed with 75 mL of water and 75 mL of saturated NaCl solution, dried over 25 g of $MgSO_4$, filtered through a 250-mL medium porosity sintered glass funnel, and concentrated by rotary evaporation (25 °C, 20 mmHg) to afford 3.25 g of a yellow oil."

> "The solution is transferred to a 250-mL, round-bottomed flask equipped with a magnetic stirbar and a 15-cm Vigreux column fitted with a short path distillation head, and then concentrated by careful distillation at 50 mmHg (bath temperature gradually increased from 25 to 75 °C)."

In cases where solid products are filtered, describe the type of filter funnel used and the amount and composition of solvents used for washes.

> " … and the resulting pale yellow solid is collected by filtration on a Büchner funnel and washed with 100 mL of cold (0 °C) hexane."

When solid or liquid compounds are dried under vacuum, indicate the pressure employed (rather than stating "reduced pressure" or "dried *in vacuo*").

> " … and concentrated at room temperature by rotary evaporation (20 mmHg) and then at 0.01 mmHg to provide … "

> "The resulting colorless crystals are transferred to a 50-mL, round-bottomed flask and dried overnight in a 100 °C oil bath at 0.01 mmHg."

Purification: Distillation

Describe distillation apparatus including the size and type of distillation column. Indicate temperature (and pressure) at which all significant fractions are collected.

" … and transferred to a 100-mL, round-bottomed flask equipped with a magnetic stirbar. The product is distilled under vacuum through a 12-cm, vacuum-jacketed column of glass helices (Note 16) topped with a Perkin triangle. A forerun (ca. 2 mL) is collected and discarded, and the desired product is then obtained, distilling at 50–55 °C (0.04–0.07 mmHg) … "

Purification: Column Chromatography

Provide information on TLC analysis in a Note, including eluent, R_f values, and method of visualization.

Provide dimensions of column and amount of silica gel used; in a Note indicate source and type of silica gel.

Provide details on eluents used, and number and size of fractions.

"The product is charged on a column (5 × 10 cm) of 200 g of silica gel (Note 15) and eluted with 250 mL of hexane. At that point, fraction collection (25-mL fractions) is begun, and elution is continued with 300 mL of 2% EtOAc-hexane (49:1 hexanes:EtOAc) and then 500 mL of 5% EtOAc-hexane (19:1 hexanes:EtOAc). The desired product is obtained in fractions 24–30, which are concentrated by rotary evaporation (25 °C, 15 mmHg) … "

Purification: Recrystallization

Describe procedure in detail. Indicate solvents used (and ratio of mixed solvent systems), amount of recrystallization solvents, and temperature protocol. Describe how crystals are isolated and what they are washed with. A photograph of the crystalline product is often valuable to indicate the form and color of the crystals.

"The solid is dissolved in 100 mL of hot diethyl ether (30 °C) and filtered through a Buchner funnel. The filtrate is allowed to cool to room temperature, and 20 mL of hexane is added. The solution is cooled at –20 °C overnight and the resulting crystals are collected by suction filtration on a Buchner funnel, washed with 50 mL of ice-cold hexane, and then transferred to a 50-mL, round-bottomed flask and dried overnight at 0.01 mmHg to provide … "

Characterization

Physical properties of the product such as color, appearance, crystal forms, melting point, etc. should be included in the text of the procedure. Comments on the stability of the product to storage, etc. should be provided in a Note.

In a Note, provide data establishing the identity of the product. This will generally include IR, MS, ^{1}H-NMR, and ^{13}C-NMR data, and in some cases UV data. Copies of the proton and carbon NMR spectra for the products of each step in the procedure should be submitted showing integration for all resonances. Submission of copies of the NMR spectra for other nuclei are encouraged as appropriate.

In the same Note, provide analytical data establishing that the purity of the **isolated** product is at least 97%. **Note that this data should be obtained for the material on which the yield of the reaction is based**, not for a sample that has been subjected to additional purification by chromatography, distillation, or crystallization. Elemental analysis for carbon and hydrogen (and nitrogen if present) agreeing with calculated values within 0.4% is preferred. However, **quantitative** NMR, GC, or HPLC analyses involving measurements versus an internal standard will also be accepted. See *Instructions for Authors* at orgsyn.org for procedures for quantitative analysis of purity by NMR and chromatographic methods. Provide details on equipment and conditions for GC and HPLC analyses.

In procedures involving non-racemic, enantiomerically enriched products, optical rotations should generally be provided, but **enantiomeric purity must be determined by another method** such as chiral HPLC or GC analysis.

In cases where the product of one step is used without purification in the next step, a Note should be included describing how a sample of the product can be purified and providing characterization data for the pure material. Copies of the proton NMR spectra of both the product both *before* and *after* purification should be submitted.

Hazard Warnings

Particularly significant hazards should be indicated in a statement within a box at the beginning of the procedure in italicized, red type. Hazard warnings should only be included in the case of procedures that involve unusual hazards such as the use of pyrophoric or explosive substances, and substances with a high degree of acute or chronic toxicity. Instructions are provided in the Article Template. For other procedures, it is not necessary to include a special caution note that refers to standard operating procedures such as working in a hood, avoiding skin contact, etc., since this is referenced in the "Working with

Hazardous Chemicals" statement within each article. Efforts should be made to avoid the use of toxic and hazardous solvents and reagents when less hazardous alternatives are available.

Discussion Section

The style and content of the discussion section will depend on the nature of the procedure.

For procedures that provide an improved method for the preparation of an important reagent or synthetic building block, the discussion should focus on the advantages of the new approach and should describe and reference all of the earlier methods used to prepare the title compound.

In the case of procedures that illustrate an important synthetic method or strategy, the discussion section should provide a mini-review on the new methodology. The scope and limitations of the method should be discussed, and it is generally desirable to include a table of examples. Please be sure each table is numbered and has a title. Competing methods for accomplishing the same overall transformation should be described and referenced. A brief discussion of mechanism may be included if this is useful for understanding the scope and limitations of the method.

Titles of Articles

In cases where the main thrust of the article is the illustration of a synthetic method of general utility, the title of the article should incorporate reference to that method. Inclusion of the name of the final product is acceptable but not required. In the case of articles where the objective is the preparation of a specific compound of importance (such as a chiral ligand), then the name of that compound should be part of the title.

Examples

Title without name of product:

"Stereoselective Synthesis of 3-Arylacrylates by Copper-Catalyzed Syn Hydroarylation" (*Org. Synth.* **2010**, *87*, 53).

Title including name of final product (*note name of product is not required*):

"Catalytic Enantioselective Borane Reduction of Benzyl Oximes: Preparation of (S)-1-Pyridin-3-yl-ethylamine Bis Hydrochloride" (*Org. Synth.* **2010**, *87*, 36).

Title where preparation of specific compound is the subject:

"Preparation of (S)-3,3′-Bis-Morpholinomethyl-5,5′,6,6′,7,7′,8,8′-octahydro-1,1′-bi-2-naphthol" (*Org. Synth.* **2010**, *87*, 59).

Style and Format for Text

Articles should follow the style guidelines used for organic chemistry articles published in the ACS journals such as *J. Am. Chem. Soc.*, *J. Org. Chem.*, *Org. Lett.*, etc. as described in the ACS Style Guide (3rd Ed.). The text of the procedure should be created using the Word template available on the *Organic Syntheses* website. Specific instructions with regard to the manuscript format (font, spacing, margins) is available on the website in the "Instructions for Article Template" and embedded within the Article Template itself.

Style and Format for Tables and Schemes

Chemical structures and schemes should be drawn using the standard ACS drawing parameters (in ChemDraw, the parameters are found in the "ACS Document 1996" option) with a maximum full size width of 15 cm (5.9 inches). The graphics files should then be pasted into the Word document at the correct location and the size reduced to 75% using "Format Picture" (Mac) or "Size and Position" (Windows). Graphics files must also be submitted separately. All Tables that include structures should be entirely prepared in the graphics (ChemDraw) program and inserted into the word processing file at the appropriate location. Tables that include multiple, separate graphics files prepared in the word processing program will require modification.

Tables and schemes should be numbered and should have titles. The title for a Table should be included within the ChemDraw graphic and placed immediately above the table. The title for a scheme should be included within the ChemDraw graphic and placed immediately below the scheme. Use 12 point Palatino Bold font in the ChemDraw file for all titles. For footnotes in Tables use Helvetica (or Arial) 9 point font and place these immediately below the Table.

Acknowledgments and Author's Contact Information

Contact information (institution where the work was carried out and mailing address for the principal author) should be included as footnote 1. This footnote should also include the email address for the principal author. Acknowledgment of financial support should be included in footnote 1.

Biographies and Photographs of Authors

Photographs and 100-word biographies of all authors should be submitted as separate files at the time of the submission of the procedure. The format of the biographies should be similar to those in the Volume 84 procedures found at the orgsyn.org website. Photographs can be accepted in a number of electronic formats, including tiff and jpeg formats.

DISPOSAL OF CHEMICAL WASTE

General Reference: *Prudent Practices in the Laboratory* National Academy Press, Washington, D.C. 2011.

Effluents from synthetic organic chemistry fall into the following categories:

1. **Gases**
   1a. Gaseous materials either used or generated in an organic reaction.
   1b. Solvent vapors generated in reactions swept with an inert gas and during solvent stripping operations.
   1c. Vapors from volatile reagents, intermediates and products.
2. **Liquids**
   2a. Waste solvents and solvent solutions of organic solids (see item 3b).
   2b. Aqueous layers from reaction work-up containing volatile organic solvents.
   2c. Aqueous waste containing non-volatile organic materials.
   2d. Aqueous waste containing inorganic materials.
3. **Solids**
   3a. Metal salts and other inorganic materials.
   3b. Organic residues (tars) and other unwanted organic materials.
   3c. Used silica gel, charcoal, filter aids, spent catalysts and the like.

   The operation of industrial scale synthetic organic chemistry in an environmentally acceptable manner* requires that all these effluent categories be dealt with properly. In small scale operations in a research or academic setting, provision should be made for dealing with the more environmentally offensive categories.

1a. Gaseous materials that are toxic or noxious, e.g., halogens, hydrogen halides, hydrogen sulfide, ammonia, hydrogen cyanide, phosphine, nitrogen oxides, metal carbonyls, and the like.
1c. Vapors from noxious volatile organic compounds, e.g., mercaptans, sulfides, volatile amines, acrolein, acrylates, and the like.

---

*An environmentally acceptable manner may be defined as being both in compliance with all relevant state and federal environmental regulations *and* in accord with the common sense and good judgment of an environmentally aware professional.

2a. All waste solvents and solvent solutions of organic waste.
2c. Aqueous waste containing dissolved organic material known to be toxic.
2d. Aqueous waste containing dissolved inorganic material known to be toxic, particularly compounds of metals such as arsenic, beryllium, chromium, lead, manganese, mercury, nickel, and selenium.
3. All types of solid chemical waste.

Statutory procedures for waste and effluent management take precedence over any other methods. However, for operations in which compliance with statutory regulations is exempt or inapplicable because of scale or other circumstances, the following suggestions may be helpful.

Gases

Noxious gases and vapors from volatile compounds are best dealt with at the point of generation by "scrubbing" the effluent gas. The gas being swept from a reaction set-up is led through tubing to a large trap to prevent suck-back and into a sintered glass gas dispersion tube immersed in the scrubbing fluid. A bleach container can be conveniently used as a vessel for the scrubbing fluid. The nature of the effluent determines which of four common fluids should be used: dilute sulfuric acid, dilute alkali or sodium carbonate solution, laundry bleach when an oxidizing scrubber is needed, and sodium thiosulfate solution or diluted alkaline sodium borohydride when a reducing scrubber is needed. Ice should be added if an exotherm is anticipated.

Larger scale operations may require the use of a pH meter or starch/iodide test paper to ensure that the scrubbing capacity is not being exceeded.

When the operation is complete, the contents of the scrubber can be poured down the laboratory sink with a large excess (10–100 volumes) of water. If the solution is a large volume of dilute acid or base, it should be neutralized before being poured down the sink.

Liquids

Every laboratory should be equipped with a waste solvent container in which *all* waste organic solvents and solutions are collected. The contents of these containers should be periodically transferred to properly labeled waste solvent drums and arrangements made for contracted disposal in a regulated and licensed incineration facility.**

**If arrangements for incineration of waste solvent and disposal of solid chemical waste by licensed contract disposal services are not in place, a list of providers of such services should be available from a state or local office of environmental protection.

Aqueous waste containing dissolved toxic organic material should be decomposed *in situ*, when feasible, by adding acid, base, oxidant, or reductant. Otherwise, the material should be concentrated to a minimum volume and added to the contents of a waste solvent drum.

Aqueous waste containing dissolved toxic inorganic material should be evaporated to dryness and the residue handled as a solid chemical waste.

Solids

Soluble organic solid waste can usually be transferred into a waste solvent drum, provided near-term incineration of the contents is assured.

Inorganic solid wastes, particularly those containing toxic metals and toxic metal compounds, used Raney nickel, manganese dioxide, etc. should be placed in glass bottles or lined fiber drums, sealed, properly labeled, and arrangements made for disposal in a secure landfill.** Used mercury is particularly pernicious and small amounts should first be amalgamated with zinc or combined with excess sulfur to solidify the material.

Other types of solid laboratory waste including used silica gel and charcoal should also be packed, labeled, and sent for disposal in a secure landfill.

Special Note

Since local ordinances may vary widely from one locale to another, one should always check with appropriate authorities. Also, professional disposal services differ in their requirements for segregating and packaging waste.

Herbert Otis House
1929–2013

Herbert Otis House, an emeritus Professor in the Department of Chemistry and Biochemistry at the Georgia Institute of Technology, died in his home in Alpharetta, Georgia on October 2, 2013, after an extended illness. He is survived by his wife, Mary; his two children, Michael Alan Hammer of Berkeley, California, and Mary Moon of Navarre, Florida; his sister, Elaine, of Denton, Texas; and six stepchildren. He was born to Otis and Lorraine House of Willoughby, Ohio on December 5, 1929.

Herb was predestined for a successful career in chemistry. By sixth grade, he had constructed a fully-equipped chemistry laboratory in his basement. At seventeen years of age, he was one of two awardees in Ohio of the Westinghouse Science Talent Award for Excellence in Chemistry.

Herb obtained his Bachelor of Science degree from the Miami University of Ohio, and went on to receive his Doctor of Philosophy degree from the University of Illinois in 1953, under the tutelage of Reynold C. Fuson. He then joined the faculty of the Massachusetts Institute of Technology, where he rapidly advanced through the ranks and became a full professor in 1964. It was at MIT where he made his indelible mark on chemistry that influenced so many research groups with his seminal work on enolate and organocopper chemistry, and their utilities in preparing specific carbon-carbon bonds. His work with generating enolates caused the chemistry community to refer to lithium diisopropylamide (LDA) as "House's" base." The reaction of lithium dialkyl cuprates with alkyl halides to produce a new alkane and an organocopper compound is termed the Corey-House synthesis. While at MIT, Herb published two editions of the useful textbook entitled "Modern Synthetic Chemistry," which became a reference book of choice for more than a generation of chemists. In 1970, Herb left MIT to accept a chaired position at the Georgia Institute of Technology, where he unselfishly focused on undergraduate programs and mentorship, until he retired in 1990 as the Vasser Woolley Emeritus Professor.

Herb received a number of coveted awards during his career, and served in key positions in the scientific industry in addition to his professorial appointments. He was an Alfred P. Sloan fellow while at MIT, from 1953 to 1959. He was also a visiting professor at both the University of California at Berkeley and at Oxford University in England. He received diverse awards from the American Chemical Society, namely, for "Creative Work in Synthetic Chemistry" in 1975 and for "Chemical Health and Safety" in 1983. He was a longstanding member of the Board of Directors for *Organic Synthesis*, and served as its *Editor-in-Chief*. He and his students provided many robust preparations for this compendium and also checked many syntheses that were submitted and subsequently accepted for publication. Herb served on the Executive Board of the American Chemical Society and was affiliated with the Swiss Chemical Society. He also served as a consultant to several companies including Union Carbide Corporation.

Herb was a very pragmatic person in all aspects of his life. His publications reflect syntheses and methodologies that were designed to be useful to others and to be consistently reproducible. Yields for synthetic steps were never "enhanced." He provided practical projects for his students, and also taught them to be thorough with laboratory techniques and transcription, and to learn things from the "ground-up". All students kept a detailed notebook, recorded in duplicate using the MIT notebook format, which he helped to design. In addition, all students submitted a monthly progress report, with experimental sections that were written in *Journal of Organic Chemistry* style. Every new compound that was prepared was accompanied by a folder that contained all of the original spectral data gathered for that compound. A sample of every new compound was submitted to the group library of compounds. All commercial organometallic reagents were titrated before use, which was no small task. The group gas chromatography laboratory at MIT contained four instruments, which were used for analytical and preparative work. Students packed their own columns, up to thirty feet in length, using the open stairwells in the Dreyfus building that accommodated filling the aluminum tubes with adsorbent, and the subsequent tamping and packing. All yields calculated using gas chromatography methods resulted from first calculating a response factor for the collected product, which was related to an internal standard. Students who matriculated through the House group were very well-versed with the basics of organic chemistry laboratory methods and they left with a good sense of how to use these methods to gather and organize data and perform credible research.

An outside of work event that produced an anecdote that underscored Herb's practical nature was a party in Cambridge at the apartment of one of his postdoctoral students. It was a festive occasion, and after the meal, Herb was given the honor of pouring the brandy over the cherry desert and igniting the mixture to create "Cherries Jubilee." Herb performed these duties to everyone's delight, but afterword remarked that "this surely seems to be a waste of perfectly good brandy."

A typical example of Herb's organization and forward planning was evident at a lecture he presented at Dartmouth College in Hanover, New Hampshire in the fall of 1970. This lecture was not well-publicized at MIT, but one of his graduate students found out that Herb would speak and all of his students decided to drive to Dartmouth and attend this lecture. When Herb arrived at the seminar room he was quite surprised, and pleasantly so, to see all of his students in the audience. After delivering a very informative lecture, the first question came from a new faculty member, seeking to make an impression, who asked a very detailed two-part question. Herb's response was that "both of these questions are answered on the first of the extra slides." Of course, after this comment the audience wondered how many extra slides there were and what additional questions they should elicit.

Herb did enjoy specific activities outside of work. He was proficient with electronics, and liked working with his hands, and he used these skills to create his own sound system from component parts. Other recreational activities included gardening and hiking. Herb especially enjoyed hiking in the Appalachian Mountains. A favorite pastime was ice skating. Herb enjoyed this activity immensely, and he became an avid pair skating dancer with his wife Mary.

Only a few people are blessed with what is referred to as a "photographic memory." Herb was certainly one of these people. It was not unusual for a student to enter Herb's office to discuss a particular laboratory procedure and for Herb to then recite the needed reference to that procedure, which included the exact reference with the page numbers. If that particular reference was contained in one of the neatly bound journal volumes in Herb's office, where he housed an extensive collection, the student was given that volume to photocopy. Herb had an exceptional and encyclopedic memory with respect to chemistry publications.

In his career, Herb published 180 research papers, which provided him, and importantly his students, visibility in the scientific community. Herb cared deeply for his graduate students and postdoctoral students, and worked to place them in positions that were appropriate to their skill sets and in environments where they could thrive. These students represent the most important products of his research efforts, and they went on to populate the next generation of academic and industrial scientists. These students are his greatest legacy.

NORTON P. PEET

Jeremiah P. Freeman
1929–2014

Jeremiah P. Freeman, Emeritus Professor of Chemistry and Biochemistry at the University of Notre Dame, passed away on February 14, 2014, at the age of 84 from complications due to cancer. Jerry's legacy is one of altruism and service to his beloved profession of organic chemistry and as an outstanding example as a husband, father, teacher, friend and mentor.

Born in Detroit, Michigan on August 3, 1929, to Bartholomew Joseph and Agnes (Ryan) Freeman, Jerry delivered newspapers as a boy and was an avid Detroit Tigers fan. He also worked in his father's drug store where he developed a lifelong love of chemistry when he discovered he could order chemicals through the store's suppliers. After graduating from Detroit Catholic Central High School in 1946, he attended the University of Notre Dame where he earned his bachelor's degree in chemistry in 1950. While at Notre Dame he participated in undergraduate research with Professor Ernest Eliel, then a young member of the faculty. Together Jerry and Ernest studied the stereochemistry of reductions of optically active halides with the then new reducing agent, lithium aluminum hydride. The results were published in *J. Am. Chem. Soc.* and Volume 33 of *Organic Syntheses*, thus beginning a life long relationship between Jerry and *OrgSyn*. Prof. Eliel and other faculty at Notre Dame recognized Jerry's genuine love of and talent in chemistry and encouraged him to move on to graduate work at the University of Illinois where he joined the research group of Professor R. C. Fuson. At Illinois, he also overlapped projects with and formed a strong friendship with Bill Emmons. He received his Ph.D. in 1953, just three years after finishing his undergraduate studies at Notre Dame. While a teaching assistant, he met a young geology major, Mary Mifflin, in one of his chemistry lab sections. Mary also graduated in 1953. She and Jerry were married at St. John's Chapel on the University of Illinois campus shortly before graduation. Together, they moved to Huntsville, Alabama, where Jerry had accepted a position at the Redstone Arsenal Research Division of the Rohm and Haas Co. where Bill Emmons was already the head of the organic chemistry group.

At the Huntsville laboratories, Jerry's fascination with multi-nitrogen containing small molecules grew, and he made significant contributions to the fundamental chemistry of nitrate esters, nitramines, poly nitrogen and fluoronitrogen compounds. Many of these were highly reactive and, in one instance, there was an explosion that resulted in serious injury to one of Jerry's hands. When Emmons moved to the Philadelphia based laboratories of Rohm and Haas in 1957, Jerry was immediately appointed organic chemistry group leader of the Huntsville section. He thought his career path was clear. Then, in the fall of 1963, he was contacted by Prof. Eliel at Notre Dame. Ernest told Jerry that he had just accepted an offer to become chair of the Department of Chemistry at Notre Dame starting in January 1964 and that he wanted Jerry to return to Notre Dame to help with teaching in the spring semester. Rohm and Haas granted Jerry a leave of absence and he expected to return to Huntsville and continue his work there after the spring semester. However, shortly after he did return to Huntsville, Ernest contacted Jerry and asked if he would be interested in a permanent faculty position at Notre Dame. Jerry quickly accepted and moved his growing family back to South Bend.

Jerry was an associate professor at Notre Dame from 1964 to 1968, then professor until his official "retirement" in 1995. After only one year back at Notre Dame, Jerry became the assistant chair and then served as chairman of the Department of Chemistry from 1970 to 1979. His chairmanship was characterized by his altruism and genuine interest in each individual faculty member on the professional and personal level, as well as his ability to politely encourage the upper administration to provide the resources needed for departmental growth. He gained admiration and respect from colleagues at all levels. At Notre Dame, he continued research on nitrogen rich compounds and heterocyclic chemistry while mentoring undergraduate and graduate students, postdoctoral fellows and visiting researchers from around the world. Despite being an active researcher and administrator, Jerry always carried a full and often extra teaching load. He became famous and highly respected for teaching multiple sections of organic chemistry to pre-professional students. Many of the students not only considered him a teacher, but mentor and, over the years, a valued friend. Many former students who became accomplished physicians and other professionals visited with Jerry over the years and several attended his memorial service. He was particularly supportive of young faculty and took an active role in encouraging incoming graduate students to join new groups, though most came to Notre Dame because of Jerry's reputation. He worked tirelessly for the department, university and for his colleagues. Often, he would nominate and organize support for early awards for his colleagues without informing the colleague - most were successful. He spent significant time meeting with each new faculty throughout their early careers providing direct, extremely valuable and much appreciated mentorship. Many consider Jerry to have been their "professional father".

Although he retired from teaching in 1995 when he became Professor Emeritus, he continued to pursue his research interests until 2004. His publications spanned five decades. He considered the later years of his research the most satisfying due to the remarkable chemists with whom he collaborated. His extensive collaborations with Prof. Makhlouf Haddadin at the American University in Beruit and Dr. Jacob Szmuszkovicz at Pharmacia-Upjohn were notable. During his career, Jerry particularly valued the relationships he developed with his students and colleagues and academia's ability to break down barriers of politics, ethnicity and religion, allowing such relationships to develop.

Academic professionals are asked to fulfill three major roles–teaching, research and service. Jerry was an award winning teacher as well as a fantastic and caring administrator. His service to our profession was exemplary. He often took on important professional tasks that were not simply honorary, but required extensive time and effort. He served as secretary-treasurer of the Organic Division of the ACS from 1969–1973 and was chairman of the Organic Division from 1974–1975. He was on the editorial advisory board of *Chemical Reviews* and on the advisory board of the Army Research Office. In 1979, he began 25 years of service to *Organic Syntheses* as Secretary and a member of the Board of Directors. His role as "Secretary" was extensive. He attended and organized every *OrgSyn* meeting. He personally interacted with each author and checker, ran the meetings like clockwork, and performed all the correspondence and extensive editing with passion and joy. He impressed everyone involved with his attention to detail and dedication to making *Organic Syntheses* a practical and useful "bible" of organic procedures. He did everything with a wonderfully positive and collegial attitude. When Jerry indicated that he would retire from *OrgSyn*, a wave of concern flowed through the editorial board that was followed by an extensive and intense search for a successor. Fortunately, Rick Danheiser (MIT) and Charles "Chuck" Zercher (New Hampshire) agreed to carry on and have done an admirable job. Jerry often expressed deep gratitude for their willingness to serve and to promote the web-based developments that have continued to make *Organic Syntheses* such a valuable resource to the research community. As an expression of thanks and deep appreciation for Jerry's extensive service, Organic Syntheses, Inc. helped the University of Notre Dame to endow the "Jeremiah P. Freeman Organic Syntheses Lectureship" in 2006. The lectureship has been extremely well received. Lectures are given annually and have included presentations by K. B. Sharpless, S. J. Danishefsky, J. Bäckvall, K. Houk, J. Macor, K. Scheidt, E. M. Carreira, A. Holmes and M. J. Miller (for the Jeremiah Freeman Organic Syntheses memorial lecture). Jerry attended the lectures and subsequent dinners of all except the most recent. Each lecture included spontaneous statements of appreciation to Jerry for his work with *Organic Syntheses*. Volume 83 of *Organic Syntheses* also published an autobiography that is the source of many of the comments in this document. The autobiography includes a few pictures of Jerry keeping busy with *OrgSyn* duties.

Jerry was an ardent scholar of history and loved to travel. He enjoyed family vacations to Revolutionary War era sites, Civil War battlefields and national parks. He particularly liked foreign travel because he could combine his love of fine dining and history. Destinations were often based upon a restaurant review he had read. No recommended restaurant was too far out of the way or unworthy of a digression from a travel route. He had a particular weakness for bakeries and was never heard to turn down dessert. As a boy, he loved sundaes and ice cream from the drug store fountain, and he enjoyed these simple pleasures to the very end.

Of all his passions, it was his family that he considered his greatest joy. In addition to his wife, Mary, Jerry is survived by his six children: Thomas M. Freeman (m. Margaret Miller) of Piedmont, CA; Christopher R. Freeman (m. Sandy Neulip Freeman) of Garland, TX; John A. Freeman (m. Pamela Corson Freeman) of Reno, NV; Susan Freeman McCortney (m. Ryan McCortney) of Huntington Beach, CA; James A. Freeman of South Bend; and Kathleen A. Freeman (m. James D. Childs) of San Luis Obispo, CA; seventeen grandchildren, two great-grandchildren and seventeen nieces and nephews. He was preceded in death by his parents, Bartholomew and Agnes Freeman, and his sister, Mary Freeman Matthew, of Palatine, Illinois.

A memorial mass, held at 12 p.m., March 15, 2014 at Little Flower Catholic Church in South Bend, was attended by a few hundred relatives, friends and colleagues. Jerry's long time friend, Rev. Richard Conyers, C.S.C. officiated. He and Jerry's oldest son reflected on the wonderful example of a life well lived—a life of exceptional teaching, research, service and friendship. Jerry was a friend and mentor to many of us. His example lives on for all of us to appreciate.

MARVIN J. MILLER
South Bend, Indiana

PREFACE

I would like to begin by expressing my great satisfaction in serving as a member of the *Organic Syntheses* Editorial Board. This organization, through its annual publication has provided the scientific community with detailed, reliable and carefully checked procedures for the preparation of organic compounds since 1921. *Organic Syntheses* was the outgrowth of an effort where graduate students at universities were paid to supply research chemicals at a time when they were not available in the United States, a consequence of WWI. This activity afforded the scientific community with an additional resource – the creation of carefully documented experimental procedures for the preparation of known compounds, a goal that drives *Organic Syntheses* today. For more information regarding the fascinating history of *Organic Syntheses*, see *http://www.orgsyn.org*.

The demand for synthetic procedures that can be duplicated may be less egregious today than in 1921, but there is still a need. To date, *Organic Syntheses* has filled this need with 90 volumes of "checked" experimentals; and with the culmination of my term as a member of the editorial board, the addition of Volume 91. I have thoroughly enjoyed working closely with the graduate students, postdocs and the occasional undergraduate to ensure the reproducibility of the procedures checked in my lab. In addition, it has been an honor and enlightening experience to work together with the other board members during my eight-year term. I would like to acknowledge Chuck Zercher, Associate Editor, for his editorial assistance crucial to the completion of Volume 91, and Carl Johnson, Treasurer and Bob Boeckman, President for their contributions to the success of this organization.

Finally, I would like to give a special tribute to Rick Danheiser, Editor-in-Chief, whose phenomenal efforts have increased the quality, reproducibility and safety of the published experimental procedures. Rick was also instrumental in converting this publication from paper to electronic volumes freely available to everyone. And while I miss receiving annual volumes in the mail, the impact that this has had on *Organic Syntheses* and the scientific community is evidenced not only by the thousands of visitors to the *Organic Syntheses* website each month, but by the increased number of researchers that work to have their chemistry published as a checked procedure in *Organic Syntheses*.

K<small>AY</small> M. B<small>RUMMOND</small>
Pittsburgh, Pennsylvania

CONTENTS

DABCO-*bis*(sulfur dioxide), DABSO, as an easy-to-handle source of SO$_2$: Sulfonamide preparation

Edward J. Emmett and Michael C. Willis

One-pot Preparation of (*S*)-*N*-[(*S*)-1-Hydroxy-4-methyl-1,1-diphenylpentan-2-yl]pyrrolidine-2-carboxamide from L-Proline

Wacharee Harnying, Nongnaphat Duangdee, and Albrecht Berkessel

Enantioselective Rhodium-Catalyzed [2+2+2] Cycloaddition of Pentenyl Isocyanate and 4-Ethynylanisole: Preparation and Use of Taddol-pyrrolidine Phosphoramidite

Kevin M. Oberg, Timothy J. Martin, Mark Emil Oinen, Derek M. Dalton, Rebecca Keller Friedman, Jamie M. Neely, and Tomislav Rovis

Synthesis and Use of a Trifluoromethylated Azomethine Ylide Precursor: Ethyl 1-Benzyl-*trans*-5-(trifluoromethyl)pyrrolidine-3-carboxylate

Daniel M. Allwood, Duncan L. Browne and Steven V. Ley

Enantioselective Organocatalytic α-Arylation of Aldehydes

Pernille H. Poulsen, Mette Overgaard, Kim L. Jensen and Karl Anker Jørgensen

Synthesis of 4,5-Disubstituted 2-aminothiazoles from α,β-Unsaturated Ketones: Preparation of 5-benzyl-4-methyl-2-aminothiazolium Hydrochloride salt

Antonio Bermejo Gómez, Nanna Ahlsten, Ana E. Platero-Prats and Belén Martín-Matute

Discussion Addendum for: Formation of γ-Keto Esters from β-Keto Esters: Methyl 5,5-dimethyl-4-oxohexanoate

Yashoda M. D. Bhogadhi and Charles K. Zercher

Ni-catalyzed Reductive Cleavage of Methyl 3-Methoxy-2-Naphthoate

Josep Cornella, Cayetana Zarate, and Ruben Martin

Indium-Catalyzed Heteroaryl–Heteroaryl Bond Formation through Nucleophilic Aromatic Substitution: Preparation of 2-Methyl-3-(thien-2-yl)-1H-indole

Yuta Nagase and Teruhisa Tsuchimoto

Synthesis of 2,3-Disubstituted Benzofurans by the Palladium-Catalyzed Coupling of 2-Iodoanisoles and Terminal Alkynes, Followed by Electrophilic Cyclization: 3-Iodo-2-phenylbenzofuran

Tuanli Yao, Dawei Yue, and Richard C. Larock

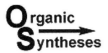

Discussion Addendum for:

Resolution of 1,1'-Bi-2-Naphthol; (R)-(+)- and (S)-(-)-2,2'-Bis(diphenylphosphino)-1,1'-binaphthyl (BINAP)

David L. Hughes*[1]

Department of Process Chemistry, Merck and Co., Inc., Rahway, NJ 07065

Original Articles: Cai, D.; Hughes, D. L.; Verhoeven, T. R.; Reider, P. J. Org. Synth. **1999**, *76, 1-5; Cai, D.; Payack, J. F.; Bender, D. R.; Hughes, D. L.; Verhoeven, T. R.; Reider, P. J. Org. Synth.* **1999**, *76, 6-11.*

A.

N-benzylcinchonidinium
chloride **4**

CH₃CN

(R,S)-BINOL

(R)-BINOL **1**

(S)-BINOL **2**

B.

(R)-BINOL **1**

Tf₂O, pyridine

3

C.

3

Ph₂PH (2.4 equiv)

10% NiCl₂(dppe)

DABCO, DMF, 100 °C

4 (R)-(+)-BINAP

The discovery of BINAP by Noyori and co-workers in 1980 ushered in an era of fertile research and practical applications of asymmetric catalysis using axially-chiral phosphine ligands.[2] The first practical syntheses of (*R*)- and (*S*)-BINAP (**4**) were reported by Takaya *et al* in 1986 (Scheme 1), involving synthesis of the racemic bis-phosphine oxide, resolution with (*R,R*)-dibenzoyl tartaric acid, followed by reduction to the bis-phosphine.[3] While this route supported production of kilogram quantities of BINAP, some drawbacks were the harsh conditions (320 °C) and low yield (45%) for conversion of 1,1'-bi-2-naphthol (BINOL) to the dibromide and a resolution in which the (*S*)-isomer could be isolated in high ee with (*R,R*)-dibenzoyl tartaric acid while the opposite enantiomer required an upgrade with (*S,S*)-dibenzoyl tartaric acid after recovery from the supernatant.

Scheme 1. First practical synthesis of BINAP

BINOL Resolution

A more concise approach to BINAP would be to start with resolved BINOL and develop mild conditions for conversion to BINAP such that the chiral integrity is not compromised. A number of viable methods for resolving BINOL have been developed over the past 40 years, including enzymatic hydrolysis of its diester, resolution of its phosphoric acid, and formation of a variety of diastereomeric salts and co-crystals with chiral bases.[4] Given the wide availability of the *Cinchona* alkaloids, we decided to further investigate the resolution of BINOL discovered by Toda[5] using *N*-benzylcinchonidinium chloride **5** (**Figure 1**). This alkaloid forms a salt co-crystal with (*R*)-BINOL with the chloride serving as a bridge between the alkaloid and BINOL with three hydrogen bonds: one to an OH of BINOL, one to the OH of **5**, and a third to an OH of a second molecule of BINOL.[6]

For most applications that employ *Cinchona* alkaloids, the two "pseudo-enantiomeric" pairs, cinchonine/cinchonidine and quinine/quinidine, lead to opposite product enantiomers since the pairs have opposite stereochemistry at the "working" part of the molecule, C8 and C9. For example, in the first practical resolution of BINOL, the phosphoric acid of BINOL is resolved using cinchonine and cinchonidine to access the (+)- and (-)-enantiomers, respectively.[7] Surprisingly, however, *N*-benzylcinchonidinium chloride **5** and *N*-benzylcinchoninium chloride **6** both form salt co-crystals only with (*R*)-BINOL.[8] Thus, to efficiently isolate both pure enantiomers of BINOL using either **5** or **6**, a near perfect resolution is required, with one diastereomeric complex crystallizing in very high yield, leaving the other enantiomer with high enantiomeric purity in the supernatant.

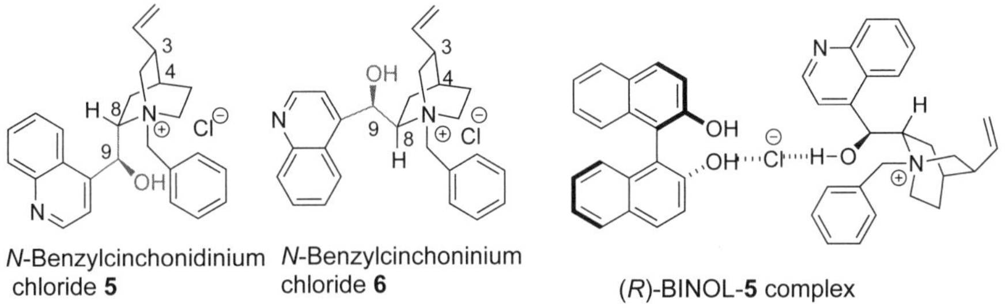

N-Benzylcinchonidinium chloride **5**   *N*-Benzylcinchoninium chloride **6**   (*R*)-BINOL-**5** complex

Figure 1. N-Benzylcinchonidinium salts used for resolution

In the published procedure of Toda,[6] MeOH was used as the crystallization solvent, affording (*R*)-BINOL in high yield and ee (after breaking the complex) but leaving the (*S*)-enantiomer in the supernatant in only 42% ee. Thus, the pure (*S*)-enantiomer could not be isolated via this method. We initiated our studies of this resolution by undertaking a survey of solvents and identified MeCN as the most promising: **5** had reasonable solubility, the crystalline diastereomeric complex (*R*)-BINOL•**5** had very low solubility at 0 °C, and no evidence was found for crystallization of (*S*)-BINOL•**5** in this solvent. A straightforward resolution for isolating both enantiomers of BINOL in high yield was thus developed.[9]

The resolution was carried out with 0.55 equiv of **5** at a concentration of 80 g of BINOL per L of solvent. The mixture was refluxed for 4 h to ensure

BINOL and **5** were fully dissolved, during which time crystallization of the (*R*)-BINOL•**5** complex occurred. After cooling to 0 °C, the solids were filtered and washed with MeCN to afford (*R*)-BINOL•**5** co-crystals with 96% de. The complex was upgraded to >99% de by slurrying in MeOH, then (*R*)-BINOL was recovered by salt break in EtOAc/aq. HCl. Concentration of the EtOAc layer to dryness afforded (*R*)-BINOL (85-88% recovery, >99% ee).

The supernatant from the original crystallization, which contained (*S*)-BINOL, was concentrated to dryness, then dissolved in EtOAc and washed with dilute HCl to remove any remaining **5**. Concentration of the EtOAc layer to dryness afforded (*S*)-BINOL (89-93%, recovery, 99% ee). This straightforward resolution thus provides access to each enantiomer of BINOL in ≥99% ee with recoveries in the 85-93% range for each enantiomer.

Nearly concurrent with our 1995 publication, Pu reported a resolution with **5**, isolating each enantiomer of BINOL in 70-75% yield.[10] This group carried out the initial resolution in MeOH to crystallize (*R*)-BINOL•**5**. The supernatant was evaporated and the resulting solids slurried in EtOAc, affording additional crystalline (*R*)-BINOL•**5** adduct, with the supernatant containing (*S*)-BINOL with 84% ee. The material from the supernatant was concentrated then further upgraded via crystallization from benzene.

Three analogs of BINOL have been similarly resolved, 6,6'-dimethyl and 7,7'-dimethoxy using **5**[11] and 7,7'-bis(benzyloxy) using quinine.[12]

A 2005 review indicates the resolution with *N*-benzylcinchonidinium chloride described in our 1999 *Organic Syntheses* procedure remains among the most effective and straightforward methods of accessing both enantiomers of BINOL.[4]

BINAP Synthesis

In parallel to the development of the BINOL resolution, a more efficient route to BINAP was explored. Five papers in the early 1990's provided the starting point for our efforts. Activation of BINOL as its bis-triflate **3** was reported by Mattay in 1990,[13] then Morgans[14a] and Hayashi[14b] discovered that mono-phosphorylation of bis-triflate **3** could be accomplished using Ph$_2$P(O)H and Pd catalysis, noting the reaction occurred stereospecifically with no loss of ee. In addition, cyanation could also be achieved using nickel catalysis, affording primarily mono-substituted product with up to 10% of the bis-nitrile.[14a] Snieckus reported conversion of bis-triflate **3** to the bis-methyl derivative using MeMgBr with Ni catalysis.[15] While no aryl triflate to aryl phosphine couplings were known at the time we began our work, cross-couplings of aryl iodides with phosphine-boranes, prepared from the corresponding phosphine oxide and $LiAlH_4$-$NaBH_4$, were reported by Imamoto using Pd catalysis.[16]

In considering the direct cross-coupling of diphenylphosphine with bis-triflate **3**, a transition-metal catalyst must be chosen that will not be poisoned by ligation to the phosphine ligands present in the starting material and product. The 2nd and 3rd row transition metals such as Pd, Rh, Ru, and Pt were thus ruled out, with Ni identified as having the best probability of success given its "hardness" and likely only weak ligation to phosphines. Screening identified a number of Ni catalysts that gave some conversion to product, with $NiCl_2$(dppe) selected for further development. The optimized procedure used 10 mol % catalyst, 4 equiv DABCO, 2.4 equiv Ph_2PH and DMF as solvent. The reaction required 2-3 days at 100 °C to reach completion. BINAP crystallized directly from the reaction mixture, providing product in 77% yield (97 % purity (HPLC area), >99% ee).[17]

Since our original report in 1994, applications of this methodology to several BINAP analogs for various applications have been reported (Table 1). The 6-Me, 6-MeO_2C, and 7-MeO analogs (Table 1, entries 1-3) were prepared according to the same procedure except the cross-coupling reactions were completed overnight rather than 2-3 days.[11] Several fluorous analogs have been prepared in 48-87% yields (Table 1, entries 4-7).[18-20] The bis-steroid ligand (Table 1, entry 8) was readily prepared from the natural product equilenine. The resulting diastereomers could be separated by column chromatography, eliminating the need for a resolution.[21] The electron-rich BINAP analog (Table 1, entry 11) was prepared in a modest 26% yield, but was found to afford superior ee's in the Rh-catalyzed asymmetric 1,4-addition of arylboron reagents to 5,6-dihydro-2(1H)-pyridones to prepare 4-aryl-2-piperidones.[23] While most of the coupling partners in Table 1 are arylphosphines, Wills reported preparation of the bis(dicyclohexylphosphine) analog in 45% yield from Cy_2PH by a modified procedure utilizing Zn dust as an additional reagent (Table 1, entry 13).[25] Attempts to prepare the bis(di-*t*-butylphosphine) analog were unsuccessful, with only the mono-substitution product formed.[25] Gilheany also reported formation of the mono-substituted product using *t*-BuPhPH in 40% yield.[26] Ni- and Pd-catalyzed cross-couplings of Ar_2PH with triflates to prepare mono-phosphine BINAP analogs, as well as non-BINAP structures, have also been reported but are outside the scope of this addendum.[27]

Table 1. Applications of Ph₂PH Cross-Coupling to Prepare BINAP Analogs

Entry	G	% Yield	Ref	Entry	R	% Yield	Ref
1	6-Me-	--	11	9	4-MeC₆H₄-	52	22
2	6-MeO₂C-	--	11	10	3,5-Me₂C₆H₃-	46	22
3	7-MeO-	--	11	11	3,5-Me₂-4-MeOC₆H₂-	26	23
4	6-C₆F₁₃-	48	18	12	Ph(CH₂)₃C₆H₄-	34	24
5	6-C₆F₁₃CH₂CH₂-	85	18	13	Cyclohexyl-	45	25
6	6-(C₆F₁₃CH₂CH₂)₃Si-	87	19				
7	6-(C₈F₁₇CH₂CH₂)₃Si-	59	20				
8		70	21				

Alternate Methods to Prepare BINAP and Analogs[27]

A number of alternate methods of preparing BINAP and analogs have been reported in the past two decades.

Cross Coupling of bis-Triflate 3 with Other Phosphorus Partners (Scheme 2). Takasago scientists reported a synthesis of BINAP via cross-coupling of bis-triflate **3** with Ph₂P(O)H, citing the ease of handling the phosphine oxide vs the phosphine (pyrophoric, stench) as the principle driver.[28] Employing the same NiCl₂(dppe) catalyst that was used in the Merck procedure, the cross-coupling affords a mixture of BINAP analog **9** (35%), the mono-oxide **8** (60%), and the bis-oxide (4%). The mixture is then treated with HSiCl₃/PhNMe₂ to reduce all species to the BINAP analog. This method has been applied to a number of analogs, including 2-napthyl, 4-tolyl, 3,5-dimethylphenyl, 4-methoxy-3,5-dimethylphenyl, and 4-*t*-butoxy-3,5-dimethylphenyl. Yields from BINOL range from 57-84%.[28]

Scheme 2. Ni-Catalyzed Cross-Coupling of Bis-Triflate 3 with Alternate Phosphorus Partners

Cross-coupling of bis-triflate **3** with $Ar_2P(O)H$ and $R_2P(O)H$ using Pd-catalysis stop at the mono-phosphine oxide **7**, leaving an unreacted triflate that can be further functionalized using Pd- or Ni-catalysis to prepare unsymmetrical BINAP analogs.[27]

Also using $NiCl_2(dppe)$ catalyst, Laneman at Monsanto reported coupling of the bis-triflate **3** with Ph_2PCl, mediated by 1.2-1.5 equiv Zn, affording a yield of BINAP of 52%.[29] The authors proposed that Zn plays two roles, reducing Ni(II) to Ni(0) as well as generating Ph_2PZnCl as the active species. This process has been reportedly licensed to Rhodia for commercial production.[27b] In a slight variation of this process, Waechtler used the $-SO_2C_4F_9$ analog for the cross-coupling with Ph_2PCl.[30]

Yu found that BINAP analogs with electron-rich aromatic groups could be prepared by the Merck process with bis-triflate **3** in moderate yields (Table 1, entries 9, 10), but that electron-withdrawing analogs were unreactive. However, using Ar_2PBr **11** instead of the chloride **10**, the method could be extended to 4-CF_3 (33% yield) and 3,5-$(CF_3)_2$ (28% yield) analogs and yields of the electron-donating analogs could be increased 10-15%.[22]

A wide-variety of electron-rich phosphines have been prepared by cross-coupling phosphine borane species **12** with bis-triflate **3** in yields ranging from 24-79%.[31-32] The use of conditions and reaction times very similar to the Merck process suggest the reaction may be proceeding via a common intermediate.

Cross Coupling of bis-Bromide with Ar$_2$PBr. Yu reported the racemic bis-bromide of BINOL could be converted to the corresponding Grignard species and cross-coupled with Ar$_2$PBr, requiring no catalyst. Five examples were reported, affording yields of 68-78% for both electron-withdrawing and donating groups. Under these conditions, Ar$_2$PCl was unreactive.[22]

Oxidative Homo-Coupling. Takasago scientists designed and prepared a family of ligands termed SEGPHOS, which have a smaller dihedral angle than BINAP and show superior performance in a number of applications. The preparation of this ligand is shown in Scheme 3, featuring an iron-mediated homo-coupling to generate the bis-phosphine oxide (*R,S*)-**15**, followed by resolution to (*R*)-**15**, and reduction to (R)-SEGPHOS (**16**).[33]

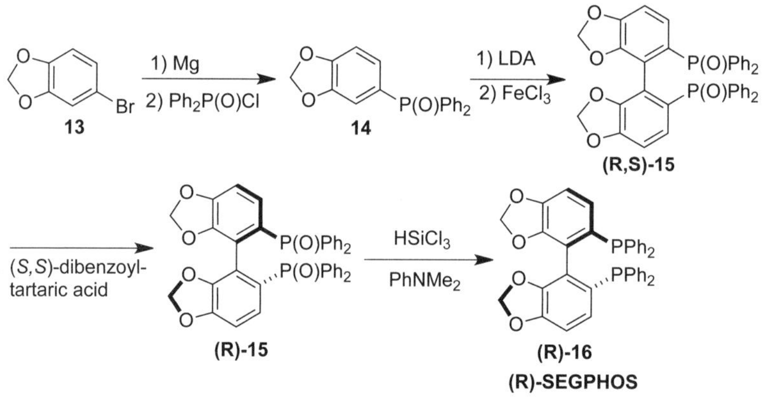

Scheme 3. Preparation of SEGPHOS via Oxidative Coupling

In summary, axially-chiral ligands, introduced by Noyori and co-workers in 1980, have played an important role in the development of asymmetric catalysis for organic synthesis. A number of practical routes to BINAP and analogs have made a wide variety of these ligands available for application in academic settings as well as industrial processes.

References

1. Department of Process Chemistry, Merck and Co., Inc., Rahway, NJ 07065. Email: dave_hughes@merck.com
2. (a) Miyashita, A.; Yasuda, A.; Takaya, H.; Toriumi, K.; Ito, T.; Souchi, T.; Noyori, R. *J. Am. Chem. Soc.* **1980**, *102*, 7932-7934. (b) A concise, excellent review: Shimizu, H.; Nagasaki, I.; Matsumura, K.; Sayo, N.; Saito, T. *Acc. Chem. Res.* **2007**, *40*, 1385-1393.
3. Takaya, H.; Mashima, K.; Koyano, K.; Yagi, M.; Kumobayashi, H.; Taketomi, T.; Akutagawa, S.; Noyori, R. *J. Org. Chem.* **1986**, *51*, 629-635; Takaya, H.; Akutagawa, S. *Org. Synth.* **1989**, *67*, 20-32.
4. For a review of approaches to each enantiomer of BINOL, see: Brunel, J. M. *Chem. Rev.* **2005**, *105*, 857-897.
5. Tanaka, K.; Okada, T.; Toda, F. *Angew. Chem. Int. Ed. Engl.* **1993**, *32*, 1147-1148.
6. Toda, F.; Tanaka, K.; Stein, Z.; Goldberg, I. *J. Org. Chem.* **1994**, *59*, 5748-5751.
7. Jacques, J.; Fouquey, C.; Viterbo, R. *Tetrahedron Lett.* **1971**, *12*, 4617-4620.
8. Wang, Y.; Sun, J.; Ding, K. *Tetrahedron* **2000**, *56*, 4447-4451.
9. Cai, D.; Hughes, D. L.; Verhoeven, T. R.; Reider, P. J. *Tetrahedron Lett.* **1995**, *36*, 7991-7994.
10. Hu, Q.-S.; Vitharana, D.; Pu, L. *Tetrahedron: Asymmetry* **1995**, *6*, 2123-2126.
11. Cai, D.; Hughes, D. L.; Levac, S.; Verhoeven, T. R. US Patent 6,333,435, December 25, 2001.
12. Yokozawa, T.; Saito, K. European Patent Appl. EP1371655, **2003**.
13. Vondenhof, M.; Mattay, J. *Tetrahedron Lett.* **1990**, *31*, 985-988.
14. (a) Kurz, L.; Lee, G.; Morgans, D., Jr.; Walkyke, M. J.; Ward, T. *Tetrahedron Lett.* **1990**, *31*, 6321-6324; (b) Uozumi, Y.; Tanahashi, A.; Lee, S.-Y.; Hayashi, T. *J. Org. Chem.* **1993**, *58*, 1945-1948.
15. Sengupta, S.; Leite, M., Raslan, D. S.; Quesnelle, C.; Snieckus, V. *J. Org. Chem.* **1992**, *57*, 4066-4068.
16. (a) Imamoto, T.; Oshiki, T.; Onozawa, T.; Kusumoto, T.; Sato, K. *J. Am. Chem. Soc.* **1990**, *112*, 5244-5252; (b) Imamoto, T. *Pure Appl. Chem.* **1993**, *65*, 655-660.
17. (a) Cai, D.; Payack, J. F.; Bender, D. R.; Hughes, D. L.; Verhoeven, T. R.; Reider, P. J. *J. Org. Chem.* **1994**, *59*, 7180-7181; (b) Cai, D.; Payack, J. F.; Verhoeven, T. R. US Patent 5,399,771, March 21, 1995.

18. Birdsall, D. J.; Hope, E. G.; Stuart, A. M.; Chen, W.; Hu, Y.; Xiao, J. *Tetrahedron Lett.* **2001**, *42*, 8551-8553.

19. Nakamura, Y.; Takeuchi, S.; Zhang, S.; Okamura, K.; Ohgo, Y. *Tetrahedron Lett.* **2002**, *42*, 3053-3056.

20. Andrushko, V.; Schwinn, D.; Tzschucke, C. C. Michalek, F.; Horn, J.; Mossner, C.; Bannwarth, W. *Helv. Chim. Acta* **2005**, *88*, 936-949.

21. Enev, V.; Ewers, C. L. J.; Harre, M.; Nickisch, K.; Mohr, J. T. *J. Org. Chem.* **1997**, *62*, 7092-7093.

22. Liu, L.; Wu, H.-C.; Yu, J.-Q. *Chem. Eur. J.* **2011**, *17*, 10828-10831.

23. Senda, T.; Ogasawara, M.; Hayashi, T. *J. Org. Chem.* **2001**, *66*, 6852-6856.

24. Ding, H.; Kang, J.; Hanson, B. E.; Kohlpaintner, C. W. *J. Mol. Catal. A: Chem.* **1997**, *124*, 21-28.

25. Morris, D. J.; Docherty, G.; Woodward, G.; Wills, M. *Tetrahedron Lett.* **2007**, *48*, 949-953.

26. Kerrigan, N. J.; Dunne, E. C.; Cunningham, D.; McArdle, P.; Gilligan, K.; Gilheany, D. C. *Tetrahedron Lett.* **2003**, *44*, 8461-8465.

27. Four reviews extensively cover the preparation and use of BINAP analogs. (a) Pereira, M. M.; Calvete, M. J. F.; Carrilho, R. M. B.; Abreu, A. R. *Chem. Soc. Rev.* **2013**, *42*, 6990-7027; (b) Berthod, M.; Mignani, G.; Woodward, G.; Lemaire, M. *Chem. Rev.* **2005**, 1801-1836; (c) Tappe, F. M. J.; Trepohl, V. T.; Oestreich, M. *Synthesis* **2010**, 3037-3062; (d) Li, W.; Zhang, X. in *Phosphorus(III) Ligands in Homogeneous Catalysis*; Kramer, P. C. J.; Piet, W. N. M., Eds; Wiley & Sons: Hoboken, NJ; 2012, pp. 27-80.

28. (a) Kumobayashi, H.; Miura, T.; Sayo, N.; Saito, T.; Zhang, X. *Synlett* **2001**, 1055-1064; (b) Sayo, N.; Zhang, X.; Ohmoto, T.; Yoshida, A.; Yokozawa, T. US Patent 5,693,868, December 2, 1997.

29. (a) Ager, D. J.; East, M. B.; Eisenstadt, A.; Laneman, S. A. *Chem. Commun* **1997**, 2359-2360; (b) Laneman, S. A.; Eisenstadt, A.; Ager, D. J. US Patent 5,902,904, May 11, 1999.

30. Waechtler, A.; Derwenskus, K.-H.; Meudt, A. WO 99/36397, July 22, 1999

31. Goto, M.; Yamano, M. European Patent Application 1452537, September 1, 2004.

32. (a) Jahjah, M.; Alame, M.; Pellet-Rostaing, S.; Lemaire, M. *Tetrahedron: Asymm.* **2007**, *18*, 2305-2312; (b) Alame, M.; Jahjah, M.; Pellet-Rostaing, S.; Lemaire, M.; Meille, V.; de Bellefon, C. *J. Mol. Catal. A: Chem.* **2007**, *271*, 18-24.

33. Saito, T.; Yokozawa, T.; Ishizaki, T.; Moroi, T.; Sayo, N.; Miura, T.; Kumobayashi, H. *Adv. Synth. Catal.* **2001**, *343*, 264-267.

David Hughes received his PhD in physical organic chemistry with Professor Bordwell at Northwestern University in 1981. After postdoctoral studies with Professor Arnett at Duke University, Dave joined the Process Chemistry group at Merck and currently holds the position of Distinguished Scientist. He served on the Board of Editors of *Organic Syntheses* from 2008-2013.

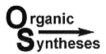

Discussion Addendum for:

Practical Synthesis of Novel Chiral Allenamides: (R)-4-Phenyl-3-(1,2-propadienyl)oxazolidin-2-one

Yong-Chua Teo[*1] and Richard P. Hsung

Natural Sciences and Science Education, National Institute of Education, Nanyang Technological University, 1 Nanyang Walk, Singapore 637616

Original article: Xiong, H.; Tracey, M. R.; Grebe, T.; Mulder, J. A.; Hsung, R. P. Org. Synth. 2005, 81, 147–156.

A.

B.

Allenamides, especially chiral allenamides, are important synthetic intermediates and exist as key structural motifs in many natural products. We recently reported an efficient route to the formation of chiral allenamides that relied upon the isomerization of chiral propargylic oxazolidinones for the introduction of the chiral allenamide functionality.[2] More recently, the stereospecific amidation of optically enriched allenyl iodides using catalytic copper(I) salt and *N,N*-dimethylethylene-diamine as ligands has emerged as a facile strategy to access chiral allenamides[3] (Scheme 1) In recent years, major advances in allenamide chemistry have been concerned with the applications of chiral allenamides

Org. Synth. **2014**, *91*, 12-26
DOI: 10.15227/orgsyn.091.0012

Published on the Web 10/15/2013
© 2014 Organic Syntheses, Inc.

Scheme 1. Synthesis of chiral allenamides promoted by CuCN/dmeda catalyst

to stereoselective synthesis. These advances will be summarized in this article.

(I)   CYCLIZATION REACTIONS

The ability of γ-substituted allenamides to undergo cyclization reactions[4] using either stoichiometric TBAF or catalytic PPTS affords stereodivergent syntheses of 2,5-disubstituted dihydrofurans (Scheme 2). The resulting products were then subjected to stereoselective dihydroxylations demonstrating the synthetic utility of chiral allenamides in the stereoselective synthesis of complex molecules.

Scheme 2. Intramolecular ring cyclization reactions of γ-substituted allenamides

Cyclization of allenamides can also proceed via a radical mechanism[5] using a combination of AIBN and *n*-Bu₃SnH (Scheme 3). The radical addition reactions are highly selective for the *central* carbon of the allene, leading to an efficient preparation of nitrogen heterocycles such as isoquinolines, and carbocycles such as indane and naphthalene derivatives. The *exo*-cyclization mode could also be achieved in some cases, leading to the synthesis of isoindoles.

Scheme 3. Radical mediated intramolecular cyclization reactions

(II) CYCLOADDITION REACTIONS

A) [2π + 1π] CYCLOADDITIONS

The Simmons-Smith cyclopropanations[6] of chiral allenamides provides a viable strategy to obtain optically enriched amido-spiro [2.2] pentanes (Scheme 4). This reaction was efficient with good substrate generality, and represents the most direct synthesis of both chemically and biologically interesting amido-spiro [2.2] pentane systems. However, it suffers from poor diastereoselectivities for unsubstituted chiral allenamides. For □-substituted allenamides, the diastereoselectivity was improved, but both mono- and bis-cyclopropanation products were observed.

Scheme 4. Cyclopropanation reactions of chiral allenamides

B) [4π + 2π] CYCLOADDITIONS

An inverse electron-demand *aza* [4π + 2π] cycloaddition[7] was observed in the reaction of chiral allenamides with 1-azadienes (Scheme 5). In addition, it was also possible to promote a stereoselective intramolecular normal demand [4π + 2π] cycloaddition[8] under thermal conditions without the need for metal catalysts (Scheme 6). Both of these works are applicable for the construction of highly functionalized *aza*-sugars and related nitrogen heterocycles synthesis.

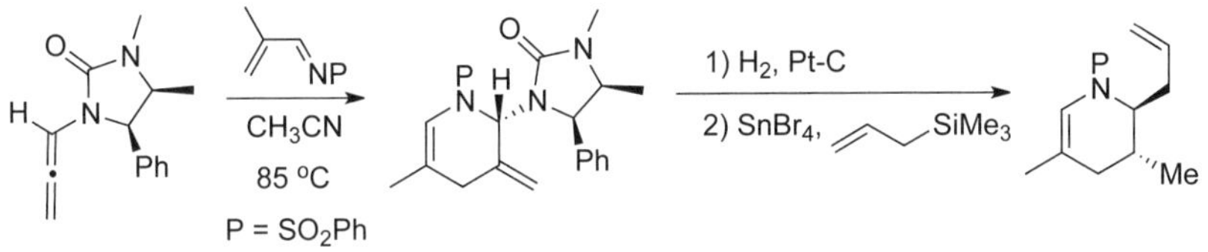

Scheme 5. Inverse electron-demand aza-[4π + 2π] cycloaddition

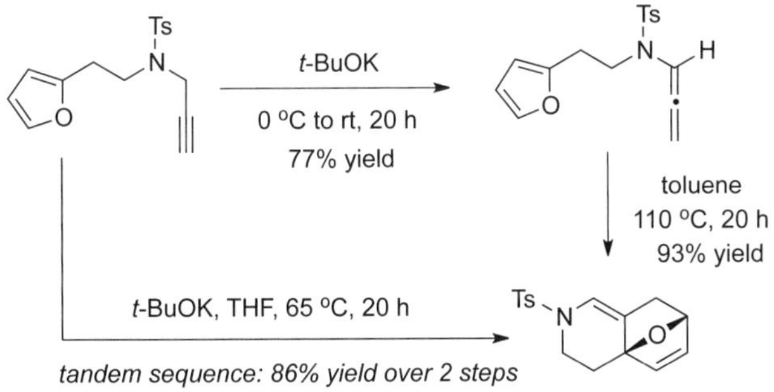

Scheme 6. Intramolecular [4π + 2π] cycloaddition

C) [4π + 3π] CYCLOADDITIONS

The first intramolecular [4π + 3π] cycloaddition[9] using nitrogen-stabilized chiral oxyallyl cations[10] was observed in the epoxidation of *N*-tethered allenamides (Scheme 7). The origin of the high selectivities for the oxyallyl cycloadditions was postulated to originate from structure of the nitrogen-stabilized chiral oxyallyl cation intermediate. This strategy was expanded to include the chemoselective epoxidation of allenamides tethered to either ☐ or ☐ carbon atom of dienes[11] followed by the tandem [4π + 3π] cycloaddition step (Scheme 8).

Scheme 7. Intramolecular [4π + 3π] cycloaddition via a nitrogen-stabilized chiral oxyallyl cation

Scheme 8. Epoxidation of α or γ tethered allenamides in tandem with intramolecular [4π + 3π] cyclocycloaddition

A highly enantioselective version of this [4π + 3π] cycloaddition was promoted by catalytic amounts of a chiral Lewis acid complex generated from Cu(OTf)$_2$ and C$_2$-symmetric bisoxazolines as ligands.[12] High enantioselectivities were obtained when either unsubstituted or substituted furans were employed as the dienes (Scheme 9).

Scheme 9. Enantioselective [4p + 3p] cycloaddition promoted by chiral copper catalysts

Mechanistically, it had been rationalized that the observed stereoselectivity for the intermolecular [4π + 3π] cycloaddition between nitrogen-stabilized oxyallyl cation and furan is a result of the furan approaching in a favorable *endo* manner from the less hindered bottom or *endo-I* face of the oxyallyl cation. This preference can be further enhanced with a bidentate metal cation such as Zn that can chelate to both the oxyallyl oxygen atom and the oxazolidinone carbonyl oxygen (Scheme 10).

	endo-I	endo-II
2.0 equiv ZnCl₂:	95	5
without ZnCl₂:	75	25

Scheme 10. Intermolecular [4p + 3p] cycloaddition in favour of *endo*-I facial attack with furans as dienes

However, an unexpected reversal of diastereoselectivity was observed when methyl 2-furoate was employed as the diene with nitrogen-stabilized oxyallyl cation.[13] This intriguing reversal in favor of the *endo-II* cycloaddition pathway is likely a result of the transition state minimizing the dipole interaction between the oxyallyl cation and ester carbonyl of methyl 2-furoate (Scheme 11).

	endo-I	endo-II
2.0 equiv ZnCl₂:	30	70
without ZnCl₂:	5	95

Scheme 11. Reversal of diastereochemistry in preference for *endo*-II facial attack with methyl-2-furoate as diene

Based on density functional theory calculations, it was further established that the stereo-induction for the [4π + 3π] cycloaddition between oxyallyl cation and unsubstituted furan is due to the stabilizing CH-π interactions between the incoming furan and the Ph group on the oxyallyl isomer (Scheme 12). These CH-π interactions cause the cycloaddition to take place preferentially via the more crowded face of the oxazolidinone to afford the product with the stereochemistry resulting from the addition via *endo* I-face.[14] In the case of methyl-2-furoate, the reversal of stereoselectivity can be ascribed to the repulsive interactions between the Ph group and the 2-COOMe group which outweighs the stabilizing effect of the CH-π interactions. As such, cycloaddition takes place preferentially through the less crowded transition state affording a product with stereochemistry that can be explained by *endo* II-facial attack.

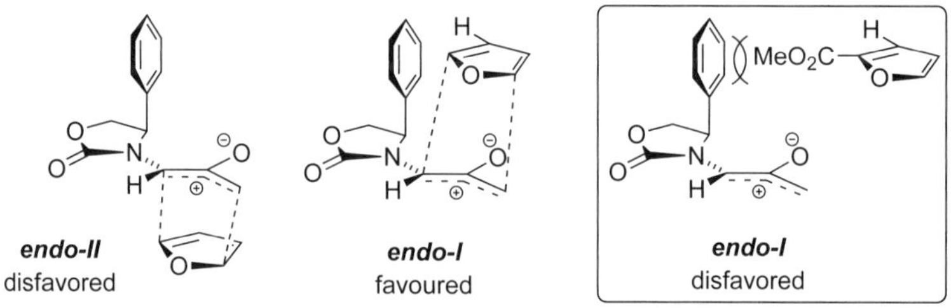

endo-II	endo-I	endo-I
disfavored	favoured	disfavored

Scheme 12. CH- p interactions between furan and Ph group on the oxyallyl cation favours product with *endo* I-stereochemistry

This [4π + 3π] cycloaddition reaction was further expanded to include the reactions between oxazolidinone-substituted oxyallyl groups and unsymmetrically substituted furans.[15] *Syn* regioselectivity was observed when the furan has a 2-Me or 2-COOR substituent, while *anti* regioselectivity is obtained with a 3-Me or 3-COOR group (Scheme 13).

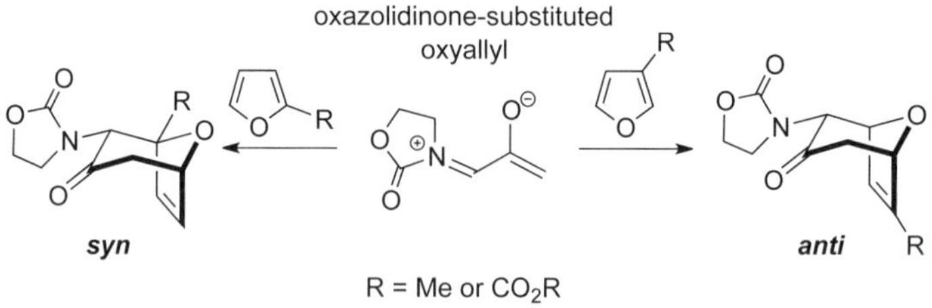

oxazolidinone-substituted
oxyallyl

syn   anti   R

R = Me or CO₂R

Scheme 13. Intermolecular [4π + 3π] cycloadditions between oxazolidinone-substituted oxyallyls and unsymmetrically-substituted furans

Finally, a practical and diastereoselective intramolecular $[4\pi + 3\pi]$ cycloaddition of *N*-sulfonyl substituted oxyallyl cations with furans was also reported (Scheme 14). Selectivity is found to depend on the tethering length as well as the stability of the oxyallyl cation intermediate, whether generated from *N*-carbamoyl- or *N*-sulfonyl-substituted allenamides.[16] The use of chiral *N*-sulfonyl-substituted allenamide provided minimal diastereoselectivity in the cycloaddition, while high diastereoselectivity can be achieved with a stereocenter present on the tether.

Scheme 14. Intramolecular $[4\pi + 3\pi]$ cycloaddition of *N*-sulfonyl substituted oxyallyl cations with furans as dienes

III. ISOMERIZATION REACTIONS

The ability of allenamides to undergo regio- and stereoselective 1,3-hydrogen shift under both acidic and thermal conditions, leads to the *de novo* preparations of 2-amido-dienes.[17] This process could be rendered in tandem with a 6π -electron pericyclic ring closure to access cyclic 2-amido-dienes in good overall yields directly from the respective allenamides (Scheme 15). Further broadening of the scope of this isomerization reaction leads to the development of a new torquoselective ring-closure of chiral amide-substituted 1,3,5-hexatrienes and its application in tandem with $[4\pi + 2\pi]$ cycloaddition[18] (Scheme 16). The trienes were derived via either a 1,3-hydrogen or 1,3-H–1,7-hydrogen shift of a-substituted allenamides, and the entire sequence through the $[4\pi + 2\pi]$ cycloaddition could be promoted in tandem.

Scheme 15. Stereoselective isomerizations of chiral allenemides in tandem with a 6π -electron pericyclic ring closure

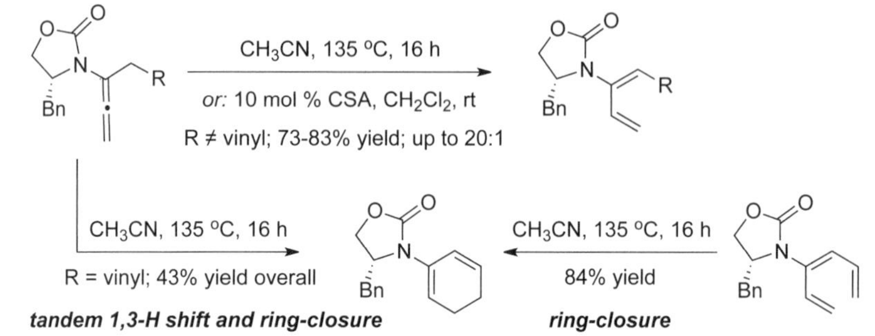

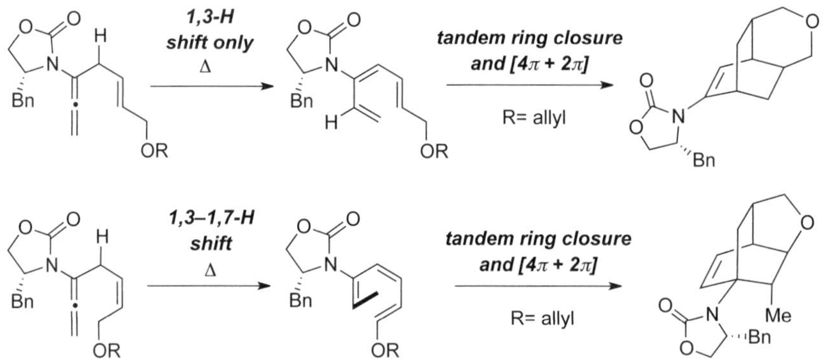

Scheme 16. Isomerizations of allenamides to chiral amide-substituted 1,3,5-hexatrienes in tandem with [4π + 2π] cycloaddition

The efforts reported thus far unveiled an invaluable opportunity not only to develop a new and attractive template for conducting stereoselective 6π-electrocyclic ring-closures, but also to achieve a highly challenging 1,6-asymmetric induction. Indeed, a diastereoselective 6π-electrocyclic ring-closure employing halogen-substituted 3-amidotrienes via a 1,6-remote asymmetric induction was subsequently reported[19] (Scheme 17). This new asymmetric manifold for pericyclic ring-closure further underscores the significance of the allenamide chemistry.

Scheme 17. Isomerizations of allenamides to chiral halogen-substituted 3-amido-triene in tandem with 6π-electron electrocyclic ring-closure

In addition, 1,3-hydrogen shifts using allenamide[20] is also applicable to the preparation of acyclic 2-amido-dienes and 3-amido-trienes. Additionally, 6π-electron electrocyclic ring-closure could be carried out using 3-amido-trienes to afford cyclic 2-amido-dienes, and such electrocyclic ring-closure could be rendered in tandem with the 1,3-hydrogen shift, thereby constituting a facile construction of synthetically rare cyclic 2-amido-dienes (Scheme 18).

Scheme 18. Isomerizations of chiral allenamides to either 2- or 3-amido-trienes

Finally, a new approach to Oppolzer's intramolecular Diels-Alder cycloaddition [IMDA] through □-isomerization[21] of readily available N-tethered allenamides is described. These IMDA reactions are carried out in tandem with the allenamide isomerization or 1,3-hydrogen shift, leading to complex nitrogen heterocycles in a highly stereoselective manner (Scheme 19).

Scheme 19. Oppolzer's intramolecular Diels-Alder cycloaddition [IMDA] via γ-isomerization of N-tethered allenamides

(IV)   NATURAL PRODUCT SYNTHESIS

The synthetic usefulness of the stereoselective inverse electron-demand [4π + 2π] cycloaddition using chiral allenamide was demonstrated in the formal synthesis of (+)-zincophorin. This effort describes the preparation of the Miyashita's intermediate which features the first synthetic application of a stereoselective inverse electron demand hetero [4π + 2π] cycloaddition of a chiral allenamide and an interesting urea directed Stork-Crabtree hydrogenation (Scheme 20).[22] This approach was subsequently applied for the construction of the Cossy's C1-C9 subunit of (+)-zincophorin[23], which also led to the observation of an unusual urea directed Stork-Crabtree hydrogenation (Scheme 21). It is noteworthy that these works represent the first application of chiral allenamides in natural product synthesis.

Scheme 20. Stereoselective inverse electron demand hetero [4p + 2p] cyclo-addition for the synthesis of (+)-Zincophorin via the Miyashita's intermediate

Scheme 21. Stereoselective inverse electron demand hetero [4p + 2p] cycloaddition for the synthesis of (+)-Zincophorin via the Cossy's intermediate

Finally, a highly stereoselective [4π + 3π] cycloaddition of *N*-substituted pyrroles with allenamide-derived nitrogen-stabilized chiral oxyallyl cations was reported which could serve as a useful approach towards the construction of the *aza*-tricyclic core of parvineostemonine[24] (Scheme 22).

Scheme 22. Stereoselective [4p + 3p] cycloaddition of *N*-substituted pyrroles with allenamides for the synthesis of parvineostemonine

References

1. Natural Sciences and Science Education, National Institute of Education, Nanyang Technological University, 1 Nanyang Walk, Singapore 637616. Email address: yongchua.teo@nie.edu.sg
2. Xiong, H.; Tracey, M. R.; Grebe, T.; Mulder, J. A.; Hsung, R. P. *Org. Synth.* **2005**, *81*, 147-156.
3. Shen, L.; Hsung, R. P.; Zhang, Y.; Antoline, J. E.; Zhang, X. *Org. Lett.* **2005**, *7*, 3081-3084.
4. Berry, C. R.; Hsung, R. P.; Antoline, J. E.; Petersen, M. E.; Rameshkumar, C.; Nielson, J. A. *J. Org. Chem.* **2005**, *70*, 4038-4042.
5. Shen, L.; Hsung, R. P. *Org. Lett.* **2005**, *7*, 775-778.
6. Lu, T.; Hayashi, R.; Hsung, R. P. DeKorver, K. A.; Lohse, A. G.; Song, Z.; Tang, Y. *Org. Biomol. Chem.* **2009**, *9*, 3331-3337.
7. Berry, C. R.; Hsung, R. P. *Tetrahedron* **2004**, *60*, 7629-7636.
8. Lohse, A. G.; Hsung, R. P. *Org. Lett.* **2009**, *11*, 3433-3430.
9. For a recent review on [4+2] cycloaddition reactions, see: Lohse, A. G.; Hsung, R. P. *Chem. Eur. J.* **2011**, *17*, 3812-3822.
10. Xiong, H.; Huang, J.; Ghosh, S. K.; Hsung, R. P. *J. Am. Chem. Soc.* **2003**, *125*, 12694-12695.

11. Rameshkumar, C.; Hsung, R. P. *Angew. Chem. Int. Ed.* **2004**, *43*, 615-618; *Angew. Chem.* **2004**, *116*, 625-628.

12. Huang, J.; Hsung, R. P. *J. Am. Chem. Soc.* **2005**, *127*, 50-51.

13. Antoline, J. E.; Hsung, R. P. *Synlett* **2008**, 739-744.

14. Krenske, E. K.; Houk, K. N.; Lohse, A. G.; Antoline, J. E.; Hsung, R. P. *Chem. Science* **2010**, *1*, 387-392.

15. Lohse, A. G.; Krenske, E. K.; Antoline, J. E.; Hsung, R. P.; Houk, K. N. *Org. Lett.* **2010**, *12*, 5506-5509.

16. Lohse, A. G.; Hsung, R. P.; Leider, M. D.; Ghosh, S. K. *J. Org. Chem.* **2011**, *76*, 3246–3257.

17. Hayashi, R.; Hsung, R. P.; Feltenberger, J. B.; Lohse, A. G. *Org. Lett.* **2009**, *11*, 2125-2128.

18. Hayashi, R.; Feltenberger, J. B.; Hsung, R. P. *Org. Lett.* **2010**, *12*, 1152-1155.

19. Hayashi, R.; Walton, M. C.; Hsung, R. P.; Schwab, J.; Yu, X. *Org. Lett.* **2010**. *12*, 5768–5771.

20. Hayashi, R.; Feltenberger, J. B.; Lohse, A. G.; Walton, M. C.; Hsung, R. P. *Beil. J. Org Chem.* **2011**, *7*, 410-420.

21. Feltenberger, J. B.; Hsung, R. P. *Org. Lett.* **2011**, *13*, 3114-3117.

22. Song, Z.; Hsung, R. P. *Org. Lett.* **2007**, *9*, 2199-2202.

23. Song, Z.; Hsung, R. P.; Lu, T.; Lohse, A. G. *J. Org. Chem.* **2007**, *72*, 9722-9731.

24. Antoline, J. E.; Hsung, R. P.; Huang, J.; Song, Z.; Li, G. *Org. Lett.* **2007**, *9*, 1275-1278.

Richard P. Hsung went to Calvin College for his B.S. in Chemistry and Mathematics, and attended The University of Chicago for his M.S. and Ph.D. degrees in Organic Chemistry, respectively, under supervisions of Professors Jeff Winkler and Bill Wulff. After post-doctoral research at Chicago with Professor Larry Sita, and as an NIH post-doctoral fellow with Professor Gilbert Stork at Columbia University, he began his independent career at University of Minnesota before moving to University of Wisconsin in 2006. He is the recipient of an NSF Career Award and The Camille Dreyfus Teacher-Scholar Award. He has co-authored over 250 publications, delivered over 200 invited lectures, and supervised over 160 students and post-doctoral fellows with research interests in developing cycloadditions and annulations to natural product total syntheses and stereoselective methods using allenamides, ynamides, enamides, and cyclic acetals.

Teo Yong Chua received both his B.Sc. and Ph.D. degrees at the National University of Singapore. He pursued his graduate studies on the design of chiral indium complexes for asymmetric catalysis, under the supervision of Professor Loh Teck Peng. In 2006, he joined the faculty at the National Institute of Education, Nanyang Technological University as an Assistant Professor. His current research interests include the development of new asymmetric catalytic systems and transition-metal-catalyzed methodologies for applications in organic synthesis.

N-Carboxylated-2-substituted Indoles and 2,3-Disubstituted-2,3-dihydro-4-quinolones from 2-Alkynylbenzamides

Noriko Okamoto,[1a] Kei Takeda,[1b] and Reiko Yanada[1a*]

(a) Faculty of Pharmaceutical Sciences, Hiroshima International University, 5-1-1 Hirokoshingai, Kure, Hiroshima 737-0112, Japan; (b) Department of Synthetic Organic Chemistry, Graduate School of Medical Sciences, Hiroshima University1-2-3 Kasumi, Minami-Ku, Hiroshima 734-8553, Japan

Checked by Hang Chu and Viresh H. Rawal

Procedure

A. *2-(1-Hexynyl)benzamide (1).*[2] A 500-mL round-bottomed flask equipped with a Teflon coated magnetic stirring bar (3.8 × 0.9 cm) is charged successively with 2-iodobenzamide (19.8 g, 80 mmol), PPh$_3$ (419 mg, 1.60 mmol, 0.02 equiv), CuI (152 mg, 0.80 mmol, 0.01 equiv),

Published on the Web 10/30/2013
© 2014 Organic Syntheses, Inc.

Pd(OAc)$_2$ (180 mg, 0.80 mmol, 0.01 equiv), N,N-dimethylformamide (DMF, 40 mL, anhydrous), Et$_3$N (120 mL), and 1-hexyne (12.0 mL, 104 mmol, 1.3 equiv) (Note 1). A glass inlet adapter (24/40) is fitted on the flask, which is then purged by two cycles of evacuation (2-5 sec.) and back-filling with nitrogen. A slight positive pressure of nitrogen is maintained. The flask is then placed in a preheated oil bath (oil bath temperature 60 °C) and allowed to stir for 16 h, during which time the color of the slurry darkens progressively to brown. Completion of the reaction is confirmed by TLC monitoring (hexanes:EtOAc, 2:1, R$_f$ = 0.35), the reaction mixture is diluted with 200 mL of EtOAc and transferred to a 1 L separatory funnel. The organic phase is sequentially washed with saturated NH$_4$Cl solution (2 x 120 mL) and saturated brine (1 x 100 mL), and the aqueous layer is back-extracted with EtOAc (1 x 100 mL). The combined organic phase is dried over anhydrous Na$_2$SO$_4$ (15 g, 10 min), filtered, and concentrated by rotary evaporation (15–20 mmHg, 23 °C) to give a dark orange solid. The solid is purified by column chromatography on silica gel (Note 2) to afford 2-(1-hexynyl)benzamide (1) as a yellowish powder (14.4 g, 90%) (Note 3).

B. *Ethyl 2-butyl-1H-indole-1-carboxylate (2).* A 500-mL, round-bottomed flask equipped with a Teflon coated magnetic stirring bar (3.8 × 0.9 cm) is successively charged with 2-(1-hexynyl)benzamide 1 (9.02 g, 45 mmol), PhI(OAc)$_2$ (15.2 g, 47 mmol, 1.04 equiv), 1,2-dichloroethane (150 mL), and ethanol (7.85 mL, 134 mmol). The flask is fitted with a Dimroth condenser topped with a glass inlet adapter (24/40) and purged by two cycles of evacuation (2-5 sec.) and back-filling with nitrogen. The solution is maintained under a slight positive pressure of nitrogen. The reaction flask is placed in an oil bath (90 °C, bath temp.) and allowed to stir for 2 h. During this period, the reaction mixture turns dark red. The oil bath is removed, and the reaction mixture is allowed to cool to room temperature. The condenser is removed and PtCl$_2$ (596 mg, 2.2 mmol, 0.05 equiv) (Note 4) is quickly added in one portion. After replacing the condenser, the mixture is put back in the 90 °C oil bath and stirred for further 3 h (Note 5), during which time the color of the reaction mixture turns dark brown. After completion of the reaction is confirmed by TLC monitoring (hexane:EtOAc, 25:1, R$_f$ = 0.30), the reaction mixture is filtered through a pad of Florisil (20 g). The Florisil filter pad is washed with an additional 40 mL of 1,2-dichloroethane. The filtrate is concentrated by rotary evaporation (15–20 mmHg, 23 °C), and the residual brown oil is purified by column chromatography on silica gel (Note 6) to afford ethyl 2-butyl-1H-indole-1-carboxylate 2 (9.59 g, 87 %) as colorless prisms (Note 7).

C. *Ethyl 3-butyl-4-oxo-2-p-tolyl-3,4-dihydroquinoline-1(2H)-carboxylate* (3). A 500-mL round-bottomed flask equipped with a Teflon coated magnetic stirring bar (3.8 × 0.9 cm) is charged with 2-(1-hexynyl)benzamide **1** (8.04 g, 40 mmol), PhI(OAc)$_2$ (14.2 g, 44 mmol, 1.1 equiv), 1,2-dichloroethane (100 mL), and ethanol (4.70 mL, 80 mmol). The flask containing the reagents is fitted with a Dimroth condenser topped with a glass inlet adapter (14/20) and purged by two cycles of evacuation (2-5 sec.) and back-filling with nitrogen. The solution is then maintained under a slight positive pressure of nitrogen. The stirred mixture is placed in a preheated oil bath (90 °C, bath temp.) for 2 h, during which period the reaction mixture turns dark red. The oil bath is removed, and the reaction mixture is allowed to cool to room temperature. The condenser is replaced with a rubber septum, into which is inserted a needle connected to a nitrogen line, and a positive nitrogen pressure is maintained. Neat *p*-tolualdehyde (7.1 mL, 60 mmol, 1.5 equiv) (Note 8) is added by syringe, dropwise over 3 min, followed by BF$_3$•Et$_2$O (8.25 mL, 40 mmol, 1.0 equiv) (Note 9), which is also added by syringe, dropwise over 5 min. The reaction mixture turns black upon the addition of BF$_3$•Et$_2$O,[3] and a slight exotherm is observed. The rubber septum is replaced with the Dimroth condenser and the system is maintained under a slight positive pressure of nitrogen. The black reaction mixture is placed in a preheated oil bath (90 °C, bath temp.) and allowed to stir for 24 h. After confirming the completion of the reaction by TLC monitoring (hexanes:EtOAc, 20:1, R$_f$ = 0.30 distinct blue spot under a UV lamp (254 nm)), the reaction mixture is diluted with 100 mL of chloroform and transferred to a 1 L round-bottomed flask. Saturated aqueous NaHCO$_3$ solution (150 mL) is added slowly via a glass funnel and the reaction mixture is stirred at room temperature until no gas evolution is visible. The mixture is then transferred to a 1 L separatory funnel and washed with saturated NaHCO$_3$ solution (2 x 100 mL) (Note 10) and brine (1 x 100 mL). The aqueous layer is back-extracted with chloroform (1 x 150 mL). The combined organic layer is dried over Na$_2$SO$_4$ (20 g, approx. 15 min). The drying agent is removed by filtration and the filtrate concentrated by rotary evaporation (15–20 mmHg, 23 °C) to give a brown oil, which is purified by column chromatography on silica gel (Note 11) to afford ethyl 3-butyl-4-oxo-2-*p*-tolyl-3,4-dihydroquinoline-1(2H)-carboxylate **3** (*trans:cis* = 20:1, based on integral ratio of ^{1}H NMR) as a slightly yellow oil (11.39 g, 78%) (Note 12).

Notes

1. 2-Iodobenzamide (98.0%) was purchased from Tokyo Chemical Industry Co., Ltd. and used as received. Pd(OAc)$_2$ (98.0%), 1-hexyne (98.0%), CuI (99.5%), PPh$_3$ (97.0%), and Et$_3$N (99.0%) were purchased from Sigma Aldrich, Co. and used as received. DMF (Fischer Optimum Grade) was dried through a molecular sieves based solvent drying system (Innovative Technologies).

2. Column chromatography was carried out using SILICYCLE SiliaFlash P60 silica gel. A glass column (5 x 40 cm) was slurry-packed with 200 g of silica gel in hexane. The compound was loaded on the column as a solution in a small amount of CH$_2$Cl$_2$, and the column was eluted first with hexane, then hexane:EtOAc, 50:1 (ca. 1 L), then hexane:EtOAc, 10:1 (ca. 0.5 L), then hexane:EtOAc, 5:1 (ca. 0.5 L), and then hexane:EtOAc, 2:1 until all the product had eluted. Fractions containing the desired product **1** were combined and concentrated using a rotary evaporator at 30 °C (15–20 mmHg) and dried under high vacuum at 23 °C (5–10 mmHg).

3. The procedure was performed at half-scale in 91% by the checkers. Characterization data for compound **1**: TLC R$_f$ = 0.35 (hexane:EtOAc, 2:1); mp 105–106 °C (slightly orange powder from hexane-EtOAc); ^{1}H NMR (500 MHz, CDCl$_3$) δ: 0.96 (t, J = 7.2 Hz, 3 H), 1.45–1.50 (m, 2 H), 1.60–1.64 (m, 2 H), 2.50 (t, J = 7.0 Hz, 2 H), 6.15 (br s, 1 H), 7.35–7.42 (m, 2H), 7.48 (td, J = 2.4, 5.6 Hz, 1H), 7.69 (br s, 1H), 8.11 (dd, J = 2.4, 7.2 Hz, 1 H); ^{13}C NMR (CDCl$_3$, 126 MHz) δ: 13.6, 19.3, 22.1, 30.5, 79.7, 97.8, 121.0, 128.1, 130.3, 131.0, 133.8, 134.0, 168.2. IR (CHCl$_3$, cm^{-1}) 3372, 3181, 2956, 2930, 2871, 1648, 1594, 1489, 1452, 1398, 1124, 817, 759, 634. HRMS (EI): m/z calcd for C$_{13}$H$_{15}$NO: 201.1154; found: 201.1151. Anal. Calcd for C$_{13}$H$_{15}$NO: C, 77.58; H, 7.51; N, 6.96. Found: C, 77.84; H, 7.51; N, 7.12.[4]

4. PhI(OAc)$_2$ (97.0%) and 1,2-dichloroethane (99.5%) were purchased from Sigma Aldrich, Co. and used as received. Ethanol (99.5%) was purchased from Acros Organic and used as received. PtCl$_2$ (98.0%) was purchased from Strem Chemicals and used as received.

5. The present procedure represents a modification of a previously published procedure.[5]

6. Column chromatography was carried out using SILICYCLE SiliaFlash P60 silica gel. A glass column (5 x 40 cm) was slurry-packed with 200 g of silica gel in hexane. The compound was loaded as a solution in a

small amount of CH$_2$Cl$_2$, and eluted first with hexane, then hexane:EtOAc, 50:1 (ca. 0.5 L), and then hexane:EtOAc, 25:1 (ca. 2 L). The fractions containing the desired product **2** were combined and concentrated by rotary evaporation at 30 °C (15–20 mmHg) and dried under high vacuum at 23 °C (5–10 mmHg).

7. The procedure was performed at half-scale in 88% by the checkers. Characterization data for compound **2**: TLC R$_f$ = 0.30 (hexanes:EtOAc, 25:1); mp 34–35 °C (colorless prisms from hexane, extensive drying is required); ^{1}H NMR (CDCl$_3$, 500 MHz) δ: 0.97 (t, J = 7.5 Hz, 3 H), 1.45 (sex, J = 7.5 Hz, 2H), 1.48 (t, J = 7.5 Hz, 3 H), 1.69 (quint, J = 7.5 Hz, 2 H), 3.01 (t, J = 7.5 Hz, 2H), 4.49 (q, J = 7.0 Hz, 2 H), 6.34 (s, 1 H), 7.18–7.23 (m, 2 H), 7.44 (dd, J = 7.4, 1.8 Hz, 1 H), 8.09 (d, J = 7.4 Hz, 1 H); ^{13}C NMR (CDCl$_3$, 126 MHz) δ: 14.0, 14.4, 22.6, 29.8, 31.1, 63.0, 107.5, 115.7, 119.8, 122.9, 123.4, 129.6, 136.6, 142.6, 152.1. IR (CHCl$_3$, cm^{-1}) 2958, 2871, 1736, 1593, 1568, 1456, 1398, 1378, 1323, 1258, 1211, 1118, 1081, 807, 766, 746. HRMS (EI): m/z calcd for C$_{15}$H$_{19}$NO$_2$: 245.1416; found: 245.1413. Anal. Calcd for C$_{15}$H$_{19}$NO$_2$: C, 73.44; H, 7.81; N, 5.71. Found: C, 73.78; H, 7.89; N, 5.87.[6]

8. *p*-Tolualdehyde (98.0%) was purchased from SigmaAldrich, Co. and used as received.

9. BF$_3$•Et$_2$O (95.0%) was purchased from Sigma Aldrich, Co. and used as received.

10. **CAUTION**: A large amount of carbon dioxide is generated in this extraction and appropriate care should be taken to release pressure from the funnel.

11. Column chromatography was carried out using SILICYCLE SiliaFlash P60 silica gel. A glass column (6 x 40 cm) was slurry-packed with 300 g of silica gel. The compound was loaded as a solution in a small amount of CH$_2$Cl$_2$, and the column eluted with hexane, then hexane:EtOAc, 50:1 (ca. 0.75 L), and then hexane:EtOAc, 30:1 (ca. 1 L). The R$_f$ value of *p*-tolualdehyde is very close to that of the desired product. It is useful to take note of the distinct UV response of the product (i.e. a blue spot). The fractions containing the desired product **3** were combined and concentrated by rotary evaporation at 30 °C (15–20 mmHg) and dried under high vacuum at 23 °C (5–10 mmHg).

12. The procedure was performed at half-scale in 77% by the checkers. Characterization data for compound **3**: TLC R$_f$ = 0.30 (hexanes:EtOAc, 20:1); slightly yellow oil; ^{1}H NMR (CDCl$_3$, 500 MHz) δ: 0.92 (t, J = 7.5 Hz, 3 H), 1.34–1.41 (m, 5 H), 1.43–1.50 (m, 1 H), 1.54–1.62 (m, 1 H), 1.71–

1.81 (m, 2 H), 2.22 (s, 3 H), 3.12 (t, J = 6.5 Hz, 1 H), 4.32–4.46 (m, 2 H), 5.98 (s , 1 H), 7.00 (d, J = 7.0 Hz, 2 H), 7.04–7.09 (m, 3 H), 7.46 (t, J = 7.0 Hz, 1 H), 7.89–7.90 (m, 2 H); ^{13}C NMR (CDCl$_3$, 126 MHz) δ: 13.9, 14.5, 20.9, 22.5, 29.3, 29.8, 51.1, 59.7, 62.8, 123.5, 123.8, 126.6, 127.4, 129.3, 134.4, 135.5, 137.1, 141.2, 155.1, 195.8. IR (CHCl$_3$, cm^{-1}) 3036, 3007, 2961, 2932, 1706, 1682, 1601, 1479, 1460, 1396, 1381, 1321, 1298, 1269, 1242, 1196, 1049. MS (EI): m/z = 365 (M$^+$). HRMS (EI): m/z calcd for C$_{23}$H$_{27}$NO$_3$: 365.1991; found: 365.1985. Anal. Calcd for C$_{23}$H$_{27}$NO$_3$: C, 75.59; H, 7.45; N, 3.83. Found: C, 75.89; H, 7.36; N, 3.75.

Handling and Disposal of Hazardous Chemicals

The procedures in this article are intended for use only by persons with prior training in experimental organic chemistry. All hazardous materials should be handled using the standard procedures for work with chemicals described in references such as "Prudent Practices in the Laboratory" (The National Academies Press, Washington, D.C., 2011 www.nap.edu). All chemical waste should be disposed of in accordance with local regulations. For general guidelines for the management of chemical waste, see Chapter 8 of Prudent Practices.

These procedures must be conducted at one's own risk. *Organic Syntheses, Inc.*, its Editors, and its Board of Directors do not warrant or guarantee the safety of individuals using these procedures and hereby disclaim any liability for any injuries or damages claimed to have resulted from or related in any way to the procedures herein.

Discussion

Synthesis of heterocyclic compounds has attracted a great deal of attention due to their biological activities. Metal-catalyzed ring closure of 2-alkynylaniline derivatives is one of the most efficient approaches for the construction of benzo-fused *N*-containing heterocyclic compounds;[7] however, one drawback of the method is the air instability of the amines. We became interested in using 2-alkynylbenzamide **1** for heterocycle formation, since alkynylbenzamides can be converted to amine derivatives via a Hofmann-type rearrangement.[8] The strategy described in this manuscript for the synthesis of indole **2** involves (1) Hofmann-type rearrangement of 2-alkynylbenzamides **1** using hypervalent iodine reagent PhI(OAc)$_2$ followed by a nucleophilic addition of an alcohol to an isocyanate intermediate,[9] and (2) platinum(II)-catalyzed 5-*endo* cyclization of carbamate nitrogen atom toward an alkyne functionality (Scheme 1). Similarly, the strategy for the synthesis of quinolone[10] **3** involves (1) Hofmann-type rearrangement of 2-alkynylbenzamides **1** followed by a nucleophilic addition of an alcohol to an isocyanate intermediate, (2) acid-catalyzed intermolecular [2+2]-cycloaddition between the carbon-carbon triple bonds of carbamates and aldehydes, and (3) acid-catalyzed intramolecular aminocyclization to the α,β-unsaturated ketones.[11]

Scheme 1. Heterocycle Formation

The present procedure provides easy access to *N*-carboxylated-2-substituted indoles **2** via a one-pot tandem reaction (Table 1).[5] The electronic nature of the substituents on the aromatic ring does not affect the reaction; in the presence of electron-withdrawing groups or electron-donating groups, the yields of indoles are within the range of 82–91 % (entries 1–3). Alkynylbenzamide, bearing a phenyl group on the acetylene terminus, also provides the corresponding indole (entry 4). The terminal alkyne is not suitable for this reaction (entry 8).

Table 1. One-pot indole synthesis from benzamide

entry	R^1	R^2	R^3	yield (%)
1	F	n-Bu	Et	84
2	NO_2	n-Bu	Et	91
3	OMe	n-Bu	Et	82
4	H	Ph	Et	84
5	H	p-Tol	Et	66
6	H	$-(CH_2)_3OTs$	Et	100
7	H	n-Bu	Bn	90
8	H	H	Et	33

In 2,3-dihydro-4-quinolone **3** synthesis, moderate yields and high *trans*-selectivities are observed (Table 2).[11] In the presence of either electron-donating or moderately electron-withdrawing substituents R^1 on the aromatic ring of 2-alkynylbenzamides **1**, the yields of desired products range from 72–80% (Table 2, entries 1–3). However, the presence of the strongly electron-withdrawing nitro group on the aromatic ring hinders the reaction (entry 4). For *p*-cyanobenzaldehyde or *p*-nitrobenzaldehyde, a higher temperature of 90 °C is required (entries 10 and 11). The reaction can tolerate aliphatic aldehydes (entries 12 and 13). Terminal alkyne is also suitable for this reaction (entry 14). The use of benzophenone instead of aldehydes fails to give the desired product (entry 15).

Table 2. One-pot 4-quinolone synthesis from benzamide

entry	R^1	R^2	R^3	R^4	temp (°C)	yield (%) (trans:cis)
1	H	n-Bu	Et	p-Tol	60	78 (20:1)
2	OMe	n-Bu	Et	p-Tol	60	72 (7:1)
3	F	n-Bu	Et	p-Tol	60	80 (38:1)
4	NO_2	n-Bu	Et	p-Tol	60	0
5	H	n-Bu	Bn	p-Tol	60	73 (8:1)
6	H	n-Bu	Me	Ph	60	72 (28:1)
7	OMe	n-Bu	Et	Ph	60	52 (11:1)
8	H	Ph	Me	p-Tol	60	73 (trans)
9	H	Ph	Et	p-Tol	60	66 (trans)
10	H	n-Bu	Et	p-CNC_6H_4	90	82 (trans)
11	H	n-Bu	Et	p-$NO_2C_6H_4$	90	80 (trans)
12	H	n-Bu	Et	hexyl	60	61 (20:1)
13	H	n-Bu	Et	cyclohexyl	60	87 (99:1)
14	H	H	Et	p-Tol	60	85
15	H	n-Bu	Et	(PhCOPh)	90	0

References

1. (a) Faculty of Pharmaceutical Sciences, Hiroshima International University, 5-1-1 Hirokoshingai, Kure, Hiroshima 737-0112, Japan; ryanada@ps.hirokoku-u.ac.jp; (b) Department of Synthetic Organic Chemistry, Graduate School of Medical Sciences, Hiroshima University1-2-3 Kasumi, Minami-Ku, Hiroshima 734-8553, Japan.
We gratefully acknowledge a Grant-in-Aid for Scientific Research (C) from JSPS KAKENHI (23590032). Dr. N. O. is grateful for the Sasakawa Scientific Research Grant from The Japan Science Society. We also thank the Natural Science Center for Basic Research and Development (N-BARD), Hiroshima University for the use of the facilities.

2. Sonogashira, K.; Tohda, Y.; Hagiwara, N. *Tetrahedron Lett.* **1975**, *50*, 4467–4470.
3. Kurtz, K. C. M.; Hsung, R. P.; Zhang, Y. *Org. Lett.* **2006**, *8*, 231–234.
4. Wu, M.-J.; Chang, L.-J.; Wei, L.-M.; Lin, C.-F. *Tetrahedron* **1999**, 55, 13193–13200.
5. (a) Okamoto, N.; Miwa, Y.; Minami, H.; Takeda, K.; Yanada, R. *Angew. Chem. Int. Ed.* **2009**, *48*, 9693–9696. (b) Okamoto, N.; Takeda, K.; Yanada, R. *J. Org. Chem.* **2010**, *75*, 7615–7625.
6. Hiroya, K.; Itoh, S.; Sakamoto, T. *J. Org. Chem.* **2004**, *69*, 1126–1136.
7. Indole synthesis: (a) Humphrey, G. R.; Kuethe, J. T. *Chem. Rev.* **2006**, *106*, 2875–2911. (b) Fürstner, A.; Davies, P.W. *Angew. Chem. Int. Ed.* **2007**, *46*, 3410–3449. (c) Huang, N.-Y.; Liu, M.-G.; Ding, M.-W. *J. Org. Chem.* **2009**, *74*, 6874–6877. Isoquinoline synthesis: (a) Fischer, D.; Tomeba, H.; Pahadi, N. K.; Patil, N. T.; Huo, Z.; Yamamoto, Y. *J. Am. Chem. Soc.* **2008**, *130*, 15720–15725. (b) Enomoto, T. Anne-Lise Girard, A.-L.; Yasui, Y.; Takemoto, Y. *J. Org. Chem.* **2009**, *74*, 9158–9164. (c) Sperger, C.; Fiksdahl, A. *J. Org. Chem.* **2010**, *75*, 4542–4553.
8. Hofmann, A. W. *Ber.* **1881**, *14*, 2725–2736.
9. Isocyanates prepared with hypervalent iodine reagents: (a) Liu, W.; Buck, M.; N. Chen, Shang, M.; Taylor, N. J.; Asoud, J.; Wu, X.; Hasinoff, B. B.; Dmitrienko, G. I. *Org. Lett.* **2007**, *9*, 2915–2918. (b) Zagulyaeva, A. A.; Banek, C. T.; Yusubov, M. S.; Zhdankin, V. V. *Org. Lett.* **2010**, *12*, 4644–4647.
10. 2,3-Dihydro-4-quinolone synthesis: (a) Saito, A.; Kasai, J.; Odaira, Y.; Fukaya, H.; Hanzawa, Y. *J. Org. Chem.* **2009**, *74*, 5644–5647. (b) Lei, B.-L.; Ding, C.-H.; Yang, X.-F.; Wan, X.-L.; Hou, X.-L. *J. Am. Chem. Soc.* **2009**, *131*, 18250–18251. (c) Liu, X.; Lu, Y. *Org. Lett.* **2010**, *12*, 5592–5595.
11. Okamoto, N.; Takeda, K.; Ishikura, M.; Yanada, R. *J. Org. Chem.* **2011**, *76*, 9139–9143.

Appendix
Chemical Abstracts Nomenclature (Registry Number)

2-(1-Hexynyl)benzamide: Benzamide, 2-(1-hexyn-1-yl)-; (110166-74-0)
Ethyl 2-butyl-1H-indole-1-carboxylate: 1H-Indole-1-carboxylic acid, 2-butyl-, ethyl ester; (221353-60-2)
Ethyl 3-butyl-4-oxo-2-p-tolyl-3,4-dihydroquinoline-1(2H)-carboxylate: 1(2H)-Quinolinecarboxylic acid, 3-butyl-3,4-dihydro-2-(4-methylphenyl)-4-oxo-, ethyl ester, (2R,3S)-rel-; (1337988-00-7)

2-Iodobenzamide; (3930-83-4)
Triphenylphosphine; (603-35-0)
Copper(I) iodide; (7681-65-4)
Palladium(II) acetate; (3375-31-3)
Triethylamine; (121-44-8)
Hexyne; (928-49-4)
(Diacetoxyiodo)benzene; (3240-34-4)
Platinum(II) chloride; (10025-65-7)
p-Tolualdehyde; (104-87-0)
Boron trifluoride diethyl ethereate; (109-63-7)

Reiko Yanada received both her B.S. degree (1977) and M.S. degree (1979) from Toyama University (with Eiichi Yoshii) and her Ph.D. degree (1988) from Kyoto University (with Fumio Yoneda). After working at the Dyson Perrins Laboratory of University of Oxford with Professor S. G. Davies, she was promoted to assistant professor, lecturer, and then to associate professor at Kyoto University. She became a professor of organic chemistry at Hiroshima International University in 2006. Her current research interest is in the development of new synthetic organic methodologies utilizing tandem reaction.

Noriko Okamoto obtained her B.S. degree (2004), M.S. degree (2006), and Ph.D. degree (2011) from Hiroshima University (with Kei Takeda). She joined Professor Reiko Yanada's research group at Hiroshima International University in 2008 as an assistant professor. Her current research interest is in the development of new synthetic reactions. She was the recipient of the Chugoku-Shikoku Branch of Pharmaceutical Society of Japan Award for Young Scientists (2012).

Kei Takeda is professor of organic chemistry at Hiroshima University. He was born in 1952 and received both his B.S. degree (1975) and M.S. degree (1977) from Toyama University (with Eiichi Yoshii) and his Ph.D. degree (1980) from the University of Tokyo (with Toshihiko Okamoto). In 1980, he joined the faculty at Toyama Medical and Pharmaceutical University (Prof. Yoshii's group). After working at MIT with Professor Rick L. Danheiser (1988-1989), he was promoted to Lecturer (1989) and then to Associate Professor (1996). He became a professor of Hiroshima University in 2000. His research interests are in the invention of new synthetic reactions and chiral carbanion chemistry. He was the recipient of the Sato Memorial Award (1998) and the 41st Senji Miyata Foundation Award.

Hang Chu was born in China in 1991. He is currently pursuing a joint BS/MS degree in chemistry at the University of Chicago. He began his research in synthetic organic chemistry under the supervision of Dr. Viresh Rawal in 2011. Currently, he is working on developing a diene to achieve "meta"-selective Diels Alder reactions. He plans to continue his studies in organic chemistry – particularly asymmetric catalysis - by pursuing a doctoral degree.

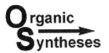

Pd-Catalyzed External-CO-Free Carbonylation: Preparation of 2,4,6-Trichlorophenyl 3,4-Dihydronaphthalene-2-Carboxylate

Hideyuki Konishi, Tsuyoshi Ueda, and Kei Manabe[1]*

School of Pharmaceutical Sciences, University of Shizuoka, 52-1 Yada, Suruga-ku, Shizuoka 422–8526, Japan

Checked by Jennifer Ciesielski and Erick Carreira

Procedure

Caution! The carbonylation (part C) should be performed in a well-ventilated hood in case of a leak of carbon monoxide.

Org. Synth. **2014**, *90*, 39-51
DOI: 10.15227/orgsyn.091.0039

Published on the Web 10/31/2013
© 2014 Organic Syntheses, Inc.

A. *3,4-Dihydronaphthalen-2-yl trifluoromethanesulfonate (2)*. An oven-dried, 500-mL, three-necked, round-bottomed flask equipped with a Teflon-coated magnetic stir bar (40 x 20 mm), rubber septa (all three necks), and an argon inlet needle (center neck) is charged with β-tetralone (**1**) (4.16 mL, 4.60 g, 31.5 mmol, 1.05 equiv) (Note 1) and tetrahydrofuran (THF) (120 mL) (Note 2) via syringes through a septum. The flask is cooled to –20 °C (bath temperature) in a cooling bath (*i*-PrOH/dry-ice). Potassium *tert*-butoxide (3.53 g, 31.5 mmol, 1.05 equiv) is added slowly to the solution of β-tetralone over 10 min (Note 3). After completing the addition, the mixture is warmed to 0 °C in an ice-water bath and stirred for 1 h. Afterward, the dark-blue solution is cooled to –20 °C (bath temperature) in a cooling bath (*i*-PrOH/dry-ice). After the septum is removed, *N*-phenylbis(trifluoromethanesulfonimide) (10.7 g, 30.0 mmol, 1.00 equiv) (Note 1) is added over 1 min to the solution under a positive pressure of argon. Argon is immediately flushed into the flask and a new rubber septum is inserted. The mixture is warmed to 0 °C in an ice-water bath and stirred for 4 h. At this point, full conversion to the product is confirmed by TLC analysis (Note 4), and the mixture is concentrated under reduced pressure (40 °C, *ca.* 100 mmHg) to approximately one-fourth of the original volume in a rotary evaporator. The resulting mixture is diluted with EtOAc (80 mL) and H$_2$O (80 mL), and the layers are then partitioned in a 500-mL separatory funnel (Note 5). The aqueous layer is extracted with EtOAc (80 mL × 2). The combined organic layers are washed with brine (80 mL), dried over anhydrous Na$_2$SO$_4$ (50 g) (Note 1), filtered through a 75 mL medium-porosity fritted funnel, and concentrated on a rotary evaporator under reduced pressure (40 °C, *ca.* 60 mmHg) to afford a black solid. The obtained residue is purified by column chromatography (SiO$_2$, hexane/EtOAc 100/1) (Note 6) to afford 7.75-8.09 g (93-97%, 27.8-29.0 mmol) of product **2** (Notes 7 and 8).

B. *2,4,6-Trichlorophenyl formate (6)*. A 1-L, three-necked, round-bottomed flask is equipped with a Teflon-coated magnetic stir bar (40 x 20 mm), rubber septa (both side necks) and a reflux-condenser (center neck). The reflux-condenser is fitted with a rubber septum and an argon inlet needle. The flask is charged with formic acid (18.9 mL, 500 mmol, 5.00 equiv) and acetic anhydride (37.8 mL, 400 mmol, 4.00 equiv) (Note 1). The mixture in the flask is heated to 60 °C (bath temperature) in an oil bath and stirred for 1 h. Subsequently, the mixture is cooled to 0 °C in an ice-water bath. Toluene (300 mL) (Note 2), 2,4,6-trichlorophenol (**5**) (19.7 g, 100 mmol, 1.00 equiv), and sodium acetate (8.20 g, 100 mmol, 1.00 equiv) are added to the solution (Notes 1 and 9). After 10 min, the ice-water bath is removed and the reaction is warmed to room temperature. During this time, a white precipitate gradually appears. The mixture is stirred for 30 min at room temperature. At this point, full conversion to the product is confirmed by TLC analysis (Note 4), and H_2O (100 mL) is added to the solution, resulting in the dissolution of the white precipitate. The layers are partitioned in a 1-L separatory funnel. The organic layer is washed with H_2O (100 mL × 3) and brine (100 mL × 2), dried over anhydrous Na_2SO_4 (50 g) (Note 1), filtered through a 75 mL medium-porosity fritted funnel, and concentrated on a rotary evaporator under reduced pressure (40 °C, *ca.* 20 mmHg) to afford a pale yellow oil, which crystallizes upon standing at ambient temperature. The obtained residue is recrystallized from hexane/EtOAc (50/1 v/v, *ca.* 150 mL) (Note 10) to afford 21.0 g (93% yield, 93.1 mmol) of product **6** (Notes 11 and 12).

C. *2,4,6-Trichlorophenyl 3,4-dihydronaphthalene-2-carboxylate (7).* An oven-dried, 300-mL, three-necked, round-bottomed flask equipped with a Teflon-coated magnetic stir bar (40 x 20 mm), rubber septa (all three necks), an argon inlet (center neck), and a bubbler (side neck) is charged with palladium acetate (135 mg, 0.600 mmol, 0.03 equiv) (Note 1), Xantphos (694 mg, 1.20 mmol, 0.06 equiv) (Note 1), and 2,4,6-trichlorophenyl formate (**6**) (5.41 g, 24.0 mmol, 1.20 equiv). The flask is evacuated and backfilled with argon three times. An oven-dried, 50-mL, single-necked, round-bottomed flask is charged with 3,4-dihydronaphthalen-2-yl trifluoromethanesulfonate (**2**) (5.57 g, 20.0 mmol, 1.00 equiv) and equipped with a rubber septa and an argon inlet. The flask is evacuated and backfilled with argon three times. Then, toluene (20 mL) (Note 2) is added to the flask via a syringe. The triflate solution is added to the 300-mL flask via a gastight syringe. Toluene (10 mL x 2) (Note 2) is added to the 50-mL flask to wash its interior, and the washing is transferred to the 300-mL flask via the same syringe. After the mixture is stirred for 5 min at ambient temperature, triethylamine (3.33 mL, 24.0 mmol, 1.20 equiv) (Note 1) is added over 15 min via a syringe (Note 13). During the addition of triethylamine, a slightly exothermic reaction ensues with gas evolution, and the brown color changes to black. After completing the addition, the mixture is stirred for another 2 h. At this point, full conversion to the product is confirmed by TLC analysis (Note 4), and the mixture is diluted with Et_2O (100 mL) and H_2O (100 mL). The layers are partitioned in a 500-mL separatory funnel. The aqueous layer is extracted with Et_2O (50 mL x 2). The combined organic layers are washed with aqueous NaOH solution (0.5 M, 60 mL) and brine (100 mL), dried over anhydrous Na_2SO_4 (40 g) (Note 1), filtered through a 75 mL medium-porosity fritted funnel, and concentrated under reduced pressure (40 °C, *ca.* 60 mmHg) in a rotary evaporator to afford a brown oil. The obtained residue is purified by column chromatography (SiO_2, hexane/EtOAc 100/1) (Note 14) to afford 6.73 g (96%, 19.1 mmol) of product **7** as a white solid (Notes 15 and 16).

Notes

1. The checkers purchased β-tetralone (98%) and $Pd(OAc)_2$ (≥99.9% trace metal basis) from Sigma-Aldrich and used as received. Formic acid (98+%) and acetic anhydride (97+%) were purchased from Merck and used as received. Xantphos (98%), sodium acetate (98.5+%), anhydrous sodium sulfate (99+%), and potassium *tert*-butoxide (98+%) were purchased from Acros Organics and used as received. N-Phenylbis(trifluoromethanesulfonimide) (>98.0%) and 2,4,6-trichlorophenol (>96.0%) were purchased from Tokyo Chemical Industry Co., Ltd. and used as received. Triethylamine (99+%) was purchased from Sigma-Aldrich and distilled from CaH_2 prior to use. The submitters purchased β-tetralone (97+%), anhydrous sodium sulfate (99+%), formic acid (98+%), acetic anhydride (97+%), sodium acetate (98.5+%), and Xantphos (98+%) from Wako Pure Chemical Industries, Ltd. and used as received. Potassium *tert*-butoxide solution in THF (12%, *ca.* 1 M), N-phenylbis(trifluoromethanesulfonimide) (>98.0%), and 2,4,6-trichlorophenol (>96.0%) were purchased from Tokyo Chemical Industry Co., Ltd. and used as received. $Pd(OAc)_2$ (≥99.9% trace metal basis) was purchased from Sigma-Aldrich and used as received. Triethylamine (99+%) was purchased from Wako and distilled from CaH_2 prior to use.
2. The checkers purchased THF (>99.5%) and toluene (>99.5%) from Sigma-Aldrich and passed it through a column of alumina before use. The submitters purchased dehydrated, stabilizer-free THF (>99.5% "Super Plus") and dehydrated toluene (>99.5% "Super Plus") from Kanto, which was purified by using a Glass Contour Solvent Purification Systems prior to use.

3. The submitters equipped the 500-mL, three-necked round-bottomed flask with rubber septa (both side necks) and a 100 mL pressure-equalizing addition funnel (center neck). After the 500-mL flask was charged with β-tetralone (1) (4.16 mL, 4.60 g, 31.5 mmol, 1.05 equiv) and THF (120 mL), the addition funnel was charged with a solution of potassium *tert*-butoxide in THF (*ca.* 1.0 M, 31.5 mL, 31.5 mmol, 1.05 equiv) via a syringe that was then added dropwise to the solution of β-tetralone over 10 min.

4. TLC analysis was performed on Merck glass plates coated with 0.25-mm 230–400 mesh silica gel containing a fluorescent indicator. The plate was eluted with hexane/EtOAc (9/1) and visualized by ultraviolet lamp at 254 nm. The following R_f values were obtained: 3,4-dihydronaphthalen-2-yl trifluoromethanesulfonate (0.63), 2,4,6-trichlorophenyl formate (0.57), and 2,4,6-trichlorophenyl 3,4-dihydronaphthalene-2-carboxylate (0.57). As for 2,4,6-trichlorophenyl formate, partial decomposition resulting in the appearance of a spot of 2,4,6-trichlorophenol was observed.

5. The interface of the two phases is difficult to recognize because of the deep color (organic phase: dark blue-black, aqueous phase: dark yellow-black). Care should be taken in the separation of the two phases. Additional amounts of EtOAc and H_2O may lead to easier separation.

6. Column chromatography is performed using a 7.5-cm wide, 22-cm high column of 270 g of Fluka Silica gel (high purity grade, 60 Å pore size, 230-400 mesh) (The submitters used Kanto Silica Gel 60 N (spherical, neutral, 63–210 μm)) packed by slurring the silica gel with hexane. The residue is dissolved with a minimum amount of CH_2Cl_2 (30 mL), and loaded onto the column. Elution with 400 mL of hexane and then hexane/EtOAc (100/1) (50 mL initial collection followed by 200 mL fractions) afforded the production in fractions 2-24. The combined fractions containing the desired product are concentrated on a rotary evaporator under reduced pressure (40 °C, *ca.* 120 mmHg).

7. The submitters reported two runs. Starting with compound 1, triflate 2 was obtained in 97–98% yield (8.06-8.16 g, 29.0-29.3 mmol).

8. 3,4-Dihydronaphthalen-2-yl trifluoromethanesulfonate (2) showed the following characterization data: $R_f = 0.63$ (hexane/EtOAc 9/1); ^{1}H NMR (400 MHz, CDCl$_3$) δ: 2.70 (t, J = 8.4 Hz, 2 H), 3.06 (t, J = 8.2 Hz, 2 H), 6.48 (s, 1 H), 7.05 – 7.10 (m, 1 H), 7.12 – 7.16 (m, 1 H), 7.17 – 7.23 (m, 2 H). ^{13}C NMR (100 MHz, CDCl$_3$) δ: 26.5, 28.5, 118.5, 118.5 (q, $^1J_{cf}$ = 320.7 Hz), 127.3, 127.5, 128.4, 131.1, 132.9, 149.9. ^{19}F NMR (376 MHz, CDCl$_3$) δ:

–73.6. IR (ATR, neat) cm^{-1}: 1664, 1416, 1248, 1202, 1137, 1062, 985, 895, 824, 753, 610. HMRS (EI): m/z calcd for $C_{11}H_9F_3O_3S$ [M$^+$] 278.0219, found 278.0222. Elemental Analysis: Anal. Calcd. for $C_{11}H_9F_3O_3S$: C, 47.48; H, 3.26. Found: C, 47.27; H, 3.19. This compound tends to develop a brown color; storage by refrigeration is recommended.

9. The submitters reported that an exothermic reaction occurred after addition of sodium acetate. The checkers avoided this exotherm by cooling the reaction to 0 °C prior to addition of sodium acetate.

10. Recrystallization is performed as follows. The residue is transferred to a 500-mL, round-bottomed flask equipped with a Teflon-coated magnetic stir bar (30 x 6 mm). Hexane/EtOAc (50/1 v/v, *ca.* 150 mL) is added to the flask and the flask is heated to 60 °C by an oil bath with stirring. After dissolution of the solid, magnetic stirring is turned off and the solution was allowed to cool to ambient temperature and crystallize for 5 h. The resulting crystals were collected by suction filtration on a 75 mL medium-porosity fritted funnel with a 250 mL round-bottomed receiving flask, and washed with ice-cold hexane (5 mL x 2). The mother liquor is concentrated on a rotary evaporator under reduced pressure (40 °C, *ca.* 20 mmHg) to give a yellow solid. A Teflon-coated magnetic stir bar (30 x 6 mm) and hexane/EtOAc (50/1 v/v, *ca.* 50 mL) were added to the flask and the flask was heated to 60 °C by an oil bath with stirring. Crystallization was carried out as described for the first crop. After filtration, the mother liquor, which was collected in a 100 mL round-bottomed flask, was concentrated on a rotary evaporator under reduced pressure (40 °C, *ca.* 20 mmHg) to give a yellow solid. A Teflon-coated magnetic stir bar (30 x 6 mm) and hexane/EtOAc (50/1 v/v, *ca.* 10 mL) were added to the flask and the flask was heated to 60 °C by an oil bath with stirring. Crystallization was carried out as described for the first crop. The three crops are dried overnight in a desiccator at ambient temperature under vacuum (3 mmHg).

11. The checkers completed two runs. The first run was completed on half the scale. Compound **6** was obtained in 95% yield (10.7 g, 47.4 mmol). The quantities and yields obtained from the three crops after recrystallization were as follows: first crop (6.90 g, 30.6 mmol, 61%), second crop (3.38 g, 15.0 mmol, 30%), and third crop (0.440 g, 1.95 mmol, 4%). For the second (full-scale) run, the quantities and yields obtained from the three crops after recrystallization were as follows: first crop (13.8 g, 61.2 mmol, 61%), second crop (6.40 g, 28.3 mmol, 28%), and third crop (0.88 g, 3.90 mmol, 4%).

12. 2,4,6-Trichlorophenyl formate (**6**) showed the following characterization data: mp 72–73 °C; $R_f = 0.57$ (hexane/EtOAc 9/1); ¹H NMR (400 MHz, CDCl₃) δ: 7.41 (s, 2 H), 8.28 (s, 1 H). ¹³C NMR (100 MHz, CDCl₃) δ: 128.7, 129.2, 132.6, 141.9, 156.2. IR (ATR, neat) cm⁻¹: 3078, 1732, 1563, 1447, 1385, 1227, 1085, 1057, 850, 820, 805, 678, 562. HRMS (ESI-TOF): *m/z* calcd for C₇H₃Cl₃O₂ [M⁺] 223.9194, found 223.9190. Elemental Analysis: Anal. Calcd. for C₇H₃Cl₃O₂: C, 37.29; H, 1.34. Found: C, 37.10; H, 1.40.

13. Triethylamine should be added slowly. Fast addition of triethylamine causes sudden decomposition of the formate.

14. Column chromatography is performed using a 7.5-cm wide, 22-cm high column of 270 g of Fluka Silica gel (high purity grade, 60 Å pore size, 230-400 mesh) (The submitters used Kanto Silica Gel 60 N (spherical, neutral, 63–210 µm)) packed by slurring the silica gel with hexane. The residue is dissolved with a minimum amount of CH₂Cl₂ (30 mL), and loaded onto the column. Elution with 400 mL of hexane and then hexane/ EtOAc (100/1) (2 L initial collection followed by 200 mL fractions) afforded the production in fractions 10-16. The combined fractions containing the desired product were concentrated on a rotary evaporator under reduced pressure (40 °C, *ca.* 120 mmHg).

15. The checkers completed two runs. The first run was completed on half the scale. Compound **7** was obtained in 92% yield (3.22 g, 9.14 mmol). The submitters reported two runs. Compound **7** was obtained as white needles in 94–96% yield (6.67–6.76 g, 18.9–19.1 mmol).

16. 2,4,6-Trichlorophenyl 3,4-dihydronaphthalene-2-carboxylate (**7**) showed the following characterization data: mp 80–82 °C; $R_f = 0.57$ (hexane/EtOAc 9/1); ¹H NMR (400 MHz, CDCl₃) δ: 2.76 (t, *J* = 7.9 Hz, 2 H), 2.97 (t, *J* = 8.3 Hz, 2H), 7.20 – 7.35 (m, 4 H), 7.40 (s, 2 H), 7.86 (s, 1H). ¹³C NMR (100 MHz, CDCl₃) δ: 22.2, 27.4, 126.7, 126.9, 127.8, 128.5, 129.0, 129.8, 130.3, 131.8, 132.0, 137.2, 139.9, 143.3, 163.5. IR (ATR, neat) cm⁻¹: 1733, 1624, 1564, 1448, 1380, 1275, 1238, 1201, 1184, 1171, 1022, 958, 855, 757, 737, 714. HRMS (EI): *m/z* calcd for C₁₇H₁₂Cl₃O₂ [M+H⁺] 352.9897, found 352.9896. Elemental Analysis: Anal. Calcd. for C₁₇H₁₁Cl₃O₂: C, 57.74; H, 3.14. Found: C, 57.72; H, 3.08.

Handling and Disposal of Hazardous Chemicals

The procedures in this article are intended for use only by persons with prior training in experimental organic chemistry. All hazardous materials should be handled using the standard procedures for work with chemicals described in references such as "Prudent Practices in the Laboratory" (The National Academies Press, Washington, D.C., 2011 www.nap.edu). All chemical waste should be disposed of in accordance with local regulations. For general guidelines for the management of chemical waste, see Chapter 8 of Prudent Practices.

These procedures must be conducted at one's own risk. *Organic Syntheses, Inc.*, its Editors, and its Board of Directors do not warrant or guarantee the safety of individuals using these procedures and hereby disclaim any liability for any injuries or damages claimed to have resulted from or related in any way to the procedures herein.

Discussion

Palladium-catalyzed carbonylation of organic (pseudo)halides employing carbon monoxide (CO) has received much attention because of its versatility for the synthesis of carbonyl-containing compounds.[2] One of the major drawbacks to this methodology is the use of CO in gaseous form, which is highly toxic, and, can, therefore, be problematic for production on both laboratory and multikilogram scales. A simple solution to this problem involves replacing gaseous CO with another, less toxic, carbonyl source. Given the increasing demand for CO surrogates, several carbonyl sources have been developed.[3] Aldehydes can be decomposed through transition metal catalysis to generate CO.[4] Formic anhydrides[5] and formic esters[6] are also known to produce CO in the presence of a transition metal or a strong base. However, these CO surrogates require high temperatures or harsh reaction conditions to promote the generation of CO, making them less attractive. Although metal carbonyl complexes have been reported to generate CO by thermal decomposition using a microwave,[7] the need for a large excess of the complex is unfavorable for handling and poses environmental issues. Recently, 9-methylfluorene-9-carbonyl chloride[8] and silacarboxylic acid[9] have been reported as CO precursors. However, tedious procedures are required for their synthesis, and a special two-chamber system must be used for the carbonylation reaction. It is noted that acylpalladium precatalyst works well in hydroxycarbonylation of aryl halides using potassium formate as a CO source.[10]

We have recently developed a Pd-catalyzed external-CO-free carbonylation of aryl, alkenyl, allyl halides, and sulfonates by using phenyl formate as a phenoxycarbonylating source.[11-13] This method is based on the finding that phenyl formate can undergo facile decarbonylation in the presence of weak base such as triethylamine to afford CO and phenol, which subsequently react with an electrophile.

Though aryl formates easily decompose to generate CO in the presence of tertiary amines, alkyl formates do not.[11] The rate of decomposition increases as electron-withdrawing groups are introduced into the aromatic rings of aryl formates.[14]

Among the aryl formates with electron-withdrawing groups, 2,4,6-trichlorophenyl formate was found to have the potential to promote the external-CO-free carbonylation reaction under much milder conditions, i.e., at ambient temperature.[14] 2,4,6-Trichlorophenyl formate is a stable crystalline compound and is easily accessible from 2,4,6-trichlorophenol, an inexpensive chemical feedstock.[15] It allows mild and fast aryloxycarbonylation of aryl iodides and alkenyl trifluoromethanesulfonates (triflates) to afford the corresponding 2,4,6-trichlorophenyl esters in excellent yields. In the case of large-scale synthesis, the reaction proceeds without any difficulties, although adjustment of the rate of addition of triethylamine is recommended to prevent sudden decomposition of the formate. Because the reaction does not require toxic gaseous CO or pressure-resistant apparatus, the experimental procedure disclosed herein is safer and more practical than those that use CO gas or other CO surrogates, as mentioned above.

Furthermore, 2,4,6-trichlorophenyl formate can be regarded as "weighable CO." This additional merit is ascribed to the ease of adjusting the amount of CO. Its chemical nature to produce CO under very mild conditions has potential to be applied to various organic reactions using CO as a reactant or a ligand.

Because products obtained via this method are esters with highly electrophilic nature, further derivatization can easily be achieved. As described in our previous paper,[14] trichlorophenyl esters can react with a slight excess quantity of nucleophiles, resulting in successful conversion into various compounds, including alkyl esters, thioesters, carboxylic acids, and amides. One-pot syntheses from (pseudo)halides to amides have also been demonstrated.[13] Analogous one-pot reactions have been shown to be feasible with other nucleophiles. When a nucleophile is used to release 2,4,6-trichlorophenol as a by-product, simple washing with diluted aqueous NaOH solution is sufficient for removal of 2,4,6-trichlorophenol. This characteristic further increases the practicality of the reaction protocol.

References

1. School of Pharmaceutical Sciences, University of Shizuoka, 52-1 Yada, Suruga-ku, Shizuoka 422–8526, Japan. E-mail: manabe@u-shizuoka-ken.ac.jp
2. (a) Schoenberg, A.; Bartoletti, I.; Heck, R. F. *J. Org. Chem.* **1974**, *39*, 3318–3326. (b) Brennführer, A.; Neumann, H.; Beller, M. *Angew. Chem. Int. Ed.* **2009**, *48*, 4114–4133.
3. Morimoto, T.; Kakiuchi, K. *Angew. Chem. Int. Ed.* **2004**, *43*, 5580–5588.
4. Fuji, K.; Morimoto, T.; Tsutsumi, K.; Kakiuchi, K. *Angew. Chem. Int. Ed.* **2003**, *42*, 2409–2411.
5. Cacchi, S.; Fabrizi, G.; Goggiamani, A. *Org. Lett.* **2003**, *5*, 4269–4272.
6. (a) Schareina, T.; Zapf, A.; Cotte, A.; Gotta, M.; Beller, M. *Adv. Synth. Catal.* **2010**, *352*, 1205–1209. (b) Ko, S.; Lee, C.; Choi, M. G.; Na, Y.; Chang, S. *J. Org. Chem.* **2003**, *68*, 1607–1610. (c) Carpentier, J. F.; Castanet, Y.; Brocard, J.; Mortreux, A.; Petit, F. *Tetrahedron Lett.* **1991**, *32*, 4705–4708.
7. Odell, L. R.; Russo, F.; Larhed, M. *Synlett* **2012**, 685–698.
8. Hermange, P.; Lindhardt, A. T.; Taaning, R. H.; Bjerglund, K.; Lupp, D.; Skrydstrup, T. *J. Am. Chem. Soc.* **2011**, *133*, 6061–6071.
9. Friis, S. D.; Taaning, R. H.; Lindhardt, A. T.; Skrydstrup, T. *J. Am. Chem. Soc.* **2011**, *133*, 18114–18117.
10. Korsager, S.; Taaning, R. H.; Skrydstrup, T. *J. Am. Chem. Soc.* **2013**, *135*, 2891–2894.
11. Ueda, T.; Konishi, H.; Manabe, K. *Org. Lett.* **2012**, *14*, 3100–3103.
12. Tsuji *et al.* also reported the Pd-catalyzed esterification of aryl halides using aryl formates: Fujihara, T.; Hosoki, T.; Katafuchi, Y.; Iwai, T.; Terao, J.; Tsuji, Y. *Chem. Commun.* **2012**, *48*, 8012–8014.
13. Ueda, T.; Konishi, H.; Manabe, K. *Tetrahedron. Lett.* **2012**, *53*, 5171–5175.
14. Ueda, T.; Konishi, H.; Manabe, K. *Org. Lett.* **2012**, *14*, 5370–5373.

15. Our procedure in the synthesis of 2,4,6-trichlorophenyl formate produces yields higher than that in a previous report: van Es, A.; Stevens, W. *Recl. Trav. Chim. Pays-Bas* **1965**, *84*, 1247.

Appendix
Chemical Abstracts Nomenclature (Registry Number)

β-Tetralone: 2(1*H*)-Naphthalenone, 3,4-dihydro-; (530-93-8)
Potassium *tert*-butoxide: 2-Propanol, 2-methyl-, potassium salt (1:1); (865-47-4)
N-Phenylbis(trifluoromethanesulfonimide): Methanesulfonamide, 1,1,1-trifluoro-*N*-phenyl-*N*-[(trifluoromethyl)sulfonyl]-; (37595-74-7)
3,4-Dihydronaphthalen-2-yl trifluoromethanesulfonate: Methanesulfonic acid, 1,1,1-trifluoro-, 3,4-dihydro-2-naphthalenyl ester; (143139-14-4)
Formic acid; (64-18-6)
Acetic anhydride: Acetic acid, 1,1'-anhydride; (108-24-7)
2,4,6-Trichlorophenol: Phenol, 2,4,6-trichloro-; (88-06-2)
Sodium acetate: Acetic acid, sodium salt (1:1); (127-09-3)
2,4,6-Trichlorophenyl formate: Phenol, 2,4,6-trichloro-, 1-formate; (4525-65-9)
Palladium acetate: Acetic acid, palladium(2+) salt (2:1); (3375-31-3)
Xantphos: Phosphine, 1,1'-(9,9-dimethyl-9*H*-xanthene-4,5-diyl)bis[1,1-diphenyl- ; (161265-03-8)
2,4,6-Trichlorophenyl 3,4-dihydronaphthalene-2-carboxylate: 2-Naphthalenecarboxylic acid, 3,4-dihydro-, 2,4,6-trichlorophenyl ester; (1402012-58-1)

Kei Manabe was born in Kanagawa, Japan. He completed his doctoral work in 1993 at the University of Tokyo. After working as a postdoctoral fellow at Columbia University, USA, he went back to the University of Tokyo and worked as an Assistant Professor, Lecturer, and Associate Professor. In 2005, he moved to RIKEN as an Initiative Research Scientist. He joined the faculty at the University of Shizuoka as a Professor in 2009. His research interests include development of new catalytic reactions for organic synthesis.

Hideyuki Konishi was born in Takamatsu, Japan in 1979. He obtained his Ph.D. degree in pharmaceutical sciences at the University of Tokyo in 2008 under the direction of Professor Shu Kobayashi. He carried out his postdoctoral research in Professor Viresh H. Rawal's laboratory at the University of Chicago. In 2009, he became a Research Assistant Professor in the group of Professor Kei Manabe at the University of Shizuoka. His research interests include development of practical and efficient catalytic reactions for construction of pharmaceutically and synthetically important compounds.

Tsuyoshi Ueda was born in Ishikawa prefecture, Japan, in 1978. He received his M.S. degree in 2003 in Industrial Chemistry from Meiji University and subsequently joined the Department of Process Development at Sankyo Co., Ltd. He completed his Ph.D. under the guidance of Professor Kei Manabe at University of Shizuoka in 2013. He is currently an Associate Senior Researcher in Daiichi Sankyo Co., Ltd., working on the development of practical synthetic methods and large-scale synthesis of active pharmaceutical ingredients.

Jennifer Ciesielski received her B.A. degree in Chemistry from Lake Forest College in 2007. She then moved to University of Rochester where she obtained her Ph.D. in 2012. She is currently a National Science Foundation postdoctoral fellow with Professor Erick M. Carreira at the ETH Zürich.

Rhodium-Catalyzed Direct Amination of Arene C-H Bonds Using Azides as the Nitrogen Source

Sae Hume Park,[†§] Yoonsu Park,[§†] and Sukbok Chang[§†]*

§Center for Catalytic Hydrocarbon Functionalizations, Institute of Basic Science (IBS), Daejeon 305-701, Republic of Korea
†Department of Chemistry, Korea Advanced Institute of Science and Technology (KAIST), Daejeon, 305-701, Republic of Korea

Checked by Kazuma Amaike, Kolby Lyn White, and Mohammad Movassaghi

Procedure

A. *N-(Benzo[h]quinolin-10-yl)-4-dodecylbenzenesulfonamide of compound (1).* A flame-dried, 100 mL, two-necked round-bottomed flask (Note 1) is equipped with a 3-cm football-shaped stir bar, dichloro(η^5-pentamethylcyclopentadienyl)-rhodium(III) dimer (124 mg, 0.20 mmol, 1.00 mol%), silver hexafluoroantimonate(V) (275 mg, 0.80 mmol, 4.00 mol%) (Note 2), and benzo[h]quinoline (3.58 g, 20 mmol) (Note 3). One neck of the flask is equipped with a reflux condenser (Note 1) and another neck is closed with a rubber septum. The flask is evacuated and then back-filled with dry N_2, a process that is performed three times. Distilled 1,2-dichloroethane (20 mL) (Note 4) and *p*-dodecylbenzenesulfonyl azide (7.36 mL, 22 mmol, 1.10 equiv) (Note 3) are added sequentially via syringe to the flask (Note 5). Under a continuous flow of N_2 gas by balloon (should not use balloons!), the mixture is placed in a preheated oil bath at 80 °C and stirred for 12 h (Note 6). After the reaction is complete as judged by TLC analysis (Note 7), the reaction mixture is cooled to room temperature and then filtered through a silica pad (19.0 g) washing with dichloromethane (5 x 20 mL) (Note 8) into a round-bottomed flask (Note 9). The filtrate is concentrated using a rotary evaporator (20 mmHg, water bath temperature 30 °C), and is purified by recrystallization from dichloromethane (5 mL) and *n*-hexane (100 mL) (Note 8) at 0 °C for 12 h (Note 10). Precipitate is filtered through a Büchner funnel (Whatman filter paper, 70 mm diameter) and washed with *n*-hexane (5 x 20 mL) at 25 °C to yield a yellowish solid. The filtered solid is transferred to a 20-mL glass vial with spatula and dried under vacuum at room temperature for 12 h to give an air stable yellow solid product **1** (7.74 g, 77%) (Note 11).

Notes

1. All glassware was flame-dried under vacuum and allowed to cool under an atmosphere of nitrogen.
2. Dichloro(η^5-pentamethylcyclopentadienyl)rhodium(III) dimer (99%) was purchased from Strem Chemicals Co., Ltd., and silver hexafluoroantimonate(V) (98%) was purchased from Aldrich Chemical Co., and both reagents were used as received.
3. Benzo[*h*]quinoline (99%) and *p*-dodecylbenzenesulfonyl azide (soft type, mixture) were purchased from Tokyo Chemical Industry Co., Inc and were used as received.
4. The checkers purchased 1,2-dichloroethane from Aldrich Chemical Co., and the liquid was distilled from calcium hydride before use.
5. The reflux condenser was used for heating the reaction mixture around the boiling point of 1,2-dichloroethane. However, the submitters obtained a similar range of product yields without using a reflux condenser, in which the flask was equipped with a rubber septum and kept under a N_2 atmosphere (balloon).
6. The reaction mixture was stirred at room temperature for 10 min before placing in a pre-heated oil bath (80 °C).
7. The progress of the reaction was monitored by TLC. For TLC analysis, silica gel 60 F_{254} TLC plates (EMD Chemicals, Inc.) were used.
8. The submitters purchased dichloromethane, ethyl acetate, and *n*-hexane from Junsei Chemical Co. Ltd. and used as received.

9. Silica gel 60 (0.040-0.063 mm) (230-400 mesh ASTM) was purchased from Aldrich Chemical Co. Inc., and used as received. A glass filter (60-mL) with a medium porosity fritted disc was used.

10. While a liquid product was first obtained upon rotary evaporation of solvent, crystallization of the crude residue yielded crystalline product.

11. The submitters and checkers obtained product yields in the range of 7.50~8.04 g (75~80%) without further purification. The product (a mixture of isomeric compounds in the dodecyl moiety) exhibits the following physicochemical properties, N-(benzo[h]quinolin-10-yl)-4-dodecylbenzenesulfonamide (1): R_f = 0.30 (20% ethyl acetate/n-hexane); mp 50–52 °C (decomp); ^1H NMR (500 MHz, CDCl$_3$) δ: 0.50–1.58 (m, 24H), 2.21–2.60 (m, 1H), 6.98–7.10 (m, 2H), 7.46–7.60 (m, 4H), 7.69 (d, J = 8.5 Hz, 1H), 7.72–7.82 (m, 2H), 7.84–7.93 (m, 1H), 8.18 (d, J = 8.0 Hz, 1H), 8.92 (d, J = 2.0 Hz, 1H), 15.06–15.28 (m, 1H); ^{13}C NMR (125 MHz, CDCl$_3$) δ: 11.8, 13.7, 13.8, 13.89, 13.93, 14.0 (3C), 20.3, 21.6, 22.2, 22.39, 22.42, 22.45, 22.47, 22.52, 26.8, 27.1, 27.2, 27.3, 28.9, 29.0, 29.07, 29.13, 29.2, 29.27, 29.31, 29.36, 29.4 (2C), 31.4, 31.5, 31.6, 31.67, 31.7, 31.8, 35.9, 36.0, 36.28, 36.34, 37.8, 38.6, 39.6, 45.5, 45.7, 47.5, 116.0, 116.4, 116.6, 117.1, 117.3 (2C), 121.0, 122.6, 122.7, 122.8 (2C), 125.2 (2C), 126.99, 127.01, 127.1 (2C), 127.2 (2C), 127.8 (2C), 127.9, 128.4 (2C), 128.7 (2C), 134.9 (2C), 136.5 (2C), 137.4, 137.5 (2C), 138.4 (2C), 145.5, 146.96 (2C), 146.99, 151.3, 151.5, 151.62, 152.9. IR (NaCl) v 2927, 2855, 1593, 1577, 1465, 1402, 1339 cm^{-1}; HRMS (EI) m/z calcd. for C$_{31}$H$_{38}$N$_2$O$_2$S [M+H]$^+$: 503.2727, found: 503.2728; Elemental analysis; calcd. for C$_{31}$H$_{38}$N$_2$O$_2$S: C, 74.06; H, 7.62; N, 5.57; S, 6.38; Found: C, 74.05; H, 7.52; N, 5.52; S, 6.50.

Handling and Disposal of Hazardous Chemicals

The procedures in this article are intended for use only by persons with prior training in experimental organic chemistry. All hazardous materials should be handled using the standard procedures for work with chemicals described in references such as "Prudent Practices in the Laboratory" (The National Academies Press, Washington, D.C., 2011 www.nap.edu). All chemical waste should be disposed of in accordance with local regulations. For general guidelines for the management of chemical waste, see Chapter 8 of Prudent Practices.

These procedures must be conducted at one's own risk. *Organic Syntheses, Inc.*, its Editors, and its Board of Directors do not warrant or guarantee the safety of individuals using these procedures and hereby disclaim any liability for any injuries or damages claimed to have resulted from or related in any way to the procedures herein.

Discussion

Aryl amines and aryl amides are a key synthetic building unit widely utilized in organic synthesis, coordination chemistry, materials science and pharmaceutical industry.[2] Currently, preparative routes to those compounds require either pre-functionalized reactants (aryl halides or haloamides) or external oxidants, and, therefore, the present procedures inevitably generate stoichiometric amounts of side products.[3] During the course of our efforts on developing new direct C-N bond formation reactions from arenes,[4] we found that an attractive nitrogen source are azides, which are able to be introduced into the arenes via direct C-H bond activation. We report herein a practical procedure of the direct C-H amination and amidation reaction of arenes using aryl sulfonyl azides as the source of the nitrogen functionality, which releases molecular nitrogen as the single by-product.[5]

This reaction is catalyzed under mild conditions by a cationic rhodium complex $[Cp^*Rh(III)]^{2+}(SbF_6)_2^{2-}$ (Cp^*, η^5-pentamethylcyclopentadienyl) without requiring external oxidants. A broad range of chelate group-containing arenes were selectively aminated with excellent functional group tolerance. Preliminary mechanistic studies suggest that the carbon-nitrogen (C-N) bond formation proceeds via rate-limiting rhodium-mediated C-H bond activation of arenes followed by azide insertion into a rhodacycle intermediate.[5]

This procedure offers a very efficient and practical route to aryl amines and amides. Although additional optimization processes may be required for each substrate, synthetically acceptable product yields were obtained for wide range of arenes and azides investigated as demonstrated in Table 1.

Table 1. Direct C-H amination of arenes with sulfonyl- and aryl azides [a]

Entry	Substrate	Azide (R²-N₃)	Product	Yield (%)[b]
1[c]		Me—⟨⟩—SO_2N_3		89
2[d]		F_3C—⟨⟩—N_3, CF_3		82
3	R¹ = H			83
4	Me	Me—⟨⟩—SO_2N_3		86
5	CHO		R¹—NHTs	82
6	CF₃			82
7		$AcHN$—⟨⟩—SO_2N_3	NHR²	71
8[d]		F_3C—⟨⟩—N_3, CF_3	NHR²	80
9		Me—⟨⟩—SO_2N_3		61
10	R¹ = H	Me—⟨⟩—SO_2N_3		60
11	Me		R¹—NHTs	79

[a] Reaction conditions: arene (5.0 mmol), azide (5.5 mmol), [RhCp*Cl₂]₂ (3.0 mol %), AgSbF₆ (12.0 mol %) in 1,2-dichloroethane (10 mL) at 80 °C for 24 h. [b] Isolated yield. [c] Scale of 20.0 mmol. [d] Scale of 1.0 mmol.

References

1. Center for Catalytic Hydrocarbon Functionalizations (IBS) and Department of Chemistry, Korea Advanced Institute of Science and

Technology (KAIST), Daejeon, 305-701, Republic of Korea, E-mail: sbchang@kaist.ac.kr.

2. Amino Group Chemistry, From Synthesis to the Life Sciences (Ed.: Ricci, A.), Wiley-VCH, Weinheim, **2007**.
3. (a) Kosugi, M.; Kameyama, M.; Migita, T. *Chem. Lett.* **1983**, *12*, 927. (b) Paul, F.; Patt, J.; Hartwig, J. F. *J. Am. Chem. Soc.* **1994**, *116*, 5969. (c) Guram, A. S.; Buchwald, S. L. *J. Am. Chem. Soc.* **1994**, *116*, 7901. (d) Evano, G.; Blanchard, N.; Toumi, M. *Chem. Rev.* **2008**, *108*, 3054.
4. (a) Cho, S. H.; Kim, J. Y.; Lee, S. Y.; Chang, S. *Angew. Chem., Int. Ed.* **2009**, *48*, 9127. (b) Kim, J. Y.; Cho, S. H.; Joseph, J.; Chang, S. *Angew. Chem., Int. Ed.* **2010**, *49*, 9899. (c) Cho, S. H.; Yoon J.; Chang, S. *J. Am. Chem. Soc.* **2011**, *133*, 5996. (d) Kim, H. J.; Kim, J.; Cho, S. H.; Chang, S. *J. Am. Chem. Soc.* **2011**, *133*, 16382. (e) Cho, S. H.; Kim, J. Y.; Kwak, J.; Chang, S. *Chem. Soc. Rev.* **2011**, *40*, 5068.
5. (a) Kim. J. Y.; Park, S. H.; Ryu, J.; Cho, S. H.; Kim, S. H.; Chang, S. *J. Am. Chem. Soc.* **2012**, *134*, 9110. (b) Ryu, J.; Shin, K.; Park, S. H.; Kim, J. Y.; Chang, S. *Angew. Chem., Int. Ed.* **2012**, *51*, 9904.

Appendix
Chemical Abstracts Nomenclature (Registry Number)

Silver hexafluoroantimonate(V) (26042-64-8)

Dichloro(η^5-pentamethylcyclopentadienyl)rhodium(III) dimer (12354-85-7)

Benzo[*h*]quinoline (230-37-3)

p-Dodecylbenzenesulfonyl azide (79791-38-1)

Professor Sukbok Chang received his B.S. degree from Korea University (1985), M.S. from KAIST (1987) and Ph.D. from Harvard University (1996) under the guidance of Professor Eric N. Jacobsen. He was a postdoctoral fellow in the lab of Professor Robert H. Grubbs at California Institute of Technology from 1996 to 1998, and became an assistant professor at Ewha Womans University (Seoul, Korea) in 1998. In 2002, he moved to the present position at KAIST and was promoted to professor 2007. He received the Award for Young Chemists (KCS-Wiley) of the Korea Chemical Society (2002), Organic Chemistry Division Award (2005), and Korean Chemical Society Academic Award (2010). His recent research interests include the development of new synthetic methods via metal catalysis.

Sae Hume Park was born in Princeton, New Jersey, United States of America, in 1985. He received his B.S. degree in chemistry from Sungkyunkwan University (2008). He is now in M.S.-Ph.D. joint course at Korea Advanced Institute of Science and Technology (KAIST) under the guidance of Professor Sukbok Chang. Currently, he is focused on researching metal-catalyzed oxidative C-H bond activation to make C-C and C-N bond formation.

Yoonsu Park was born in Suwon, Korea, in 1992. He is an undergraduate student in Korea Advanced Institute of Science and Technology (KAIST), and now performing undergraduate research at KAIST under the guidance of Professor Sukbok Chang. Currently, he is interested in discovering C-H bond activation reactions to make biologically active compounds

Kazuma Amaike is pursuing his graduate studies in Professor Kenichiro Itami's group at Nagoya University, Nagoya, Japan. His studies focus on a range projects related to C–H activation and the synthesis of natural products. In 2013, he joined the laboratory of Professor Mohammad Movassaghi at MIT as a visiting graduate student via the National Science Foundation CCI Center for selective C–H functionalization.

Kolby Lyn White pursued her undergraduate studies at Northwestern University, where she received her B.A./M.S. degree in Chemistry in 2011. While an undergraduate, she worked in the laboratory of Professor Karl Scheidt. In 2011, she joined the laboratory of Professor Mohammad Movassaghi at MIT for her graduate studies, which is focused on the total syntheses of alkaloid natural products.

Synthesis of Alkynyliodonium Salts: Preparation of Phenyl(phenylethynyl)iodonium Trifluoroacetate

Luke I. Dixon,[1] Michael A. Carroll,[1*] George J. Ellames[2] and Thomas J. Gregson[2]

[1]School of Chemistry, Newcastle University, Newcastle upon Tyne, NE1 7RU, UK; [2]Department of Isotope Chemistry and Metabolite Synthesis, Sanofi-Aventis, Alnwick, NE66 2JH, UK.

Checked by David M. Guptill and Huw M. L. Davies

$$PhI(OAc)_2 \xrightarrow[\substack{\text{2. Room temp., 1 h} \\ \text{3. Phenylacetylene, 3.5 h}}]{\substack{\text{1. TFA, DCM, -30 °C,} \\ \text{0.5 h}}} Ph\!-\!\!\equiv\!\!-\overset{\oplus}{\underset{Ph}{I}} \quad {}^{\ominus}O_2CCF_3 \quad \mathbf{1}$$

Procedure

Caution! Hypervalent iodine compounds are potentially explosive and should be handled taking appropriate precautions.

Phenyl(phenylethynyl)iodonium trifluoroacetate (1) (Note 1). A 250-mL, two-necked, round-bottomed flask (Note 2), equipped with a 35 × 16 mm, PTFE-coated, oval magnetic stirrer bar, is charged with phenyliodo bis(acetate) (6.44 g, 20 mmol) (Note 3) followed by DCM (120 mL) (Note 4). The suspension is then stirred (Note 5) until dissolution is observed, at which point the solution is then cooled to between −30 °C and −35 °C using an acetone/dry ice bath (Note 6). Trifluoroacetic acid (3.06 mL, 40 mmol, 2.0 equiv) is added drop-wise over a period of 10 min via syringe pump and plastic syringe (Note 7) and stainless steel needle (20 gauge). The solution is then stirred for a further 0.5 h keeping the temperature between −30 °C and −35 °C before the reaction flask is removed from the acetone/dry ice bath

Org. Synth. **2014**, *91*, 60-71
DOI: 10.15227/orgsyn.091.0060

Published on the Web 11/19/2013
© 2014 Organic Syntheses, Inc.

and stirred at room temperature for 1 h. At this point the solution is clear and colorless in appearance. Phenylacetylene (2.20 mL, 20 mmol, 1.0 equiv) is then added to the solution by syringe, as above, over a period of 10 min, and the reaction vessel is wrapped in aluminium foil and stirred in darkness (Note 8) for 3.5 h (Notes 9 and 10) at room temperature. The resulting clear pale yellow solution is then transferred to a 250-mL, single-necked, round-bottomed flask and concentrated to a volume of 30–40 mL (Note 11) *in vacuo* on a rotary evaporator, keeping the water bath at 25 °C. Diethyl ether (40 mL) followed by petroleum ether 40–60 (80 mL) are then added to the solution, mixing by 'swirling' manually, to initiate crystallization of the product, visible by the appearance of a white precipitate. The suspension is then stoppered before being wrapped in foil, placed in the freezer and kept in darkness at −25 °C for 48 h (Note 12). The flask is then removed from the freezer and left for 30 min to warm to room temperature before filtration of the white crystalline solid by suction using a medium-porosity sintered glass funnel and washed with diethyl ether (2 × 15 mL), agitating the crystals quickly with a spatula before the removal of each portion of diethyl ether (Note 13). The solid is then transferred to a pre-weighed vial and dried *in vacuo* for 12–24 h in a vacuum desiccator to give the product as a white crystalline solid 3.98 – 4.59 g, (48 – 55%) (Notes 14 and 15). Purity of the product can be determined by ^{1}H NMR initially, though, due to decomposition in solution (Notes 9 and 10), elemental analysis provides a more accurate indication of purity (Note 16). The described procedure has also been shown to be suitable for larger scale syntheses (Note 17).

Notes

1. Iodonium salts have, in certain cases, been reported to be explosive.[3] In the case of **1**, TGA and DSC data (*vide infra*) suggests that the material is not explosive, but it does experience a little decomposition around its melting point. Caution should, however, be taken in the synthesis and isolation of **1**, conducting all procedures behind a blast shield (e.g. 6 mm Perspex®) wherever possible, in addition to the protection afforded by the sash of a fume hood.

2. All reaction glassware was dried in an oven overnight (16 h) at 178 °C (submitters) or flame-dried (checkers) then evacuated under vacuum and back-filled with nitrogen or argon three times while hot, then

maintained under the inert atmosphere during the course of the reaction.

3. Checkers obtained (diacetoxyiodo)benzene (98%), phenylacetylene (98%) and trifluoroacetic acid (ReagentPlus, 99%) from Aldrich Chemical Company, and the reagents were used as received. Submitters obtained phenyliodo bis(acetate) (98+%) and phenylacetylene (98+%) from Alfa Aesar and trifluoroacetic acid (99.5%) was obtained from Apollo Scientific; all chemicals were used as supplied from new, unopened bottles.

4. Checkers obtained dichloromethane (>99.5%), diethyl ether (Laboratory grade) and petroleum ether (Certified ACS) from Fisher Scientific and were used as supplied. Submitters obtained dichloromethane (99+%), diethyl ether (99+%) and petroleum ether (40-60 °C) from Fisher and were used as supplied from new, unopened bottles.

5. Magnetic stirrer hotplates, 'IKAMAG® RCT Classic', stirred at ~633 rpm, setting 6.

6. Acetone/dry ice baths were also equipped with a 15 mm, PTFE-coated, cross-shaped, stir bar to ensure even temperature distribution.

7. Checkers purchased plastic syringes (Norm-ject®, 5 mL) from VWR. Submitters purchased plastic syringes ('BD Discardit™ II', 5 mL) from Fisher Scientific.

8. Alkynyliodonium salts have previously been shown to exhibit photoinitiation behavior under irradiative conditions[4] and should therefore be regarded as photosensitive.

9. Submitters noted that compound (1) decomposes in solution to 2-phenyl-2-oxoethyl trifluoroacetate (2), the major by-product of the reaction; prolonged reaction times encourage formation of this by-product.

10. Submitters provided the following characterization data and procedure for the preparation of by-product (2) (Scheme 1). This method was not repeated by the Checkers. *2-Phenyl-2-oxoethyl trifluoroacetate (2)*. Phenyl (phenylethynyl)iodonium trifluoroacetate (1) (2.07 g, 4.96 mmol) is added to a 100-mL, single-necked, round-bottomed flask fitted with a water condenser and equipped with a 2 cm, PTFE-coated, oval magnetic stirrer bar. Chloroform (10 mL) and a few drops of water (*ca* 4 drops) are then added. The apparatus was equipped with a Schlenk line and the solution heated to reflux (70 °C, oil bath temperature) for 2 h under an atmosphere of nitrogen with stirring. The solution is then cooled,

dried with MgSO₄, filtered by suction, using a sintered glass funnel, then concentrated *in vacuo*. The crude product is then purified by flash chromatography (silica), eluting with dry hexane/dry toluene (1:1)

Ph—≡—I⁺(Ph) O₂CCF₃⁻ **1** →[H₂O, CHCl₃][2 h, 70 °C] Ph—C(O)—CH₂—O—C(O)—CF₃ **2**

Scheme 1. Preparation of 2-phenyl-2-oxoethyl trifluoroacetate

(both dried by storing over sodium wire), to give the product as a white crystalline solid (0.40 g, 1.71 mmol, 34%). TLC R_f = 0.27 (1:1, hexane: toluene), mp = 54 °C – 55 °C (DCM: diethyl ether: petroleum ether 40 – 60), IR (v[cm⁻¹]) 1785, 1700, 1600, 1451, 1433, 1384, 1356, 1304, 1287, 1228, 1206, 1154, 1078, 1037, 1027, 1001; ¹H NMR (300 MHz, CDCl₃) δ : 5.60 (s, 2 H, CH₂), 7.54 (t, 2 H, J = 7.8 Hz, H-3′ , H-5′), 7.68 (tt, 1 H, J = 7.5, 1.8 Hz, H-4′), 7.92 (dd, 2 H, J = 8.4, 1.2 Hz, H-2′ , H-6′); ¹³C NMR (75 MHz, CDCl₃) □: 68.6 (CH₂), 114.9 (q, J = 285.3 Hz, CF₃), 128.2 (C-3□, C-5□), 129.5 (C-2□, C-6□), 133.7 (C-1□), 134.9 (C-4□), 157.5 (q, J = 43.0 Hz, (CO)CF₃); ¹⁹F NMR (376 MHz, CDCl₃) □: −74.1; ACPI-MS m/z: 231(32, [M-H]⁻), 227(100), 203(14), 183(5), 134(20); HRMS (ACPI): [M-H]⁻ calcd. for C₁₀H₆F₃O₃: 231.0275. Found 231.0273; Anal. calcd. for C₁₀H₇F₃O₃: C, 51.74; H, 3.04. Found: C, 51.66; H, 2.98.

11. The volume in the flask was assessed by filling the flask with 35 mL of DCM prior to its use and marking the level with an indelible marker.

12. The total time taken from the end of the 3.5 h stirring period to placing the flask in the freezer was no longer than 30 minutes.

13. The submitters noted that the mother liquor, including all ether washings, can be 'seeded' with a small amount of the pure product and replaced in the freezer (−25 °C) for 7 days to recover a second batch of the product using method outlined above; recovery is 0.55 – 1.47 g.

14. *Phenyl(phenylethynyl)iodonium trifluoroacetate (1)*: mp = 90 – 92 °C (dec.); ¹H NMR (600 MHz, CDCl₃) δ: 7.37 (t, J = 7.5 Hz, 2 H), 7.43 (t, J = 7.5 Hz, 1 H), 7.46-7.52 (m, 4 H), 7.60 (t, J = 7.4 Hz, 1 H), 8.17 (d, J = 8.0 Hz, 2 H); ¹³C NMR (100 MHz, CDCl₃) δ: 45.2, 103.9, 115.6 (q, J = 292 Hz), 120.5, 120.6, 128.7, 130.8, 131.9, 132.1, 132.9, 133.6, 162.7 (q, J = 36.2 Hz); IR (solid): 3077, 2166, 1666, 1185, 1124 cm⁻¹; HRMS (NSI): m/z calcd. for

$C_{14}H_{10}I$ ([M - TFA]$^+$): 304.9822. Found: 304.9821; Anal. calcd. for $C_{16}H_{10}F_3IO_2$: C, 45.96; H, 2.41. Found: C, 46.14; H, 2.35.

15. Submitters reported yields of 4.68 – 5.05 g (56 – 60%). Submitters provided the following additional characterization data for product (**1**): ^{19}F NMR (376 MHz, [D$_6$]DMSO) □: –73.6.

16. Samples (30 mg) were put in small glass vials then wrapped in foil and sent by post for elemental analysis.

17. Submitters also performed the reaction in triplicate on a 100 mmol scale showing some improvements in yield, to provide 22.80 – 24.74 g from first batch followed by 1.08 – 2.30 g (total yield: 23.87 – 26.01 g, 57 – 62%).

Handling and Disposal of Hazardous Chemicals

The procedures in this article are intended for use only by persons with prior training in experimental organic chemistry. All hazardous materials should be handled using the standard procedures for work with chemicals described in references such as "Prudent Practices in the Laboratory" (The National Academies Press, Washington, D.C., 2011 www.nap.edu). All chemical waste should be disposed of in accordance with local regulations. For general guidelines for the management of chemical waste, see Chapter 8 of Prudent Practices.

These procedures must be conducted at one's own risk. *Organic Syntheses, Inc.*, its Editors, and its Board of Directors do not warrant or guarantee the safety of individuals using these procedures and hereby disclaim any liability for any injuries or damages claimed to have resulted from or related in any way to the procedures herein.

Discussion

Alkynyliodonium salts are highly versatile reactive intermediates[5] and as such, have been used in a number of cycloaddition reactions[6] including 1,3-dipolar cycloadditions with nitrile-oxides,[7] diazoketones[8] and azides[8] as well as Diels-Alder chemistry.[9] Further syntheses of several complex ring systems from alkynyliodonium salts have been reported *via* carbene insertion reactions (Scheme 2).[10]

Scheme 2. Synthesis of heteroaromatics from alkynyliodonium salts[10b, 10d-m]

In addition, alkynyliodonium salts have been exploited in terms of their photosensitivity,[4] oxidizing nature and, most commonly, their potential for alkynylation reactions.[5, 11]

The most common methods for the synthesis of alkynyliodonium salts are *via* the use of Koser's Reagent ([hydroxy(tosyloxy)iodo]benzene, **3**)[12] and Stang's Reagent ([cyano(trifloxy)iodo]benzene, **5**).[13]

Koser's reagent is commercially available and non-toxic making it a highly attractive precursor to alkynyliodonium salts, however, it has been shown that the yields can be low due to formation of the corresponding vinylic species as a result of tosylate addition.[14] It was shown that alkynes with bulky groups afforded the desired product in high yields, whereas those with less steric hindrance in the □-acetylenic position afforded high amounts of, or even exclusively, the addition products, phenyl[□-(tosyloxy)vinyl]iodonium tosylates.[14a, 15]

$$\underset{\textbf{3}}{\overset{OTs}{\underset{OH}{Ph-I}}} + R\!\!=\!\!\!=\!\!\!-H \xrightarrow[\text{-78 °C to rt}]{\text{DCM, dessicant}} \underset{\textbf{4 (12-67\%)}}{R\!\!=\!\!\!=\!\!\!-\overset{\oplus}{\underset{Ph}{I}}}\ \ ^{\ominus}OTs$$

R = Ph (**a**), *n*-Bu (**b**), *sec*-Bu (**c**), *t*-Bu (**d**), Me (**e**), 4-ClC$_6$H$_4$ (**f**), 4-MeOC$_6$H$_4$ (**g**), 4-C$_8$H$_{17}$OC$_6$H$_4$ (**h**), 4-C$_{10}$H$_{21}$OC$_6$H$_4$ (**i**), 4-C$_{12}$H$_{25}$OC$_6$H$_4$ (**j**), 4-C$_{14}$H$_{29}$OC$_6$H$_4$ (**k**).

Scheme 3. Synthesis of alkynyliodonium tosylates from Koser's Reagent[7b, 16]

The addition of a desiccant to the reaction mixture has proven advantageous allowing formation of a range of novel analogues in spite of low steric bulk in the □-acetylenic position (Scheme 3).[16a]

Though triflate anions have also been shown to add,[17] the lower nucleophilicity makes alkynliodonium triflates much less susceptible to decomposition and thus Stang's reagent has found a much wider application in the synthesis of alkynyliodonium salts.[18]

Stang demonstrated the extensive range of functionality possible in the □-alkynyl position of alkynyliodonium salts from the Stang reagent and, although compounds **6b–w** (Scheme 4) had to be isolated at low temperature, the synthetic route proved that alkynyliodonium triflates could be synthesized bearing electron-withdrawing □-alkynyl functionality, without significant triflate addition to the □-alkynyl position.[18a]

OTf
Ph—I   R——SnBu₃   ⊖ OTf
     | ————————→   R——⊕ I
CN   DCM, -40 °C, 15 min |
                                           Ph
5   **6** (42-89%)

R = Me (**a**), *t*-BuC(O) (**b**), CN (**c**), *p*-MeC₆H₄SO₂ (**d**), MeO(CO) (**e**), MeOCH₂ (**f**), PhC(O) (**g**), (CH₃)₂NC(O) (**h**), Cl (**i**), BrCH₂ (**j**), ClCH₂ (**k**), 1-cyclohexenyl (**l**), adamantylC(O) (**m**), *t*-pentylC(O) (**n**), cyclopropylC(O) (**o**), furan-2-ylC(O) (**p**), thiophen-2-ylC(O) (**q**), N-piperidinylC(O) (**r**), N-azepanylC(O) (**s**), N-morpholinylC(O) (**t**), cyclobutylC(O) (**u**), cyclohexylC(O) (**v**), N-pyrrolidinylC(O) (**w**)

Scheme 4. Synthesis of alkynyliodonium triflates bearing electron-withdrawing β-alkynyl functionality from Stang's reagent[9b, 10c 18]

It was noted in 1993[9b] that many functionalised alkynyliodonium salts could not be prepared in satisfactory manner from Koser's reagent, Zefirov's reagent, iodosylarenes or related compounds. Stang's reagent, however, has made possible the synthesis of the most comprehensive library of over forty highly functionalized alkynyliodonium salts to date.[9b, 10c, 11a, 18a, 19]

Despite their widespread use, a number of problems remain to be resolved with regard to the synthesis and isolation of alkynyliodonium salts:

• Many procedures require use of the stannous alkynyl-derivatives creating problems with isolation and toxicity issues.

- The most versatile and commonly used of the available precursors, Stang's Reagent,[13] produces the noxious asphyxiant HCN as a by-product.
- Many syntheses require use of iodosylbenzene (PhIO), which is known to disproportionate to the highly explosive iodylbenzene (PhIO$_2$).[20]

Trifluoroacetate analogues have previously been reported as too unstable and hydroscopic to be of practical interest.[21] We have found that not only are they stable isolatable crystalline solids, but they also address many of the problems associated with other syntheses, by the use of unmodified, terminal alkynes and commercially available, non-explosive starting materials with the production of no noxious by-products. The yields of the reaction are also comparable to those of other approaches and the purification is by far the simplest we have experienced.

References

1. School of Chemistry, Newcastle University, Newcastle upon Tyne, NE1 7RU, UK
2. Department of Isotope Chemistry and Metabolite Synthesis, Sanofi-Aventis, Alnwick, NE66 2JH, UK. *Current address*: Department of Isotope Chemistry, Covance Laboratories, Alnwick, Northumberland, UK, NE66 2JH.
3. (a) Dess, D. B.; Martin, J. C. *J. Am. Chem. Soc.* **1991,** *113*, 7277–7287; (b) Tohma, H.; Takizawa, S.; Maegawa, T.; Kita, Y. *Angew. Chem.* **2000,** *112*, 1362–1364; (c) Stang, P. J. *Chem. Eng. News* **1989,** *67*, 4.
4. Hofer, M.; Liska, R. *J. Polym. Sci., Part A: Polym. Chem.* **2009,** *47*, 3419–3430.
5. Wirth, T.; Ochiai, M.; Varvoglis, A.; Zhdankin, V. V.; Koser, G. F.; Tohma, H.; Kita, Y. *Hypervalent Iodine Chemistry*. Springer: London, 2003.
6. Stang, P. J. *Angew. Chem., Int. Ed. Engl.* **1992,** *31*, 274–285.
7. (a) Kitamura, T.; Mansei, Y.; Fujiwara, Y., *J. Organomet. Chem.* **2002,** *646*, 196–199; (b) Kotali, E.; Varvoglis, A.; Bozopoulos, A., *J. Chem. Soc., Perkin Trans. 1* **1989**, 827-832; (c) Stang, P. J.; Murch, P., *Tetrahedron Lett.* **1997**, *38*, 8793–8794.

8. Maas, G.; Regitz, M.; Moll, U.; Rahm, R.; Krebs, F.; Hector, R.; Stang, P. J.; Crittell, C. M.; Williamson, B. L. *Tetrahedron* **1992**, *48*, 3527–3540.

9. (a) Murch, P.; Arif, A. M.; Stang, P. J. *J. Org. Chem.* **1997**, *62*, 5959–5965; (b) Williamson, B. L.; Stang, P. J.; Arif, A. M. *J. Am. Chem. Soc.* **1993**, *115*, 2590-2597; (c) Shimizu, M.; Takeda, Y.; Hiyama, T. *Chem. Lett.* **2008**, *37*, 1304–1305.

10. (a) Ochiai, M.; Kunishima, M.; Tani, S.; Nagao, Y. *J. Am. Chem. Soc.* **1991**, *113*, 3135–3142; (b) Feldman, K. S.; Bruendl, M. M.; Schildknegt, K. *J. Org. Chem.* **1995**, *60*, 7722–7723; (c) Williamson, B. L.; Tykwinski, R. R.; Stang, P. J. *J. Am. Chem. Soc.* **1994**, *116*, 93–98; (d) Kitamura, T.; Tsuda, K.; Fujiwara, Y. *Tetrahedron Lett.* **1998**, *39*, 5375–5376; (e) Liu, Z.; Chen, Z. C.; Zheng, Q. G. *Synth. Commun.* **2004**, *34*, 361–367; (f) Wipf, P.; Venkataraman, S. *J. Org. Chem.* **1996**, *61*, 8004–8005; (g) Zhang, P. F.; Chen, Z. C. *Synthesis* **2001**, 358–360; (h) Miyamoto, K.; Nishi, Y.; Ochiai, M. *Angew. Chem., Int. Ed. Engl.* **2005**, *44*, 6896–6899; (i) Liu, Z.; Chen, Z.-C.; Zheng, Q.-G. *J. Heterocycl. Chem.* **2003**, *40*, 909–912; (j) Kitamura, T.; Zheng, L.; Fukuoka, T.; Fujiwara, Y.; Taniguchi, H.; Sakurai, M.; Tanaka, R. *J. Chem. Soc., Perkin Trans. 2* **1997**, 1511–1515; (k) Kitamura, T.; Zheng, L.; Taniguchi, H.; Sakurai, M.; Tanaka, R. *Tetrahedron Lett.* **1993**, *34*, 4055–4058; (l) Nikas, S.; Rodios, N.; Varvoglis, A. *Molecules* **2000**, *5*, 1182–1186; (m) Feldman, K. S.; Perkins, A. L. *Tetrahedron Lett.* **2001**, *42*, 6031–6033.

11. (a) Stang, P. J.; Zhdankin, V. V. *J. Am. Chem. Soc.* **1990**, *112*, 6437–6438; (b) Stang, P. J.; Crittell, C. M. *Organometallics* **1990**, *9*, 3191–3193; (c) Canty, A. J.; Rodemann, T. *Inorg. Chem. Commun.* **2003**, *6*, 1382–1384; **(d)** Stang, P. J.; Tykwinski, R. *J. Am. Chem. Soc.* **1992**, *114*, 4411–4412; (e) Ochiai, M.; Kunishima, M.; Nagao, Y.; Fuji, K.; Fujita, E. *J. Chem. Soc., Chem. Commun.* **1987**, 1708–1709; (f) Ochiai, M.; Nagaoka, T.; Sueda, T.; Yan, M.; Chen, D. W.; Miyamoto, K. *Org. Biomol. Chem.* **2003**, *1*, 1517–1521.

12. (a) Koser, G. F.; Rebrovic, L.; Wettach, R. H. *J. Org. Chem.* **1981**, *46*, 4324–4326; (b) Zhdankin, V. V.; Tykwinski, R.; Berglund, B.; Mullikin, M.; Caple, R.; Zefirov, N. S.; Kozmin, A. S. *J. Org. Chem.* **1989**, *54*, 2609–2612; (c) Neiland, O. Y.; Karele, B. Y. *Zh. Org. Khim.* **1970**, *6*, 885-886.

13. Zhdankin, V. V.; Crittell, C. M.; Stang, P. J.; Zefirov, N. S. *Tetrahedron Lett.* **1990**, *31*, 4821–4824.

14. (a) Rebrovic, L.; Koser, G. F. *J. Org. Chem.* **1984**, *49*, 4700–4702; (b) Lodaya, J. S.; Koser, G. F. *J. Org. Chem.* **1990**, *55*, 1513–1516.

15. Stang, P. J.; Kitamura, T.; Ishihara, K.; Yamamoto, H. *Org. Synth. Coll. Vol.* **1998**, *9*, 477–483.

16. (a) Stang, P. J.; Surber, B. W.; Chen, Z. C.; Roberts, K. A.; Anderson, A. G. *J. Am. Chem. Soc.* **1987**, *109*, 228-235; (b) Kitamura, T.; Lee, C. H.; Taniguchi, H.; Matsumoto, M.; Sano, Y. *J. Org. Chem.* **1994**, *59*, 8053–8057.

17. Kitamura, T.; Furuki, R.; Taniguchi, H.; Stang, P. J. *Tetrahedron* **1992**, *48*, 7149–7156.

18. (a) Stang, P. J.; Williamson, B. L.; Zhdankin, V. V. *J. Am. Chem. Soc.* **1991**, *113*, 5870–5871; (b) Feldman, K. S.; Bruendl, M. M.; Schildknegt, K.; Bohnstedt, A. C. *J. Org. Chem.* **1996**, *61*, 5440–5452.

19. (a) Stang, P. J.; Zhdankin, V. V.; Tykwinski, R.; Zefirov, N. S. *Tetrahedron Lett.* **1992**, *33*, 1419–1422; (b) Stang, P. J.; Tykwinski, R.; Zhdankin, V. V. *J. Org. Chem.* **1992**, *57*, 1861–1864.

20. (a) Lucas, H. J.; Kennedy, E. R., *Org. Synth. Coll. Vol.* **1955**, *3*, 485–487; (b) Bell, R.; Morgan, K. J. *J. Chem. Soc.* **1960**, *245*, 1209–1214; (c) McQuaid, K. M.; Pettus, T. R. R. *Synlett* **2004**, 2403–2405.

21. Merkushev, E. B.; Karpitskaya, L. G.; Novosel'tseva, G. I. *Dokl. Akad. Nauk SSSR* **1979**, *245*, 607–610.

Appendix
Chemical Abstracts Nomenclature (Registry Number)

Phenyliodo bis(acetate): Iodine, bis(acetato-□*O*)phenyl-; (3240-34-4)
Trifluoroacetic acid: Acetic acid, 2,2,2-trifluoro-; (76-05-1)
Phenylacetylene: Benzene, ethynyl-; (536-74-3)

Organic
Syntheses

Luke Dixon was born in Cuckfield, England, in 1984. He obtained his BSc in Chemistry (2005) and MSc in Drug Chemistry (2007) from the Universities of Sheffield and Newcastle respectively. In 2007, he began his PhD with Dr M. A. Carroll in the area of alkynyliodonium salts. Upon completion of his PhD (2011) he moved into the synthesis of complex diaryliodonium salts as precursors to fluorine-18 radiolabelled PET imaging agents.

Michael Carroll gained his BSc and PhD, under the supervision of Dr D. A. Widdowson, from Imperial College London. A number of research positions followed, most notably with Prof. A. B. Holmes FRS at the University of Cambridge. Michael was appointed to the staff at Newcastle University in 2001 where the focus of his research programme is on the development of novel synthetic methodology using hypervalent iodine compounds as reactive intermediates. A key application is the development of diaryliodonium salts as selective precursors to fluorine-18 radiolabelled PET imaging agents.

George Ellames was born in London, UK in 1952 and gained his BSc in Chemistry from the University of Southampton in 1973 and his PhD in Diterpene Biosynthesis under the supervision of Jim Hanson at the University of Sussex in 1976. After 8 years in Medicinal Chemistry with Searle at High Wycombe in the UK, where he led Anti-Infective Chemistry, he moved to Alnwick in Northumberland where he has built up and led Isotope Chemistry and Metabolite Synthesis since 1986 under a succession of site ownerships (Sterling Winthrop, Sanofi, Sanofi-Synthelabo, sanofi-aventis and now Covance Laboratories). He has been heavily involved with the International Isotope Society with a continuing interest in the synthesis and applications of isotopically labelled compounds.

Tom Gregson was born in Sunderland, UK in 1979 and gained his MChem with a Year in Industry, spent with Merck, Sharpe and Dohme, Harlow, from the University of Sheffield in 2001. After achieving his PhD under the supervision of E. J. Thomas at Manchester University in 2005, investigating the total synthesis of the Bryostatins, he moved into industry. He has worked the last six years at Alnwick, Northumberland as part of the Isotope Chemistry and Metabolite Synthesis group, formerly with sanofi-aventis and currently with Covance laboratories.

David M. Guptill was born in Fridley, MN in 1986. He earned his B.A. in ACS Chemistry from Gustavus Adolphus College in Saint Peter, MN in 2005. Currently a fifth-year graduate student in the laboratory of Professor Huw M. L. Davies at Emory University, David's research involves expanding the scope of selective C–H functionalization reactions of donor/acceptor rhodium-carbenoids.

Palladium-catalyzed 1,4-Addition of Terminal Alkynes to Conjugated Enones

Feng Zhou, Liang Chen and Chao-Jun Li[1]*

Department of Chemistry, McGill University, 801 Sherbrooke St. West, Montreal, Quebec H3A 0B8, Canada; Liang Chen, Department of Chemistry, Tulane University, New Orleans, LA 70118, USA

Checked by Lan-Ting Xu and Dawei Ma

Procedure

7-Phenyl-6-heptyn-3-one. An oven-dried, 100-mL, three-necked, round-bottomed flask is equipped with a nitrogen inlet adapter, a stopcock, a rubber septum and a Teflon-coated magnetic stir bar (22 mm x 8 mm) at room temperature under a flow of nitrogen. While temporarily removing the septum, the flask is charged with Pd(II) acetate (Pd(OAc)$_2$) (0.224 g, 1 mmol, 0.02 equiv) (Note 1) and trimethylphosphine (PMe$_3$) (4 mL of 1 M PMe$_3$ solution in toluene, 4 mmol, 0.08 equiv) (Note 2) under nitrogen. The mixture is stirred at room temperature for 10 min, resulting in a homogeneous brown solution (Note 3). Deionized water (50 mL) is added (Note 4); then phenylacetylene (3.07 g, 30 mmol, 0.6 equiv) (Note 5) and ethyl vinyl ketone (6.31 g, 75 mmol, 1.5 equiv) (Note 6) are added successively via syringe (Note 7). The solution is stirred vigorously (Note 8). With the flask equipped with a stopper, additional phenylacetylene (2.04 g, 20 mmol, 0.4 equiv) is added slowly at a rate of 0.4 mL/min. At the same time the flask is heated in an oil bath and kept at 60 °C (oil bath temperature) for 48 h with the color of the solution gradually becoming

Published on the Web 11/26/2013
© 2014 Organic Syntheses, Inc.

darker during the course of the reaction (Note 9). Upon completion of the reaction the color of the solution turns dark brown. The reaction mixture is allowed to cool to room temperature, transferred to a 250-mL separatory funnel, and extracted with ethyl ether (3 × 60 mL) (Note 10). The organic layer is separated and washed with saturated brine (3 × 5 mL) (Note 11). The combined organic phases are then dried over 25 g of Na_2SO_4 (Note 12) for 8 h. After the solid Na_2SO_4 is separated via gravity filtration on a funnel padded with cotton and washed with (3 × 30 mL) of ethyl ether, the filtrate is concentrated at room temperature (25 °C) by rotary evaporation (200 mmHg and then at 10 mmHg) to provide the residue (Note 13). This residue, a dark-brown oil, is adsorbed onto silica (6 g) to form a dry sample, loaded on a silica gel column, and purified by flash column chromatography (Note 14) to afford 6.85–6.94 g (75%) of the product as a brown liquid. This liquid appears to be pure by ^1H and ^{13}C NMR spectra (Note 15). Further purification is achieved by distillation at reduced pressure (120 °C/ 1 mmHg) (Note 16), which provided the product as a colorless liquid (6.20 g, 66%) (Note 17).

Notes

1. $Pd(OAc)_2$ (98%) was purchased from the Aldrich Chemical Company and used as received.
2. PMe_3 (1.0 M solution in toluene) was purchased from the Aldrich Chemical Company and used as received. This reagent should only be used under an inert atmosphere.
3. $Pd(OAc)_2$ may not dissolve completely.
4. The use of nitrogen gas is no longer necessary after the addition of deionized water.
5. Phenylacetylene (98%) was purchased from the Aldrich Chemical Company and used as received.
6. Ethyl vinyl ketone (97%, stabilized with BHT) was purchased from the Aldrich Chemical Company and used as received. It is a highly volatile reagent (bp 103-105 °C/760 mmHg) and 1.5 equiv of ethyl vinyl ketone was used.
7. The solution became a pale-yellow emulsion after the addition of phenylacetylene and ethyl vinyl ketone.
8. The vigorous stirring (1400 rpm) is required to reduce the self-coupling of the phenylacetylene.

9. The progress of the reaction was monitored by TLC analysis on silica gel with 10% EtOAc-hexanes as eluent and visualization under the UV light as well as alkaline $KMnO_4$ solution. R_f (self-coupling of phenylacetylene): 0.61; R_f (product): 0.23. The TLC plates were purchased from the EMD Chemicals Inc. (an affiliate of Merck) and were used as received.

10. Ethyl ether (99.9%) was purchased from the Fisher Scientific.

11. Sodium chloride (99.0%) was purchased from ACP Chemicals.

12. Sodium sulfate (certified anhydrous) (99.4%) was purchased from the Fisher Scientific.

13. An ethanol-cooled recirculator was used to condense the solvent and ethyl vinyl ketone in the collecting flask. The submitters used a dry-ice acetone to provide this cooling.

14. The silica gel was purchased from the Silicycle Inc. with the particle size as 40-63 μm (230-400 mesh). Flash column chromatography was performed using silica gel (3 cm diameter × 30 cm height), eluting with 500 mL of hexanes/ethyl acetate (50/1) first to afford the byproduct *1,4-diphenyl-1,3-butadiyne* (0.225 g, 4%) (^1H NMR (400 MHz, $CDCl_3$) δ: 7.56-7.53 (m, 4 H), 7.38-7.32 (m, 6 H); ^{13}C NMR (100 MHz, $CDCl_3$) δ: 132.5, 129.2, 128.4, 121.8, 81.5, 73.9 and then hexanes/ethyl acetate (15/1) to elute the product. The collected fractions (totaling 500 mL) were analyzed using TLC (hexanes/ethyl acetate = 10/1) (Caution! A small amount of the impurity may co-elute with the product). The spots were visualized using UV light and an alkaline $KMnO_4$ solution.

15. The physical properties are as follows: HRMS: [M + Na] calcd for $C_{13}H_{14}NaO$: 209.0937. Found: 209.0932; ^1H NMR (500 MHz, $CDCl_3$) δ: 0.99 (t, J = 7.0 Hz, 3 H), 2.38 (q, J = 7.5 Hz, 2 H), 2.55–2.59 (m, 2 H), 2.62–2.65 (m, 2 H), 7.16–7.19 (m, 3 H), 7.26–7.29 (m, 2 H); ^{13}C NMR (125 MHz, $CDCl_3$) δ: 7.7, 14.1, 36.0, 41.2, 80.9, 88.7, 123.7, 127.7, 128.2, 131.6, 209.3; IR (neat) cm^{-1}: 3055, 2976, 2239, 1715, 1481, 1363; GC-MS (Relative Intensity) m/z: 186 ([M$^+$], 31), 171 (12), 157 (100), 128 (65), 115 (70), 102 (13), 89 (10).

16. The product obtained by column chromatography was transferred into a 25-mL round-bottomed flask equipped with a magnetic stirbar. The product was distilled under vacuum through a water-cooled condenser topped with a short path distillation head to afford a colorless liquid (oil bath temperature gradually increased from 25 to 140 °C). This compound is stable toward air and moisture and can be stored at room temperature.

17. Purity analysis data on distilled material are as follows: Anal. Calcd. for $C_{13}H_{14}O$: C, 83.83; H, 7.56. Found: C, 83.33; H, 7.58; HPLC > 99% area % purity at 254 nm detection (HPLC conditions, Agilent™ 20RBAX RX-SiL column (4.6 × 250 mm), 5 µM particle size; 0.70 mL/min flow; eluent (hexanes/isopropanol = 98/2); product elutes at 5.42 min. (HPLC conditions in submitter's report: Supelcosil™ LC-PAH C18 column (4.6 × 250 mm), 5 µM particle size; 0.75 mL/min flow; eluent (hexanes/isopropanol = 99/1); product elutes at 5.2 min.) The sample was prepared by dissolving 5 µL of the product in 5 mL of hexanes and the injection volumes equal to 50 µL.

Handling and Disposal of Hazardous Chemicals

The procedures in this article are intended for use only by persons with prior training in experimental organic chemistry. All hazardous materials should be handled using the standard procedures for work with chemicals described in references such as "Prudent Practices in the Laboratory" (The National Academies Press, Washington, D.C., 2011 www.nap.edu). All chemical waste should be disposed of in accordance with local regulations. For general guidelines for the management of chemical waste, see Chapter 8 of Prudent Practices.

These procedures must be conducted at one's own risk. *Organic Syntheses, Inc.*, its Editors, and its Board of Directors do not warrant or guarantee the safety of individuals using these procedures and hereby disclaim any liability for any injuries or damages claimed to have resulted from or related in any way to the procedures herein.

Discussion

The development of palladium-catalyzed C–C bond formation reactions has dramatically advanced the "state-of-the-art" of organic synthesis.[2] The well known palladium catalyzed C–C bond formation reactions include the Heck reaction,[3] the Stille reaction,[4] the Suzuki reaction,[5] the Trost–Tsuji reaction,[6] and the Sonogashira-coupling,[7] to name a few. On the other hand, an addition reaction is an atom-economical way to construct more complex molecules from simpler units.[8]

Recently, increased interest has been shown in the addition of terminal alkynes to those compounds that involve sp[2] carbon, such as C=O bonds[9] or C=N bonds.[10] However, only a few examples of the addition of terminal alkynes to C=C bonds have been reported.[11] Although palladium is one of the most commonly used metals for the purpose of catalysis, palladium-catalyzed conjugate addition of alkynes to enones has not been reported prior to our work.[12]

We hypothesized that the absence of this method could be attributed to either (1) the facile homo- or heterodimerization of terminal alkynes (a well-known, synthetically useful process)[13] to form by-products or (2) a lower reactivity of the alkynyl palladium intermediate towards enones. Conceivably, such obstacles can be overcome by tuning the electronic properties of the ligands to coordinate with palladium.

As part of a continued interest in developing organic synthesis in water,[14] herein a simple and highly efficient Pd-catalyzed addition of a terminal alkyne to a C=C double bond of a conjugated enone, either in water or in acetone under an atmosphere of air, was achieved.[15]

Using the procedure described herein, various 1,4-addition products can be easily synthesized from the terminal alkynes and conjugated enones in the presence of catalytic amounts of $Pd(OAc)_2$ and PMe_3. Table 1 lists several examples of these products. Alkynes bearing silyl, alkenyl, aromatic, aliphatic or halide all reacted smoothly with vinyl ketone to afford good yields of the desired products. With diyne as a substrate, a bis-addition adduct was achieved as a major product. In addition to ethyl vinyl ketone, methyl vinyl ketone also participated in this addition reaction, albeit in a lower yield. It should be noted that both water and acetone are effective as solvents and similar results were obtained in either solvent.

Table 1. Addition of terminal alkyne to vinyl ketone catalyzed by Pd(OAc)₂/PMe₃ in water and in acetone

Entry	Vinylketone	Terminal alkyne	Conditions	Product isolated	Yield (%)
1		Ph—≡	40 h/acetone	Ph (1)	85
2			40 h/water		91
3		(cyclohexenyl)≡	43 h/water	(2)	74
4		n-C₈H₁₇—≡	43 h/acetone	n-C₈H₁₇ (3)	61
5			43 h/water		70
6		TMS—≡	42 h/acetone	TMS (4)	70
7			42 h/water		67
8		Cl~~~≡	44 h/water	Cl (5)	65
9		n-C₆H₁₃—≡	44 h/acetone	n-C₆H₁₃ (6)	63
10			44 h/water		72
11[b]		≡...≡	44 h/acetone	(7)	57
12[b]			44 h/water		62
13		n-C₈H₁₇—≡	42 h/acetone	n-C₈H₁₇ (8)	51
14			42 h/water		56
15		(cyclohexenyl)≡	45 h/water	(9)	58
16		TMS—≡	43 h/acetone	TMS (10)	49
17			43 h/water		52
18		n-C₆H₁₃—≡	45 h/water	n-C₆H₁₃ (11)	53
19		Ph—≡	39 h/acetone	Ph (12)	66
20			39 h/water		61

[a] The reactions were carried out by using 1.0 mmol of a terminal alkyne, 2 mmol of a vinylketone, 5 mol% of Pd(OAc)₂ and 20 mol% of PMe₃ at 60 °C in water or acetone. The product structures were determined by comparison with known compounds in the literature. [b] 1.0 mmol of terminal alkyne reacted with 4 mmol of vinylketone.

DOI: 10.15227/orgsyn.091.0072   *Org. Synth.* **2014**, *91*, 72-82

Scheme 1. Tentative mechanism of the reaction

In the tentative mechanism in Scheme 1, the η^2-coordination of the triple bond to the palladium center followed by a direct deprotonation of the coordinated terminal alkyne by the palladium catalyst[16] generates the alkynyl-palladium intermediate. Then, η^2-coordination of C=C double bond to the palladium center followed by the carbopalladation,[17] and the substitution of Pd with hydrogen (either from the solvent or terminal alkyne) produces the □, δ-ynone product with concomitant regeneration of the Pd catalyst.

In summary, the first palladium-catalyzed 1,4-addition of terminal alkynes to the C=C double bond of conjugated enones was developed in water and in acetone, under an atmosphere of air. The corresponding □, δ-alkynyl ketones were obtained in high yields. The process is simple and can generate a wide range of alkynyl ketones.

References

1. Feng Zhou and Chao-Jun Li, Department of Chemistry, McGill University, 801 Sherbrooke St. West, Montreal, Quebec H3A 0B8, Canada; Liang Chen, Department of Chemistry, Tulane University,

New Orleans, LA 70118, USA. Email: cj.li@mcgill.ca. We thank NSF (US) and NSERC for partial support of our research.

2. For monographs on palladium catalysis, see: (a) *Handbook of Organopalladium Chemistry for Organic Synthesis*, ed. Negishi, E. Wiley, Hoboken, NJ, **2002**; *Handbook of Palladium-Catalyzed Organic Reactions*, ed. Malleron, J. L.; Fiaud, J. C.; Legros, J. Y. Academic Press, London, **1997**; (b) *Palladium Reagents and Catalysts*, ed. Tsuji, J. Wiley, Chichester, **1995**.

3. (a) Beletskaya, I. P.; Cheprakov, A. V. *Chem. Rev.* **2000**, *100*, 3009; (b) de Meijere, A; Brase, S. *J. Organomet. Chem.* **1999**, *576*, 88; (c) Shibasaki, M.; Vogl, E. M. *J. Organomet. Chem.* **1999**, *576*, 1; (d) Heck, R. F. *Org. React.* **1982**, *27*, 345; (e) Heck, R. F. *Acc. Chem. Res.* **1979**, *12*, 146; (f) Cortese, N. A.; Ziegler, C. B.; Hrnjez, B. J.; Heck, R. F. *J. Org. Chem.*, **1978**, *43*, 2952; (g) Dieck, H. A.; Heck, R. F. *J. Organomet. Chem.* **1975**, *93*, 259.

4. (a) Kosugi, M.; Fugami, K. *J. Organomet. Chem.* **2002**, *653*, 50; (b) Farina, V.; Roth, G. P. *Adv. Met. Org. Chem.* **1996**, *5*, 1; (c) Echavarren, A. M.; Stille, J. K. *J. Am. Chem. Soc.* **1988**, *110*, 1557; (d) Labadie, J. W.; Tueting, D.; Stille, J. K. *J. Org. Chem.* **1983**, *48*, 4634.

5. (a) Nguyen, H. N.; Huang, X.; Buchwald, S. L. *J. Am. Chem. Soc.* **2003**, *125*, 11818; (b) Sato, M.; Miyaura, N.; Suzuki, A. *Chem. Lett.* **1989**, 1405; (c) Suzuki, A. *Pure Appl. Chem.* **1985**, *57*, 1749; (d) Miyaura, N.; Yanagi, T.; Suzuki, A. *Synth. Commun.* **1981**, *11*, 513.

6. (a) Jellerichs, B. G.; Kong, J. R.; Krische, M. J. *J. Am. Chem. Soc.* **2003**, *125*, 7758; (b) Ferroud, D.; Genet, J. P.; Muzart, J. *Tetrahedron Lett.* **1984**, *25*, 4379; (c) de Bellefon, C.; Pollet, E.; Grenouillet, P. *J. Mol. Catal. A.* **1999**, *145*, 121.

7. (a) Sonogashira, K. *J. Organomet. Chem.* **2002**, *653*, 46; (b) Sonogashira, K.; Tohda, Y.; Hagihara, N. *Tetrahedron Lett.* **1975**, *50*, 4467.

8. (a) Trost, B. M. *Angew. Chem. Int. Ed. Engl.* **1995**, *34*, 259; (b) Trost, B. M. *Science* **1991**, *254*, 1471.

9. (a) Boyall, D.; Frantz, D. E.; Carreira, E. M. *Org. Lett.* **2002**, *4*, 2605; (b) Sasaki, H.; Boyall, D.; Carreira, E. M. *Helv. Chim. Acta* **2001**, *84*, 964; (c) Anand, N. K.; Carreira, E. M. *J. Am. Chem. Soc.* **2001**, *123*, 9687; (d) Frantz, D. E.; Fassler, R.; Carreira, E. M. *J. Am. Chem. Soc.* **2000**, *122*, 1806.

10. (a) Li, C.-J. *Acc. Chem. Res.* **2010**, *43*, 581; (b) Wei, C.; Li, Z.; Li, C.-J. *Synlett* **2004**, 1472; (c) Zani, L.; Bolm, C. *Chem. Commun.* **2006**, 4263; (d) Yoo, W.-J.; Zhao, L.; Li, C.-J. *Aldrichimica Acta* **2011**, *44*, 43; (e) Shintani, R.; Fu, G. C. *J. Am. Chem. Soc.* **2003**, *125*, 10778; (f) Wei, C.; Li, C.-J. *J. Am.*

Chem. Soc. **2003**, *125*, 9584; (g) Li, C. J.; Wei, C. *Chem. Commun.* **2002**, 268; (h) Wei, C.; Li, C.-J. *J. Am. Chem. Soc.* **2002**, *124*, 5638; (i) Zhang, J.; Wei, C.; Li, C.-J. *Tetrahedron Lett.* **2002**, *43*, 5731; (j) Wei, C.; Li, Z.; Li, C.-J. *Org. Lett.* **2003**, *5*, 4473; (k) Gommermann, N.; Koradin, C.; Polborn, K.; Knochel, P. *Angew. Chem. Int. Ed. Engl.* **2003**, *42*, 5763; (l) Gommermann, N.; Knochel, P. *Chem. Eur. J.* **2006**, *12*, 4380; (m) Gommermann, N.; Knochel, P. *Tetrahedron* **2005**, *61*, 11418; (n) Fischer, C.; Carreira, E. M. *Org. Lett.* **2001**, *3*, 4319.

11. (a) Chang, S.; Na, Y.; Choi, E.; Kim, S. *Org. Lett.* **2001**, *3*, 2089; (b) Picquet, M.; Bruneau, C.; Dixneuf, P. H. *Tetrahedron* **1999**, *55*, 3937; (c) Nikishin, G. I.; Kovalev, I. P. *Tetrahedron Lett.* **1990**, *31*, 7063; (d) Knopfel, T. F.; Carreira, E. M. *J. Am. Chem. Soc.* **2003**, *125*, 6054; (e) Kanai, M.; Shibasaki, M. in *Catalytic Asymmetric Synthesis*, ed. Ojima, I. Wiley-VCH: New York, 2nd Edn., **2000**; pp. 569–592.

12. Ritleng, V.; Sirlin, C.; Pfeffer, M. *Chem. Rev.* **2002**, *102*, 1731.

13. (a) Trost, B. M.; Harms, A. E. *Tetrahedron Lett.* **1996**, *37*, 3971; (b) Trost, B. M.; Chan, C.; Ruhter, G. *J. Am. Chem. Soc.* **1987**, *109*, 3486; (c) Trost, B. M.; Li, C.-J. *Synthesis* **1994**, 1267.

14. Subsequently, the reaction was extended to alkenoates with the same concept via the use of N-heterocyclic carbene (NHC) ligands, see: Zhou, L.; Chen, L.; Skouta, R.; Li, C.-J. *Org. Biomol. Chem.* **2008**, *6*, 2969.

15. (a) Li, C.-J.; Chan, T.-H. *Organic Reactions in Aqueous Media*, Wiley, Chichester, **1997**; (b) *Organic Synthesis in Water*, ed. Grieco, P. A. Thomson Science: Glasgow, **1998**; (c) Cornils, B.; Herrmann, W. A.; Eds., *Aqueous-Phase Organometallic Catalysis, Concepts and Applications*, Wiley-VCH; Weinheim, **1996**.

16. For previous examples involving direct deprotonation of terminal alkynes by palladium catalysts, please see: Scheffknecht, C.; Peringer, P. *J. Organomet. Chem.* **1997**, *535*, 77.

17. For previous examples involving coordination of C=C double bond to transition metal, see: Picquet, M.; Bruneau, C.; Dixneuf, P. H. *Tetrahedron* **1999**, *55*, 3937.

Appendix
Chemical Abstracts Nomenclature (Registry Number)

Phenylacetylene: Benzene, ethynyl-: (536-74-3)
Ethyl vinyl ketone: 1-penten-3-one: (1629-58-9)
Palladium (II) acetate: (3375-31-3)

Trimethylphosphine: Phosphine, trimethyl- (594-09-2)

7-Phenyl-6-heptyn-3-one: 6-Heptyn-3-one, 7-phenyl-: (185309-04-0)

Dr. Feng Zhou was born in 1983 in Wuhan, China. He received his B.S. degree from Wuhan University in 2006 and Ph.D. at the Shanghai Institute of Organic Chemistry, Chinese Academy of Sciences in 2011 under the supervision of Prof. Xiyan Lu. He is currently working as a postdoctoral fellow with Prof. Chao-Jun Li at the McGill University in Canada, focusing on the Cross-Dehydrogenative-Coupling (CDC) reactions via C-H activations.

Dr. Liang Chen received his B.S. degree in Environmental Science (minor: Economics) in 1999 at Nanjing University. He studied chemistry at Tulane University, where he received his Ph.D. degree in 2005 under the supervision of Professor Chao-Jun Li. He then became a Postdoctoral Researcher at Stanford University, and subsequently a Research Associate at LSU Dental School. Since 2008, he has been working in Bisco (Dental) Inc. as a Research Scientist and Senior Research Scientist.

Organic
Syntheses

Chao-Jun Li received his Ph.D. at McGill University (1992). After a two year NSERC Postdoctoral position at Stanford University, he became Assistant (1994), Associate (1998) and Full Professor (2000) at Tulane University. In 2003, he became a Canada Research Chair (Tier I) in Organic/Green Chemistry and a Professor (E. B. Eddy Chair since 2010) of Chemistry at McGill University in Canada. His current research efforts are focused on developing innovative and fundamentally new organic reactions that will defy conventional reactivities and have high synthetic efficiency.

Dr. Lan-Ting Xu received her BS degree from West China School of Pharmacy, Sichuan University in 2008, and her Ph.D. degree from Fudan University in 2013, under the supervision of Dawei Ma. She is now a MSD China R&D Postdoc Research Fellow in Shanghai Institute of Organic Chemistry. Her research interests include copper-catalyzed coupling reactions, metal-catalyzed direct C-H functionalization and heterocycle synthesis.

Cross-Coupling of Alkenyl/Aryl Carboxylates with Grignard Reagents via Fe-Catalyzed C-O Bond Activation

Submitted by Bi-Jie Li,[1] Xi-Sha Zhang[1] and Zhang-Jie Shi[1]*

Beijing National Laboratory of Molecular Sciences (BNLMS) and Key Laboratory of Bioorganic Chemistry and Molecular Engineering of Ministry of Education, College of Chemistry and Green Chemistry Center, Peking University, Beijing 100871 and State Key Laboratory of Organometallic Chemistry, Chinese Academy of Sciences, Shanghai 200032, China

Checked by Heemal Dhanjee and John L. Wood

A.

$$
\text{(1-tetralone)} + \text{NaH} + \text{PivCl} \xrightarrow[\substack{15\,°C\ to\ 25\,°C \\ 40\ min}]{\text{THF}} \text{(3,4-dihydronaphthalen-2-yl pivalate)}
$$

B.

$$
\text{OPiv} + n\text{-HexylMgCl} \xrightarrow[\substack{\text{THF} \\ 0\,°C\ to\ -5\,°C,\ 2.5\ h}]{\substack{\text{FeCl}_2\ (1\ mol\%) \\ \text{(NHC ligand)} \\ (2\ mol\,\%)}} \text{(2-hexyl-3,4-dihydronaphthalene)}
$$

Procedure

A. *3,4-Dihydronaphthalen-2-yl pivalate*. A 1-L three-necked round-bottomed flask is equipped with a 4-cm x 8mm octagonal Teflon coated stir-bar (Note 1). The left neck is fitted with a glass stopper. The center neck is fitted with a Claisen adapter to which is attached a 50 mL pressure-equalizing addition funnel capped with a rubber septum on one joint and reflux condenser with a N₂ inlet on the other. The right neck is capped with

Published on the Web 12/6/2013
© 2014 Organic Syntheses, Inc.

a rubber septum (Note 2). The glass stopper is removed and the flask is charged with sodium hydride (6.00 g, 150 mmol, 1.5 equiv, 60% dispersion in mineral oil) (Note 3). The glass stopper is replaced and the flask is purged for 10 min with N_2 with a vent needle on the right rubber septum. After the addition of THF (200 mL) (Note 4) by syringe through the right rubber septum, the mixture is stirred at room temperature for 10 min to give a grey suspension. The right rubber septum is then replaced with a glass stopper. 2-Tetralone (13.4 mL, 14.8 g, 100 mmol, 1.0 equiv) (Note 5) is then transferred to the addition funnel via syringe and subsequently added dropwise from the addition funnel to the NaH suspension over a period of 8 min. During the addition of 2-tetralone, H_2 gas evolves while the reaction mixture becomes warm to the touch and turns into an opaque yellow suspension (Note 6). The addition funnel is then rinsed with an additional portion of THF (10 mL) (Note 7). After stirring the reaction mixture for 30 min, pivaloyl chloride (18.6 mL, 18.3 g, 150 mmol, 1.5 equiv) (Note 8) is added dropwise to the yellow suspension from the same addition funnel over a period of 10 min. The addition funnel is then rinsed with THF (20 mL) (Note 7). After stirring the reaction for an additional 30 min, TLC analysis indicated the consumption of starting material (Product R_f=0.35, 5% EtOAc in hexanes, visualized under 254 nm UV light (Note 10)). One of the glass stoppers is then removed and water (2.0 mL) is added dropwise using a pipette (Note 9). After H_2 release subsides, water (98 mL) is added followed by ethyl acetate (50 ml). The mixture is stirred vigorously for 30 min and then transferred to a 500 mL separatory funnel giving a yellow organic layer and a clear, colorless aqueous layer. The layers are separated. The aqueous phase is extracted with ethyl acetate (2 x 50 mL) and the organic phases combined. The combined organic phases are washed with 5% aqueous NaOH solution (2 x 100 mL of a solution prepared with 10.5 g NaOH in 200 mL water) and with saturated aqueous NaCl solution (2 x 100 mL). The yellow organic phase is dried over anhydrous Na_2SO_4 and gravity filtered through a cotton plug and the filtrate collected in a 1-L round-bottomed flask. The Na_2SO_4 is rinsed with EtOAc (2 x 25 mL). The combined organics are concentrated under reduced pressure (190 mmHg, 20 °C) to provide a dark brown-yellow oil. The crude oil is loaded onto a fritted chromatography column (8.0 cm outer diameter) that had been dry-packed with silica gel (300 g) (Note 10) and wetted with hexanes under air pressure. The crude product is eluted with 900 mL of hexanes followed by 2000 mL of CH_2Cl_2:hexanes (30:70) and the eluent is collected in 700 mL fractions in 1L Erlenmeyer flasks (Note 11). Fractions 2, 3, and 4 are combined in a 3.0 L round-bottomed flask and concentrated under reduced pressure (190 mmHg, 20 °C, 150 rpm) to give 22.5 g (98%) of 3,4-dihydronaphthalen-2-yl pivalate as light yellow-green oil (Note 12).

B. *3-Hexyl-1,2-dihydronaphthalene.* To a 500-mL three-necked round-bottomed flask is added a 3 cm x 8 mm Teflon-coated octagonal stir bar (Note 1). The left neck is equipped with a N_2 gas inlet adapter, the center neck with a rubber septum, and right neck with an alcohol thermometer attached to a rubber thermometer adapter. To the flask is added iron(II) chloride (63.5 mg, 0.5 mmol, 0.01 equiv) (Note 13) and NHC ligand (344.8 mg, 1.0 mmol, 0.02 equiv) (Note 14), after which the flask is purged with N_2 for 10 min with a vent needle in the center rubber septum. The flask is charged with 3,4-dihydronaphthalen-2-yl pivalate (11.3 mL, 11.5 g, 50 mmol, 1.0 equiv), which is added by 12 mL syringe, and THF (70.0 mL) (Note 4), which is also added by syringe. The mixture is stirred at room temperature (20.1 °C) for 20 min after which the dark brown suspension is moved to an ice-salt bath and cooled to –4.1 °C (Note 15). *n*-HexylMgCl (50.0 mL of a 2.0 M solution in THF, 100 mmol, 2 equiv) (Note 16) is added dropwise by 60 mL plastic syringe over a period of 43 min using a syringe-pump set to a rate of 1.16 mL/min so as to maintain an internal temperature between –5 °C and 0 °C. Upon addition of the Grignard reagent, the reaction mixture turned brown-yellow in color and over the course of the addition became a darker brown-yellow solution.

After the addition of Grignard reagent, the solution is allowed to stir for 1.5 h in the ice-water bath before quenching the reaction with 1.0 M aqueous HCl (3.0 mL) (Product R_f=0.46, hexanes, visualized under 254 nm UV light) (Note 10). After 1 min, an additional 70 mL of aqueous 1.0 M HCl solution is added during which time a slight exotherm is observed. The mixture is warmed to 25 °C and 20 mL EtOAc is then added (Note 17). The mixture is vigorously stirred for 15 min and then transferred to a 500 mL separatory funnel. The layers are allowed to separate over a period of 30 min into a red/brown organic layer and a faintly red aqueous layer. The layers are separated and the aqueous layer extracted with EtOAc (3 x 50 mL). The combined organic phases are washed with saturated aqueous NaCl (2 x 50 mL) and subsequently dried over $MgSO_4$. The solution is gravity filtered through a plug of cotton into a 1 L round-bottomed flask. The $MgSO_4$ is then washed with EtOAc (2 x 25 mL), and the combined filtrate concentrated under reduced pressure by rotary evaporation (190 mmHg, 25 °C, 150 rpm).

The crude product is purified by column chromatography. A fritted chromatography column (8.0 cm outer diameter) is dry-packed with 180.0 g of silica gel (Note 10) and wetted with hexanes under air pressure. The crude oil is loaded directly onto the column and the solvent front lowered to the top of the silica gel. The flask is then washed twice with 5 mL portions of hexanes and then loaded onto the column with the solvent front lowered to the top of the silica gel. A 1.0 cm layer of sand is placed on top of the column. Hexanes (950 mL) is used as an eluent, which is collected as 2 x 100 mL fractions, then a 50 mL fraction, and then 7 x 100 mL fractions. Fractions 4-10 are combined and concentrated under reduced pressure by rotary evaporation (190 mmHg, 20 °C, 150 rpm) to afford 3-hexyl-1,2-dihydronaphthalene as a clear colorless oil (10.3 g, 96%) (Note 18).

Notes

1. The flask and stir bar were flame-dried under vacuum, cooled to room temperature under vacuum, and then placed under N_2 atmosphere.
2. The submitters used a 500 mL three-necked round-bottomed flask equipped with a balloon and a stopper and proceeded to suspend 3.72 g of 96% NaH in 130 mL of THF. After stirring the suspension for 5 min, a solution of 2-tetralone (14.6 g , 13.2 mL, 100 mmol, 1.0 equiv) in 70 mL THF was added dropwise via syringe over a period of 25 min. After the completion of the addition of 2-tetralone, the system was warmed to 25 °C (oil bath temperature) and stirred for 5 h under N_2 atmosphere. A solution of PivCl (30.1 g, 31 mL, 250 mmol, 2.5 equiv) in 30 mL THF was then added by syringe over a period of 3 min. After stirring the reaction for an additional 8 h, TLC analysis showed the completion of the reaction. The checkers found that increasing the rates of addition and reducing the overall reaction time resulted in better yields of the product.
3. NaH (60% oil dispersion in mineral oil, grey in color) was purchased from Sigma Aldrich Co. and used as received.
4. The submitters distilled THF over sodium using benzophenone as indicator. The checkers used non-stabilized THF purchased from Fischer Scientific and passed through a column of activated alumina.

5. The submitters used 2-tetralone (96% purity) from a domestic chemical company Beijing Ouhe Technology Co. The checkers used 2-tetralone (99%) purchased from Sigma Aldrich Co. and used as received.

6. When the reaction was exposed to the air, the reaction turns purple/black in color.

7. THF was added to the addition funnel with a closed stopcock. The entire volume of the washing solvent was added all at once to the reaction by fully opening the stopcock.

8. Pivaloyl chloride (99%) was purchased from Sigma Aldrich Co. and used as received.

9. Caution: Strong effervescence was observed upon addition of the quenching solution.

10. Silica gel SilicaFlash® F60 (40-63 μm/230-500 mesh) was purchased from Silicycle. Glass-backed extra hard layer TLC plates, 60 Å (250 μm thickness) were also purchased from Silicycle containing F-254 indicator.

11. EtOAc, hexanes, and CH_2Cl_2 (all ACS Grade) used in chromatography columns and TLC solvents were all purchased from Fisher Scientific and used as received.

12. The procedure was performed at half-scale in 98% by the checkers. The characterization data of the product was as follows: 3,4-dihydronaphthalen-2-yl pivalate:[2] ^{1}H NMR (CDCl$_3$, 400 MHz) δ: 1.31 (s, 9 H), 2.49 (t, J = 8.4 Hz, 2 H), 3.00 (t, J = 8.4 Hz, 2 H), 6.20 (t, J = 1.2 Hz, 1 H), 6.99–7.03 (m, 1 H), 7.07–7.19 (m, 3 H). ^{13}C NMR (CDCl$_3$, 101 MHz) δ: 26.5, 27.4, 28.9, 39.3, 114.7, 126.4, 126.9, 127.0, 127.5, 133.5, 133.6, 151.6, 177.0. IR (cm^{-1}): ν 2974, 1745, 1129. HRMS (ESI): (M+H$^+$) 231.1385, Found: 231.1383. Anal. Calcd. for $C_{15}H_{18}O_2$, C, 78.23; H, 7.88; Found, C, 77.97; H, 7.86. R_f = 0.35 (5% EtOAc in hexanes as solvent for TLC and visualized under 254 nm UV light (Note 10)).

13. FeCl$_2$ (98% purity) was purchased from Strem Chemicals Inc. and used as received.

14. The checkers purchased 1,3-bis(2,4,6-trimethylphenyl)imidazolinium chloride from Sigma Aldrich Co. and used it as received. The submitters synthesized the NHC ligand using an *Organic Syntheses* procedure.[3]

15. This temperature refers to the internal temperature of the reaction.

16. A 2.0 M solution of *n*-hexylMgCl in THF was obtained from Acros Organics and used as received. The submitters observed that a 1.8 M

solution of *n*-hexylMgCl in THF from J&K Co. can also be used without difference in efficiency.

17. The submitters used NH₄Cl (40 mL) to quench the reaction and observed the formation of a white emulsion. The checkers found that quenching procedure with 1 M HCl avoids the formation of an emulsion thus allowing for a more efficient work-up.

18. The procedure was performed at half-scale in 95% by the checkers. The characterization data of the product 3-hexyl-1,2-dihydronaphthalene. ^{1}H NMR (CDCl$_3$, 400 MHz) δ: 0.87-0.98 (m, 3 H), 1.26–1.42 (m, 6 H), 1.48–1.59 (m, 2 H), 2.21 (t, J = 7.6 Hz, 2H), 2.26 (t, J = 8.3 Hz, 2 H), 2.79 (t, J = 8.3 Hz, 2 H), 6.23 (s, 1 H), 7.00 (d, J = 7.2 Hz, 1 H), 7.05–7.18 (m, 3 H). ^{13}C NMR (CDCl$_3$, 101 MHz) δ: 14.5, 23.0, 27.7, 27.9, 28.6, 29.4, 32.1, 37.8, 122.4, 125.6, 126.3, 126.7, 127.5, 134.8, 135.4, 142.9. MS (C$_{16}$H$_{22}$): 214 (M$^+$). HRMS (ESI) (M+H$^+$) 215.1800, Found: 215.1796. IR (cm^{-1}): ν 2926, 2855, 1485, 1453. Anal. Calcd. for C$_{16}$H$_{22}$, C, 89.65; H, 10.35; Found, C, 89.40; H, 10.32; The submitters analyzed their reaction by GC analysis: retention time of the product was 6.25 min, > 98% purity. GC 6820, oven: initial temp = 80 °C, initial time = 2 min, rate = 25 °C/min, final temp = 250 °C, final time = 4 min; Inlet: initial temp=250 °C; Column: capillary column, length = 30 m, diameter=320 μm; Detector (FID): temperature = 30 °C. R$_f$=0.46 (100% hexanes as solvent for TLC and visualized under 254 nm UV light (Note 10)).

Handling and Disposal of Hazardous Chemicals

The procedures in this article are intended for use only by persons with prior training in experimental organic chemistry. All hazardous materials should be handled using the standard procedures for work with chemicals described in references such as "Prudent Practices in the Laboratory" (The National Academies Press, Washington, D.C., 2011 www.nap.edu). All chemical waste should be disposed of in accordance with local regulations. For general guidelines for the management of chemical waste, see Chapter 8 of Prudent Practices.

These procedures must be conducted at one's own risk. *Organic Syntheses, Inc.*, its Editors, and its Board of Directors do not warrant or guarantee the safety of individuals using these procedures and hereby disclaim any liability for any injuries or damages claimed to have resulted from or related in any way to the procedures herein.

Discussion

Transition metal-catalyzed cross-coupling reactions are useful methods to construct C-C bonds and have been widely used in organic synthesis. However, the generally used halide-coupling partners have some disadvantages: high costs and pollution to the environment. The use of carboxylates as coupling partners can overcome these disadvantages. The coupling of carboxylates with Grignard reagents proceeds smoothly with high yield and efficiency under mild conditions.[4] The use of cheap, stable, and non-toxic iron salts as catalysts make it possible for application in industry.

Both cyclic and acyclic alkenyl pivalates give good yields (Scheme 1, **1a** and **1c**). In the presence of both an alkenyl or aryl ester, the alkenyl pivalate reacted selectively (Scheme 1, **1b**). Aryl pivalates are also capable of reacting, although in a lower yield (Scheme 1, **1d**); however, the reactions of inactivated alkenyl carboxylates such as 1-cyclohexenyl pivalate have been unsuccessful under these conditions.

$$
\begin{array}{c}
\text{Alkenyl—OPiv} \\
\text{or Aryl—OPiv} \\
\text{0.5 mmol} \\
\mathbf{1}
\end{array}
+
\begin{array}{c}
\textit{n}\text{-HexylMgCl} \\
\text{2 eq.} \\
\mathbf{2a}
\end{array}
\xrightarrow[\substack{\text{H}_2\text{IMes}\langle\text{HCl (2 mol\%)} \\ \text{THF} \\ 0\,^\circ\text{C, 1 h}}]{\text{FeCl}_2\ (1\ \text{mol\%})}
\begin{array}{c}
\text{Alkenyl—}^n\text{hexyl} \\
\text{or Aryl—}^n\text{hexyl} \\
\mathbf{3}
\end{array}
$$

1a	**1b**	**1c**	**1d**
93%	90%	91% (E/Z=2:1) (25 °C, 2 h)	40% (25 °C, 2 h, 4% ligand)

Scheme 1. Examples of different alkenyl pivalates

For the Grignard reagent, although magnesium bromides have almost no reactivity under the same conditions, the addition of LiCl can overcome this problem with no ligand necessary (Scheme 2).[4] Under these modified conditions, various Grignard reagents could be coupled and several functional groups tolerated. However, methyl Grignard reagents were unreactive. Secondary and tertiary Grignard reagents did not react either.

$$\text{1a (0.5 mmol)} + \text{RMgBr (2 eq.)} \xrightarrow[\substack{\text{LiCl (6.0 equiv)} \\ \text{THF, 0 °C, 1 h}}]{\text{FeCl}_2 \text{ (1 mol%)}} \text{3}$$

1a: OPiv, CO₂Et substituted cyclohexene

2: RMgBr

3: OPiv, R substituted cyclohexene

n-hexylMgBr

2a
94%

2b
90%

2c
74%

Scheme 2. Examples of different Grignard reagents

References

1. Beijing National Laboratory of Molecular Sciences (BNLMS) and Key Laboratory of Bioorganic Chemistry and Molecular Engineering of Ministry of Education, College of Chemistry and Green Chemistry Center, Peking University, Beijing 100871 and State Key Laboratory of Organometallic Chemistry, Chinese Academy of Sciences, Shanghai 200032, China. Support of this work by Peking University and the grant from National Sciences of Foundation of China (Nos. 20672006, 20821062) and the "973" Project from National Basic Research Program of China (2009CB825300) is gratefully acknowledged. E-mail: zshi@pku.edu.cn
2. Ruberu, S. R.; Fox, M. A. *J. Am. Chem. Soc.* **1992**, *114*, 6310–6317.
3. Hans, M.; Delaude, L. *Org. Synth.* **2010**, *87*, 77–87.
4. Li, B.-J.; Xu, L.; Wu, Z.-H.; Guan, B.-T.; Sun, C.-L; Wang, B.-Q.; Shi, Z.-J. *J. Am. Chem. Soc.* **2009**, *131*, 14656–14657.

Appendix
Chemical Abstracts Nomenclature (Registry Number)

Sodium hydride (60% in oil dispersion): Sodium Hydride; (7646-69-7)

2-Tetralone: 2(1H)-Naphthalenone, 3,4-dihydro-; (530-93-8)
Pivaloyl chloride: Pivalyl chloride; (3282-30-2)
Iron (II) chloride: iron dichloride; (7758-94-3)
1,3-Bis(2,4,6-trimethylphenyl)imidazolinium chloride; (173035-10-4)
n-HexylMgCl: *n*-Hexylmagnesium chloride; (44767-62-6)

Bi-Jie Li was born in 1985 in Hubei, China. He received his BS degree from Peking University in 2007, after which he continued his graduate studies in Professor Zhang-Jie Shi's group. He earned his PhD from Peking University in 2012 from Professor Zhang-Jie Shi's group after his research in C-H and C-O activation. He is now taking up a postdoctoral fellowship at U.C. Berkeley with Prof. J. F. Hartwig.

Xi-Sha Zhang was born in 1988 in Henan, China. He received his BS degree from Zhengzhou University in 2010. He is currently a fourth-year graduate student with Professor Shi at Peking University.

Zhang-Jie Shi obtained his BS at East China Normal University in 1996, and PhD with Professor Sheng-Ming Ma in the Shanghai Institute of Organic Chemistry (SIOC) in 2001. After his postdoctoral work with Professor Gregory L. Verdine at Harvard University and Professor Chuan He at the University of Chicago, he joined the chemistry faculty of Peking University in 2004, where he was promoted to a full Professor in 2008. His current research interests focus on transition metal catalyzed functionalization of inert chemical bonds and small molecules.

Heemal Dhanjee obtained his BA in Mathematics and Molecular and Cell Biology at the University of California, Berkeley in 2007. He subsequently moved to California State University, Northridge and performed research in organic synthesis under the supervision of Professor Thomas Minehan. He is currently pursuing graduate research at Baylor University under the guidance of Professor John L. Wood.

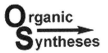

Preparation of 5-Hydroxycyclopentenones Via Conjugate Addition-Initiated Nazarov Cyclization

Joshua L. Brooks, Yu-Wen Huang, and Alison J. Frontier[*1]

Department of Chemistry, University of Rochester, Rochester, NY 14627

Checked by Raffael Vorberg and Erick M. Carreira

Procedure

> *Caution: IBX is heat- and shock-sensitive compound, showing exothermic behavior at temperatures exceeding 130 °C. All operations should be conducted behind an explosion shield.*

A. *3-Buten-1-yn-1-yl-benzene (1).* An oven-dried 1-L, three-necked (2 x 29/32 and 1 x 14.5/23 joint) round-bottomed flask containing a Teflon-coated magnetic stir bar (55 x 8 mm), equipped with three rubber septa and a contact thermometer inserted through the rubber septa in the center neck, is charged with bis(triphenylphosphine)palladium(II) dichloride (2.81 g, 4.00 mmol, 0.02 equiv) (Note 1) and copper(I) iodide (0.38 g, 2.00 mmol, 0.01 equiv.). The flask is then purged with argon, followed by the addition of anhydrous tetrahydrofuran (200 mL) (Note 2) and vinyl bromide (1 M solution in tetrahydrofuran, 200 mL, 200 mmol). The solution is then cooled to 0 °C with an ice bath. Triethylamine (70.0 mL, 500 mmol, 2.5 equiv) is added over the course of 90 sec, which is followed by the addition of phenylacetylene (23.0 mL, 210 mmol, 1.05 equiv) over the course of 30 sec, at which time the solution turns black (Note 3). The solution stirs for 1 h at 0 °C, after which the ice bath is removed and the solution is stirred at room temperature for an additional 12 h (Note 4).

The reaction is diluted with saturated aqueous NH_4Cl solution (300 mL), and the resulting solution is transferred to a 1-L separatory funnel. The aqueous phase is separated and extracted with diethyl ether (3 x 100 mL). The combined organic layers are washed sequentially with 10% aqueous HCl solution (200 mL) and saturated NaCl solution (200 mL). The organic layer is then dried over $MgSO_4$ (20 g), filtered through a 250-mL coarse porosity sintered glass funnel, and concentrated by rotary evaporation (25 °C, 20 mmHg). The resulting oil is transferred to a 100-mL round-bottomed flask equipped with Teflon-coated magnetic stir bar (25 x 6 mm). The product is distilled under vacuum through a 12 cm Vigreux column equipped with a short path distilling head with Vigreux indentations connected to a spider (3 x 14.5/23 joint). The sample is placed under vacuum (3.7 mmHg) and the temperature is slowly increased until distillation commences. A forerun (ca. 1 mL) is collected and discarded, and the desired product is then obtained, distilling at 63–65 °C (3.7 mmHg) to yield pure **1** as a colorless oil (19.4-22.2 g, 151-173 mmol, 75-86%) (Notes 5 and 6).

B. (±)-(E)-2-Hydroxy-1,3-diphenylhexa-3,5-dien-1-one (**2**).[2] An oven-dried, 100-mL one-necked (29/32 joint) round-bottomed flask containing a Teflon-coated magnetic stir bar (14 x 6 mm) is purged with argon, and bis(1,5-cyclooctadiene)rhodium(I) trifluoromethanesulfonate (0.80 g, 1.70 mmol, 0.020 equiv), (±)-BINAP (1.17 g, 1.88 mmol, 0.022 equiv), and anhydrous 1,2-dichloroethane (50 mL) (Note 7) are added. The solution is purged with argon for 15 min. A separate oven-dried 500-mL, one-necked (29/32 joint) round-bottomed flask containing a Teflon-coated magnetic stir bar (38 x 7 mm) is purged with argon, and phenylglyoxal monohydrate (13.0 g,

85.4 mmol) (Note 7), **1** (16.3 g, 127.0 mmol, 1.50 equiv), and anhydrous 1,2-dichloroethane (200 mL) (Note 8) are added. This suspension is purged with argon for 15 min, and the solution containing the rhodium catalyst is transferred to this flask via cannula over the course of 20 min (Note 9). The resulting mixture is then purged with hydrogen gas for 30 min, during which time the cloudy suspension dissolves (Notes 10 and 11). The reaction is then stirred under a hydrogen atmosphere for 40 h (Note 12), until TLC analysis indicates complete consumption of phenylglyoxal.

The solution is then purged with nitrogen for 20 min and concentrated by rotary evaporation (35 °C, 40 mmHg). The residue is purified by column chromatography (20% ethyl acetate/hexanes) to yield (±)-(*E*)-2-hydroxy-1,3-diphenylhexa-3,5-dien-1-one (**2**) (19.6 g, 74.3 mmol, 87%) as a thick orange oil (Notes 13, 14 and 15).

C. (±)-*Dimethyl-2-((5-hydroxy-4-oxo-3,5-diphenylcyclopent-2-en-1-yl)methyl)* *malonate (4)*. To an oven-dried, 100-mL one-necked (29/32 joint) round-bottomed flask containing a Teflon-coated magnetic stir bar (27 x 5 mm) is added **2** (9.0 g, 34.1 mmol) as a solution in DMSO (27 mL), followed by 2-iodoxybenzoic acid (14.3 g, 51.1 mmol, 1.5 equiv) in one portion (Notes 16 and 17). The cloudy suspension is allowed to stir while open to air. The suspension is stirred until TLC analysis indicates complete consumption of alcohol **2** (usually 15 min) (Note 18). The solution is then diluted with water (200 mL), transferred to a 500-mL separatory funnel, and extracted with diethyl ether (5 x 150 mL) (Note 19). The combined ether extracts are transferred to a 1-L separatory funnel and washed with saturated aqueous NaHCO₃ solution (150 mL), dried over MgSO₄ (30 g), filtered through a 250 mL coarse porosity sintered glass funnel, and concentrated by rotary evaporation (25 °C, 15 mmHg). (*E*)-1,3-Diphenylhexa-3,5-diene-1,2-dione (**3**) is obtained as a yellow oil, which is used in the next step without further purification (Note 20).

An oven-dried 250-mL, one-necked (29/32 joint) round-bottomed flask containing a Teflon-coated magnetic stir bar (27 x 5 mm) is charged with yttrium (III) triflate (0.18 g, 0.34 mmol, 0.01 equiv) and lithium chloride (2.89 g, 68.2 mmol, 2 equiv), after which tetrahydrofuran (68 mL), triethylamine (4.75 mL, 34.1 mmol, 1 equiv), and dimethyl malonate (4.87 mL, 42.6 mmol, 1.25 equiv) are added. The suspension is stirred at room temperature and left open to air. Diketone **3**, as a solution in 10 mL of tetrahydrofuran, is added rapidly over 10 sec (Note 21). The reaction turns from yellow to dark red and is stirred until TLC analysis indicates complete consumption of diketone (**3**), which typically requires 5 min (Note 22).

The reaction is then quenched by the addition of 1M aqueous HCl solution (50 mL) over 5 min and the resulting solution is transferred to a 250-mL separatory funnel. The aqueous phase is separated and extracted with diethyl ether (3 x 40 mL). The combined ether extracts are washed with 50 mL of saturated NaCl solution, dried over 20 g of MgSO$_4$, filtered through a 250-mL coarse porosity sintered glass funnel, and concentrated by rotary evaporation (25 °C, 20 mmHg). Purification by column chromatography (33.3% ethyl acetate/hexanes) yields **4** (9.93 g, 25.2 mmol, 74%) as a pale yellow solid (Notes 23 and 24).

Notes

1. All glassware was dried overnight at 120 °C prior to use. All reactions were run under 1 atm of argon, unless otherwise noted. The checkers purchased bis(triphenylphosphine)palladium (II) dichloride (98%) from Combi-Blocks Inc. and used as received. Vinyl bromide (1 M) was purchased from Sigma-Aldrich. Phenylacetylene (98%) and copper iodide (99+%) were purchased from abcr and were used as received. Triethylamine (99+%) was purchased from Sigma-Aldrich and distilled from CaH$_2$ prior to use. The submitters purchased bis(triphenylphosphine)palladium(II) dichloride from Strem Chemical Inc. and used it as received. All other chemicals were purchased from Sigma-Aldrich and used as received unless another vendor is specified.

2. The checkers purchased THF (>95.5%) from Sigma-Aldrich and passed it through a column of alumina before use. The submitters purchased anhydrous 99.9%, inhibitor free tetrahydrofuran from Fisher Scientific and purified it using the Glass Contour solvent purification system directly before use.

3. The checkers purged the reaction mixture with argon for 20 min while the reaction was stirring at 0 °C. This process increased the yield by 11% compared to the reactions that were not purged with argon.

4. TLC analysis was performed on Merck glass plates coated with 0.25 mm 230-400 mesh silica gel containing a fluorescent indicator. The plate was eluted with hexane/EtOAc (95/5) and visualized by ultraviolet lamp at 254 nm. The product and phenylacetylene possess similar R$_f$ values of approximately 0.7. The product stains well with cerium-ammonium-

molybdate (CAM) while phenylacetylene does not react. Vinyl bromide could not be detected by UV or CAM staining.

5. The submitters reported a yield of 79% (20.3 g, 159 mmol).

6. Compound **1** exhibits the following characteristics: ^{1}H NMR (400 MHz, CDCl$_3$) δ: 5.55 (dd, J = 11.1, 2.1 Hz, 1 H), 5.74 (dd, J = 17.5, 2.1 Hz, 1 H), 6.03 (dd, J = 17.5, 11.1 Hz, 1 H), 7.29–7.35 (m, 3 H), 7.42–7.48 (m, 2 H). ^{13}C NMR (101 MHz, CDCl$_3$) δ: 88.22, 90.10, 117.34, 123.28, 127.04, 128.42, 128.45, 131.71. IR (CDCl$_3$) (cm^{-1}) 3081-3011, 2925, 2856, 1606, 1489, 1442, 1069, 1026, 969, 916, 754, 689. HRMS (EI) m/z calc. for C$_{10}$H$_8$ (M$^+$) 128.0626, found 128.0621. Anal. Calcd. for C$_{10}$H$_8$: C, 93.71; H, 6.29. Found: C, 93.55; H, 6.43.

7. The checkers purchased bis(1,5-cyclooctadiene)rhodium(I) trifluoromethanesulfonate (99%) from Strem Chemical Inc. and used it as received. (rac)-BINAP (98%) was obtained from abcr and used as received. Phenylglyoxal monohydrate (97%) was purchased from Alfa Aesar and recrystallized twice from water prior to use. The submitters purchased bis(1,5-cyclooctadiene)rhodium(I) trifluoromethanesulfonate and (rac)-BINAP from Strem Chemical Inc. and used them as received. Phenylglyoxal monohydrate was prepared by the reaction of selenium (IV) oxide and acetophenone followed by recrystallization in water.[3]

8. The checkers purchased anhydrous 1,2-dichloroethane (99.8%) from Sigma Aldrich and used it as received. The submitters purchased anhydrous 99.9%, inhibitor free 1,2-dichloroethane from Fisher Scientific and the solvent was purified using the Glass Contour solvent purification system directly before use.

9. The solution with catalyst contained small amounts of white crystals, some of which were left behind after cannula transfer. The remaining crystals were partially dissolved with additional 1,2-dichloroethane and added to the reaction mixture. A small amount of white crystals were discarded.

10. While purging the suspension with argon, the needle got clogged and was rinsed with small amounts of 1,2-dichloroethane several times. The suspension was cleared after 20 min.

11. The needle used for purging was kept in the solution for the entire reaction time while an additional balloon filled with hydrogen and outfitted with a needle was used to provide positive pressure of H$_2$. Without keeping the needle in the solution yields dropped by 15 %.

12. The reaction was purged again with hydrogen gas for 30 min after 4 h, 19 h and 26 h.

13. In 20% ethyl acetate/hexane, enyne **1**, R_f= 0.94; phenylglyoxal R_f= 0.16; alcohol **2** R_f= 0.58.

14. The product was isolated by the checkers in 73% when the reaction was performed on a 63 mmol scale, and in 71% when the reaction was performed on a 47 mmol scale. The purging needle was not immersed in solvent for the duration of these smaller scale reactions (Note 11).

15. Alcohol **2** is purified on a column (5.5 x 33 cm) packed with 280 g of silica (high purity grade from Fluka, 60 Å pore size, 230-400 mesh) in 10% ethyl acetate/hexane, 1000 mL of 10% ethyl acetate/hexane is first flushed through the column. At this point the solvent is switched to 20% ethyl acetate/hexane and fraction collection is begun (30 mL fractions). Tubes 16-22 are pooled and contain unreacted enyne **1**. Fractions 24-46 are pooled and contain the desired product. The product has an R_f= 0.58 in 20% ethyl acetate/hexane, is UV active and stains strongly with potassium permanganate. The checkers were not able to completely separate the product. Small amounts of enyne (<2%) are contained in the product. For a better separation adding more silica is recommended. Alcohol **2** exhibits the following characteristics: ^{1}H NMR (400 MHz, CDCl$_3$) δ: 4.32 (d, *J* = 6.1 Hz, 1 H), 5.12 (dd, *J* = 10.1, 1.4 Hz, 1 H), 5.35 (dd, *J* = 16.9, 1.4 Hz, 1 H), 5.68 (d, *J* = 6.1 Hz, 1 H), 6.15–6.29 (m, 1 H), 6.47 (d, *J* = 10.9 Hz, 1 H), 7.01–7.08 (m, 2 H), 7.24–7.29 (m, 3 H), 7.38–7.43 (m, 2 H), 7.56 (t, *J* = 7.4 Hz, 1 H), 7.82–7.87 (m, 2 H). ^{13}C NMR (101 MHz, CDCl$_3$) δ: 78.83, 120.46, 127.85, 128.26, 128.44, 128.72, 129.00, 129.36, 133.36, 133.63, 134.03, 136.86, 140.84, 198.65. IR (neat) (cm^{-1}) 3451, 3058, 1679, 1597, 1492, 1448, 1260, 1087, 1001, 972, 951, 916, 774, 755, 699, 688, 670, 617. HRMS (EI) *m/z* calc for C$_{18}$H$_{16}$O (M$^+$) 264.1150, found 264.1145. Anal. calcd for C$_{18}$H$_{16}$O$_2$: C, 81.79; H, 6.10. Found: C, 81.31; H, 5.86.

16. 2-Iodoyxbenzoic acid (>95%) was prepared from 2-iodobenzoic acid by the procedure of Santagostino.[4a] DMSO (>99%) was purchased from Sigma Aldrich and used as received. The submitters prepared 2-iodoxybenzoic acid (>90%) from 2-iodobenzoic acid by the procedure of Boeckman.[4b]

17. The submitters report that other oxidation procedures were attempted (see Discussion), but they all led to formation of byproducts that could not be separated using typical purification techniques, and led to diminished yields in the subsequent (Nazarov cyclization) step.

18. The submitters report that the suspension dissolves to leave a yellow solution. The reaction was extremely exothermic, but no cooling was

provided. The submitters do, however, recommend the use of an ice bath to control the exotherm. In 20% ethyl acetate/hexane, diketone **3**, R_f = 0.65.

19. During the first extraction with ether, an emulsion is formed and the organic layer is slow to separate from the aqueous layer. The mixture is swirled gently and allowed to stand for 20 min. Any remaining emulsion is removed with the ether layer, and extraction of the homogeneous aqueous layer is continued as described.

20. Aqueous workup typically yielded the product (**3**) in high purity (< 95% as determined by NMR analysis). If diketone of higher purity is required, the product can be purified by column chromatography with 10% ethyl acetate/hexane; however, the product does show signs of decomposition on the column. Diketone **3** exhibits the following characteristics: ^{1}H NMR (400 MHz, CDCl$_3$) □: 5.56 (d, J = 10.1 Hz, 1 H), 5.71 (d, J = 16.9 Hz, 1 H), 6.48–6.61 (m, 1 H), 7.16 (d, J = 11.0 Hz, 1 H), 7.26–7.31 (m, 2 H), 7.36–7.46 (m, 3 H), 7.52 (t, J = 7.7 Hz, 2 H), 7.65 (t, J = 10.6 Hz, 1 H), 7.95 (d, J = 7.2 Hz, 2 H). ^{13}C NMR (101 MHz, CDCl$_3$) □: 128.46, 128.60, 129.13, 129.23, 129.92, 130.31, 132.78, 133.21, 133.31, 134.89, 138.45, 146.99, 195.10, 196.08. IR (neat) (cm^{-1}) 3059, 1735, 1677, 1658, 1612, 1450, 1222, 1141, 698. HRMS (ESI) m/z calc. for C$_{18}$H$_{15}$O$_2$ (M+H) 263.1072, found 263.1072.

21. The checkers purchased yttrium(III) triflate (98%) from Strem Chemicals Inc. and used it as received. Lithium chloride (99%) and THF (>95.5%) were purchased from Sigma Aldrich and both were used as received. Triethylamine (99+%) was purchased from Sigma-Aldrich and distilled from CaH$_2$ prior to use. Dimethyl malonate (>99%) was purchased from Acros Organics and used as received.

22. In 20% ethyl acetate/hexane, cyclopentenone **4**, R_f=0.14 .

23. The column (5 x 28 cm) is packed with 200 g of silica in hexane, 400 mL of 12.5% ethyl acetate/hexane are first eluted, followed of 33.3% ethyl acetate/hexane. Fractions are collected using 30 mL test tubes. Fractions 70-104 are pooled and contain the desired product. The product has an R_f= 0.14 in 20% ethyl acetate/hexane, is UV active and stains strongly with potassium permanganate. Cyclopentenone **4** exhibits the following characteristics: mp = 114 °C; ^{1}H NMR (400 MHz, CDCl$_3$) δ: 1.80 (ddd, J = 14.3, 9.2, 5.5 Hz, 1 H), 1.98 (ddd, J = 14.3, 9.2, 6.5 Hz, 1 H), 3.20 (ddd, J = 9.2, 6.5, 2.5 Hz, 1 H), 3.26 (s, 1 H), 3.41 (dd, J = 9.2, 5.5 Hz, 1 Hz), 3.62 (s, 3 H), 3.76 (s, 3 H), 7.26–7.47 (m, 8 H), 7.78 (dd, J = 8.0, 1.7 Hz, 2 H), 7.80 (d, J = 2.5 Hz, 1 H). ^{13}C NMR (101 MHz, CDCl$_3$) δ: 28.66, 49.31,

49.84, 52.75, 52.92, 84.11, 125.57, 127.09, 128.26, 128.71, 128.90, 129.40, 130.49, 139.46, 142.09, 158.40, 169.20, 169.57, 206.66. IR (neat) (cm^{-1}) 3463, 3059, 3027, 1716, 1493, 1436, 1306, 1236, 1155, 1051, 909, 832, 788, 765, 698. HRMS (ESI) m/z calc. for $C_{23}H_{23}O_6$ (M+H) 395.1489, found 395.1483. Anal. calcd for $C_{23}H_{22}O_6$: C, 70.04; H, 5.62 . Found: C, 69.75; H, 5.53.

24. The product was isolated by the checkers in 75% when the reaction was performed on a half scale.

Handling and Disposal of Hazardous Chemicals

The procedures in this article are intended for use only by persons with prior training in experimental organic chemistry. All hazardous materials should be handled using the standard procedures for work with chemicals described in references such as "Prudent Practices in the Laboratory" (The National Academies Press, Washington, D.C., 2011 www.nap.edu). All chemical waste should be disposed of in accordance with local regulations. For general guidelines for the management of chemical waste, see Chapter 8 of Prudent Practices.

These procedures must be conducted at one's own risk. *Organic Syntheses, Inc.,* its Editors, and its Board of Directors do not warrant or guarantee the safety of individuals using these procedures and hereby disclaim any liability for any injuries or damages claimed to have resulted from or related in any way to the procedures herein.

Discussion

The Nazarov electrocyclization has become a versatile method for the synthesis of cyclopentenones with a broad range of substitution patterns, with excellent potential as building blocks for complex molecule synthesis.[5] While the classical Nazarov cyclization requires stoichiometric or super amounts of Bronsted or Lewis Acid, recent advances in substrate design, Lewis acid catalysts, and methods for the synthesis of reactive intermediates have increased the utility of the reaction. Herein, we describe the synthesis and use of dienyl diketones as substrates in Nazarov cyclizations, initiated by 1,6-conjugate addition of an amine or malonate nucleophile.[6] A three-step protocol is employed to prepare the dienyl diketones, and the cyclization produces 5-hydroxycyclopentenones in high yields.

Table 1. Nucleophile Scope for Nazarov Cyclization Initiated by 1,6-Addition to Dienyl Diketones

The Nazarov cyclization of dienyl diketone **5** can be initiated with various nucleophiles, proceeding in good to excellent yield with primary, cyclic, and acyclic secondary amines (Table 1). Malonate derivatives are also effective nucleophiles, allowing the creation of a new carbon-carbon bond in the first step of the cascade. In all cases only one diastereoisomer was detected, and the *syn* relationship was confirmed by either nOe analysis or single X-ray crystallographic data.[6]

The conjugate addition-initiated cyclization is efficient for both dienyl diketone **5** and a number of dienyl diketones of type **3** (Table 2). The cyclization can be used to install a quaternary stereogenic center at the 4-position of the cyclopentenone, using either pyrrolidine or malonate to initiate the reaction (Table 2).[7] The cyclization cascade is remarkably convenient and robust: the reaction is not air- or moisture sensitive, and the reactions are complete in less than 30 minutes at ambient temperature.

Table 2. Scope of the Conjugate Addition-Initiated Nazarov Cyclization

entry	R,	nucleophile	yield (%)
1	-H	pyrrolidine	87
2	-H	dimethylmalonate	83
3	-CH$_3$	pyrrolidine	85
4	-CH$_3$	dimethylmalonate	78
5	-(CH$_2$)$_2$OAc	dimethylmalonate	89
6	-(CH$_2$)$_3$OAc	dimethylmalonate	87

The synthesis of dienyl diketones of type **3** relies upon the Rh-catalyzed reductive coupling of enynes with glyoxals, using a protocol developed by Krische and coworkers.[2] The reaction is regioselective and delivers the target dienyl alcohols of type **2** in good to excellent yield.[8] After extensive experimentation, we found that the optimal method for oxidizing the dienyl alcohols to dienyl diketones of type **3** was the use of 2-iodoxybenzoic acid (IBX) in DMSO. The target substrates were obtained in excellent yields, and they could be cyclized without the need for additional purification. A number of alternative oxidants and oxidation methods were screened, including the Dess-Martin reagent, the Jones reagent, MnO$_2$, TEMPO, TPAP- NMO, and Bi(NO$_3$)$_3$, but in all cases, the oxidation led to byproduct formation resulting in reduced yields and requiring purification of the dienyl diketones by column chromatography.[9]

References

1. Department of Chemistry, University of Rochester, Rochester, NY 14627, email frontier@chem.rochester.edu. This project has been supported by the NIGMS (R01 GM079364).
2. (a) Huddleston, R. R.; Jang, H.-Y.; Krische, M. J. *J. Am. Chem. Soc.* **2003**, *125*, 11488–11489. (b) Jang, H.-Y.; Huddleston, R. R.; Krische, M. J. *J. Am. Chem. Soc.* **2004**, *126*, 4664–4668.
3. Riley , H. A.; Gray, A. R. *Org. Synth.* **1935**, *15*, 67.
4. (a) Boeckman, R. K.; Shao, P.; Mullins, J. J. *Org. Synth.* **2000**, *77*, 141. (b) Frigerio, M.; Santagostino, M.; Sputore, S. *J. Org. Chem.* **1999**, *64*, 4537.
5. (a) Pellissier, H. *Tetrahedron* **2005**, *61*, 6479–6517. (b) Frontier, A. J.; Collison, C. *Tetrahedron* **2005**, *61*, 7577–7606. (c) Tius, M. A. *Eur. J. Org. Chem.* **2005**, *2005*, 2193–2206. (d) Grant, T. N.; Rieder, C. J.; West, F. G. *Chem. Commun.* **2009**, 5676–5688. (e) Nakanishi, W.; West, F. G. *Curr. Opin. Drug Discov. Devel.* **2009**, *12*, 732–751. (f) Vaidya, T.; Eisenberg, R.; Frontier, A. J. *Chem Cat Chem* **2011**, *3*, 1531–1548.
6. Brooks, J. L.; Caruana, P. A.; Frontier, A. J. *J. Am. Chem. Soc.* **2011**, *133*, 12454–12457.
7. Brooks, J. L.; Frontier, A. J. *J. Am. Chem. Soc.* **2012**, *134*, 16551–16553.
8. The Krische protocol can only be employed in couplings with aryl glyoxals or non-enolizable aliphatic glyoxals, because alphatic glyoxals with α-protons polymerize rapidly (Riley, H. L.; Morley, J. F.; Friend, N. A. C. *J. Chem. Soc.* **1932**, 1875–1883). Therefore, a different synthetic strategy must be used to prepare dienyl diketones like **5** (see reference 6).
9. Byproducts resulting from oxidative cleavage were observed in oxidation experiments using the Dess-Martin reagent, Jones reagent and MnO$_2$, while decomposition was observed when TEMPO, TPAP-NMO, and Bi(NO$_3$)$_3$ were tested.

Appendix
Chemical Abstracts Nomenclature (Registry Number)

Bis(triphenylphosphine)palladium(II) dichloride; (13965-03-2)
Copper(I) iodide; (7681-65-4)
Vinyl bromide solution; (593-60-2)
Triethylamine; (121-44-8)
Phenylacetylene; (536-74-3)
3-Buten-1-yn-1-yl-Benzene; (13633-26-6)
Bis(1,5-cyclooctadiene)rhodium(I) trifluoromethanesulfonate; (99326-34-8)
(±)-BINAP; 2,2'-Bis(diphenylphosphino)-1,1'-binaphthyl (98327-87-8)
Phenylglyoxal monohydrate; (1074-12-0)
(±)-(E)-2-Hydroxy-1,3-diphenylhexa-3,5-dien-1-one; (690211-20-2)
2-Iodylbenzoic acid; (64297-64-9)
(E)-1,3-Diphenylhexa-3,5-diene-1,2-dione; (1401539-00-1)
Yttrium (III) triflate; Yttrium(III) trifluoromethanesulfonate (52093-30-8)
Lithium chloride; (7447-41-8)
Dimethyl malonate; (108-59-8)
(±)-Dimethyl 2-((-5-hydroxy-4-oxo-3,5-diphenylcyclopent-2-en-1-yl)methyl)
malonate; (1401539-14-7)

Joshua L. Brooks received his B.S. The State University of New York, College of Environmental Science and Forestry (SUNY-ESF) in 2005. He continued at SUNY-ESF until 2008 where he studied under Professor José-L. Giner completing a Masters Degree. He earned his Ph.D. in 2012 from the University of Rochester where he studied with Alison J. Frontier, and is currently undertaking postdoctoral work with Derek S. Tan at Memorial Sloan-Kettering Cancer Center

Yu-Wen Huang was born in Hsinchu, Taiwan (R.O.C.) in 1982. He received his bachelor's degree from National Cheng Kung University in 2005. He then joined M.S. program at National Tsing Hua University, whereas he obtained his master degree under the supervision of Professor Shang-Cheng Huang in carbohydrate synthesis. He is currently a Ph.D. student in University of Rochester with Professor Alison J. Frontier. His research focuses on the new methodology development and its application toward total synthesis.

Alison J. Frontier received her AB from Harvard in 1992, and then took a two-year position as an Associate Chemist at the Merck Research Laboratories in Rahway, NJ. She earned her PhD in 1999 from Columbia University, where she studied with Samuel Danishefsky, and then she did postdoctoral work with Barry Trost at Stanford University. She has been on the faculty at the University of Rochester since 2002. Her research interests focus on target molecule synthesis and reaction development using novel catalytic methods, cationic and neutral pericyclic reactions and multistep cyclization cascades.

Raffael Vorberg received his B.S. and M.S degrees at the ETH Zürich, Switzerland in 2011. He then carried out an 8 month internship at Roche in Basel, Switzerland. In 2012 he joined the research group of Prof. Carreira at ETH Zürich to pursue his Ph. D. degree. His research is focused on the synthesis of small fluorinated building blocks for applications in drug design.

Synthesis of Arylboronic Pinacol Esters from Corresponding Arylamines

Di Qiu, He Meng, Liang Jin, Shengbo Tang, Shuai Wang, Fangyang Mo, Yan Zhang and Jianbo Wang[*1]

Beijing National Laboratory of Molecular Sciences (BNLMS) and Key Laboratory of Bioorganic Chemistry and Molecular Engineering of Ministry of Education, College of Chemistry, Peking University, Beijing 100871, China

Checked by Austin Smith and Margaret Faul

$$O_2N-\text{C}_6\text{H}_4-NH_2 + B_2pin_2 + tBuONO \xrightarrow[\text{MeCN, rt, 4 h}]{\text{BPO}} O_2N-\text{C}_6\text{H}_4-Bpin$$

1   78%   **2**

Procedure

A. *4,4,5,5-Tetramethyl-2-(4-nitrophenyl)-1,3,2-dioxaborolane (2)*. A 3-necked 250-mL round-bottomed flask equipped with a 2.5 cm rod-shaped, Teflon coated magnetic stir bar and fitted with a reflux condenser, glass addition funnel, and thermocouple is charged with 4-nitroaniline **1** (5.52 g, 39.6 mmol, 1.0 equiv), diboronpinacol ester (B_2pin_2, 10.66 g, 41.5 mmol, 1.05 equiv), and benzoyl peroxide (193 mg, 0.77 mmol, 0.02 equiv) under ambient atmosphere (Notes 1 and 2). The mixture is dissolved in MeCN (60 mL), and the resulting solution is stirred (700 rpm) at 22 °C with the temperature maintained by a water bath. *tert*-Butyl nitrite (*t*BuONO, 6.17 g, 53.8 mmol, 1.36 equiv) diluted by MeCN (40 mL) is then added via an addition funnel over a 1 h time period (Notes 3 and 4). The solution is allowed to stir at 23 °C for an additional 4 h, with the temperature maintained by a water bath (Note 5). Upon complete consumption of **1** as indicated by TLC analysis, the resulting solution is concentrated with a rotary evaporator to remove MeCN (40 °C, 40 mmHg) (Note 6). The crude

Org. Synth. **2014**, *91*, 106-115
DOI: 10.15227/orgsyn.091.0106

Published on the Web 1/22/2014
© 2014 Organic Syntheses, Inc.

residue is filtered over a flash silica gel plug (60 g) eluting with petroleum ether-ethyl acetate (20:1, 750 mL) until TLC analysis shows that no product **2** is still eluting from the silica gel (Note 7). The filtrate is then concentrated with a rotary evaporator (25 °C, 40 mmHg), and the residue is further purified by silica gel column chromatography, eluting with approximately 500 mL petroleum ether until product emerges as indicated by TLC, and then eluting with approximately 1000 mL of petroleum ether-ethyl acetate (100:1)(Note 8). The combined eluents are concentrated with a rotary evaporator (25 °C, 40 mmHg) and further evaporated with an oil vacuum pump for 30 min at room temperature to afford the product **2** (7.10–7.32 g, 72–74%) as a pale yellow solid (Note 9), which was >97% pure as determined by GC analysis (Note 10).

Notes

1. 4-Nitroaniline (99%) was purchased from Aldrich Chemical Company, Inc. and used as received. The submitters used 4-nitroaniline that was supplied by Sinopharm.
2. Diboronpinacol ester (B$_2$pin$_2$, 99%) and benzoyl peroxide (BPO, 97%) were purchased from Aldrich Chemical Company, Inc. and used as received. The submitters used diboronpinacol ester (B$_2$pin$_2$) from Ouhe Corporation in Beijing, China.
3. MeCN (>99%) was purchased from Aldrich Chemical Company and used as received. The submitters used MeCN (AR) from Beijing Chemical Corporation. There action is not sensitive to the presence of water. *tert*-Butyl nitrite (*t*BuONO, 90%) was purchased from Aldrich Chemical Company, Inc. and used as received.
4. The temperature of the solution rose from 22 °C to 23.5 °C during the 1h addition of *t*BuONO. No gas evolution or solvent reflux was observed.
5. A water bath was used throughout the course of the 4 h reaction, with the temperature stabilizing at 23 °C. No gas evolution or solvent reflux was observed.
6. TLC analysis was performed with silica gel plates (2.5 cm x 7.5 cm, glass backed, purchased from EMD Chemicals in Darmstadt, Germany), using petroleum ether-ethyl acetate (3:2) as eluent. The aniline **1** is yellow by visible light and dark gray under 254 nm UV lamp. R$_f$= 0.31 (petroleum ether-ethyl acetate, 3:2).

7. For silica gel column chromatography, 230-400 mesh, 0.040-0.063 mm particle size silica gel (EMD Chemicals, Darmstadt, Germany) was employed. TLC analysis was performed with silica gel plates (2.5 cm x 7.5 cm, glass backed, purchased from EMD Chemicals in Darmstadt, Germany), using petroleum ether-ethyl acetate (20:1) as eluent. The product **2** is purple under 254 nm UV lamp. R_f= 0.23 (petroleum ether-ethyl acetate, 20:1).

8. The crude residue was dissolved in 20:1 petroleum ether-ethyl acetate (15 mL), and the solution was charged onto a column (diameter = 5 cm) of silica gel (220 g). The column was eluted with approximately 500 mL petroleum ether. When TLC analysis showed that product **2** appeared from the column, the eluent was changed to petroleum ether-ethyl acetate (100:1, approximately 1000 mL). Fractions 16-35 were indicated by TLC to contain **2** and thus combined and evaporated (25 °C, 40 mmHg).

9. Characterization data of **2**: 4,4,5,5-Tetramethyl-2-(4-nitrophenyl)-1,3,2-dioxaborolane (**2**). Pale yellow solid; R_f= 0.23(petroleum ether-ethyl acetate, 20:1); mp = 108.5–109.5 °C; ^{1}H NMR (400 MHz, CDCl$_3$) δ: 1.37 (s, 12H), 7.96 (d, J = 8.0 Hz, 2H), 8.18 (d, J = 8.0 Hz, 2 H); ^{13}C NMR (100 MHz, CDCl$_3$) δ: 24.8, 84.6, 122.3, 135.6, 149.8, IR (thin film): 2961, 1600, 1519, 1365, 1343, 1141, 1086, 857, 701 cm^{-1}. EI-MS (m/z, relative intensity): 249 (M$^+$, 18), 234 (100), 163 (51), 150 (23), 149 (6), 104 (9), 85 (7), 59 (7), 43 (11), 42 (7). HRMS (ESI) calcd for C$_{12}$H$_{17}$BNO$_4$ [M+H]$^+$ 250.1251; found 250.1245. Anal. calcd for C$_{12}$H$_{16}$BNO$_4$: C, 57.87; H, 6.47; N, 5.62. Found: C, 58.12; H, 6.88; N, 5.41.

10. GC conditions: Agilent 6890N; column 19091J-413 (30 m x 0.32 mm), HP-5; temperature profile: initial temp = 61 °C, final temp = 280 °C, temperature gradient = 20 °C /min. Retention time of **2**: 11.29 min.

Handling and Disposal of Hazardous Chemicals

The procedures in this article are intended for use only by persons with prior training in experimental organic chemistry. All hazardous materials should be handled using the standard procedures for work with chemicals described in references such as "Prudent Practices in the Laboratory" (The National Academies Press, Washington, D.C., 2011 www.nap.edu). All chemical waste should be disposed of in accordance with local regulations.

For general guidelines for the management of chemical waste, see Chapter 8 of Prudent Practices.

These procedures must be conducted at one's own risk. *Organic Syntheses, Inc.,* its Editors, and its Board of Directors do not warrant or guarantee the safety of individuals using these procedures and hereby disclaim any liability for any injuries or damages claimed to have resulted from or related in any way to the procedures herein.

Discussion

The pinacol arylboronates can be accessed by Pd-catalyzed borylation of arylhalides (Miyaura borylation), or by Ir- or Rh-catalyzed direct borylation via aromatic C-H bond activation. These borylation methods have been widely applied in organic synthesis. The direct conversion of aromatic amine to arylboronates represents an entirely different method. It is metal-free, using cheap and abundant arylamines as starting materials. The reaction is carried out under mild conditions and tolerates various functional groups.

The procedure described here is based on our initial communication.[2] We have noted that the substituents on the aromatic ring of arylamine substrates significantly affect the yields of borylation products. Recently, we have reported modified reaction conditions, under which the borylation is carried out in MeCN at 80 °C without the use of BPO.[3] In most cases, the modified reaction conditions give slightly improved yields. For *p*-nitroaniline **1** and *p*-toluidine **5**, the borylation has also been carried out under modified reaction conditions (MeCN, 80 °C, without BPO). The borylation products were obtained in 59% and 58% yields, respectively. In Figure 1 and Figure 2, the substrate scope of the metal-free borylation has been summarized.

Figure 1. Substrate Scope of Borylation in the Presence of BPO at Room Temperature[2]

$$\text{ArNH}_2 + \text{B}_2\text{pin}_2 + t\text{BuONO} \xrightarrow[\text{MeCN, rt}]{\text{BPO (2 mol\%)}} \text{Ar-Bpin}$$
$$\text{(1 mmol)} \quad \text{(1 mmol)} \quad \text{(1.5 mmol)}$$

AcHN—⟨⟩—Bpin

2 h, 93%[a]

BocHN—⟨⟩—Bpin

2 h, 70%

AcO—⟨⟩—Bpin

2 h, 73%

MeO—⟨⟩—Bpin

2 h, 72%

1 h, 67%

2 h, 65%

1 h, 66%

2 h, 53%

O_2N—⟨⟩—Bpin

1 h, 91%

O_2N—⟨⟩—Bpin

1 h, 62%

NO_2 ⟨⟩—Bpin

trace[b]

Bpin—⟨⟩—⟨⟩—Bpin

2 h, 55%[c]

F_3C—⟨⟩—Bpin

2 h, 75%

F_3C ⟨⟩—Bpin

2 h, 82%

F—⟨⟩—Bpin

2 h, 54%

Cl ⟨⟩—Bpin Cl—

2 h, 54%

Cl ⟨⟩—Bpin

2 h, 56%[d]

Cl—⟨⟩—Bpin

2 h, 30%[d]

Br—⟨⟩—Bpin

1 h, 30%[d]

EtO_2C—⟨⟩—Bpin

2 h, 82%

NC—⟨⟩—Bpin

1 h, 41%

CN ⟨⟩—Bpin

1 h, 22%

[a] Isolated yields by column chromatography. [b] The reaction gave complex mixture with trace amount of product. [c] Benzidine was used as substrate. B_2pin_2 (2.2 equiv) and $t\text{BuONO}$ (3 equiv) were used. [d] Reaction was carried out at 60 °C with 10% BPO and 1.2 equiv of B_2pin_2.

Figure 2. Substrate Scope of Borylation without the use of BPO at 80 °C[3]

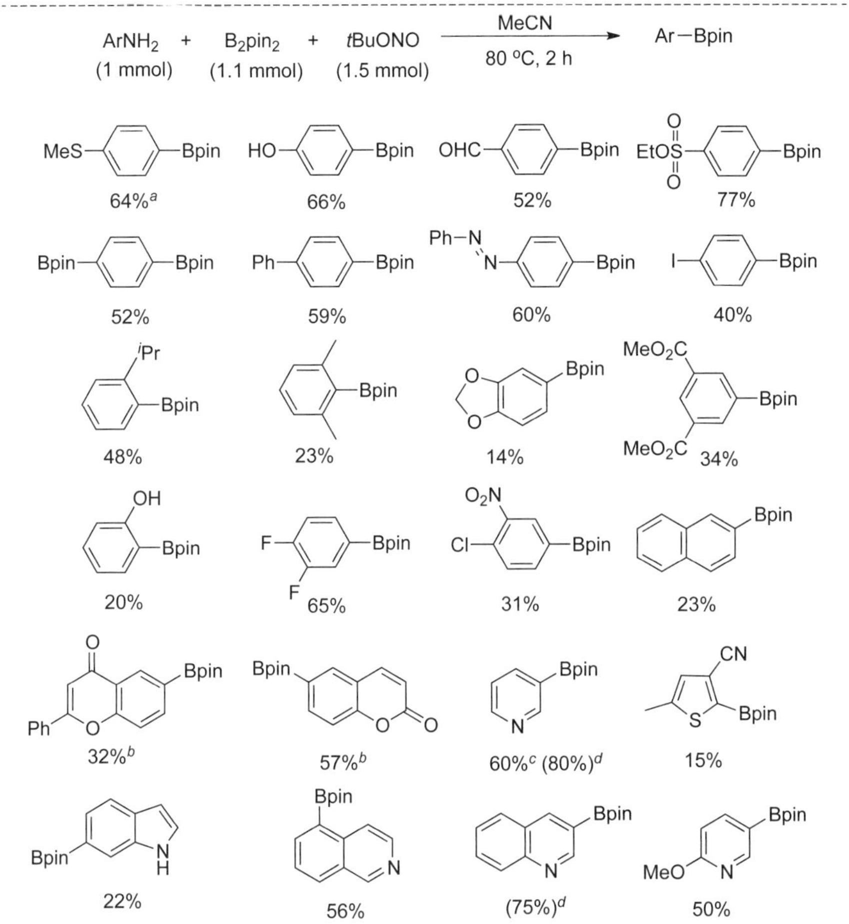

ArNH₂ (1 mmol) + B₂pin₂ (1.1 mmol) + *t*BuONO (1.5 mmol) →[MeCN, 80 °C, 2 h] Ar—Bpin

MeS—⟨⟩—Bpin 64%[a]

HO—⟨⟩—Bpin 66%

OHC—⟨⟩—Bpin 52%

EtOS(=O)(=O)—⟨⟩—Bpin 77%

Bpin—⟨⟩—Bpin 52%

Ph—⟨⟩—Bpin 59%

Ph—N=N—⟨⟩—Bpin 60%

I—⟨⟩—Bpin 40%

*i*Pr ⟨⟩—Bpin 48%

⟨⟩—Bpin (dimethyl) 23%

methylenedioxy—Bpin 14%

MeO₂C ⟨⟩—Bpin, MeO₂C 34%

OH ⟨⟩—Bpin 20%

F ⟨⟩—Bpin F 65%

O₂N, Cl—⟨⟩—Bpin 31%

naphthalene—Bpin 23%

Ph—O chromone—Bpin 32%[b]

Bpin coumarin 57%[b]

pyridine—Bpin 60%[c] (80%)[d]

thiophene CN Bpin 15%

Bpin indole 22%

isoquinoline—Bpin 56%

quinoline—Bpin (75%)[d]

MeO—N pyridine—Bpin 50%

[a] Isolated yields by column chromatography. [b] Arylamine (0.5 mmol), B₂pin₂ (1.1 equiv), *t*BuONO (1.5 equiv), MeCN (1.5 mL), 2 h. [c] Isolated by recrystallization. [d] Yields in brackets refer to GC yield measured with GC-MS instrument (mesitylene as internal standard).

References

1. Beijing National Laboratory of Molecular Sciences (BNLMS) and Key Laboratory of Bioorganic Chemistry and Molecular Engineering of

Ministry of Education, College of Chemistry, Peking University, Beijing 100871, China; Email: wangjb@pku.edu.cn. The project is supported by National Basic Research Program (973 Program, No. 2012CB821600), Natural Science Foundation of China (No. 21272010).

2. Mo, F.; Jiang, Y.; Qiu, D.; Zhang, Y.; Wang, J. *Angew. Chem. Int. Ed.* **2010**, *49*, 1846–1849.
3. Qiu, D.; Jin, L.; Zheng, Z.; Meng, H.; Mo, F.; Wang, X.; Zhang, Y.; Wang, J. *J. Org. Chem.* **2013**, *78*, 1923–1933.

Appendix
Chemical Abstracts Nomenclature (Registry Number)

4,4,4',4',5,5,5',5'-Octamethyl-2,2'-bi(1,3,2-dioxaborolane): diboronpinacol ester; (73183-34-3)

tert-Butyl nitrite; (540-80-7)

Benzoyl peroxide; (94-36-0)

4-Nitroaniline; (100-01-6)

4,4,5,5-Tetramethyl-2-(4-nitrophenyl)-1,3,2-dioxaborolane; (171364-83-3)

Di Qiu was born in Tianjin of China. He received his B.S. degree from Peking University in 2010. He is now a graduate student in Prof. Jianbo Wang's laboratory. His current research projects are focused on gold-catalyzed C-H bond functionalization and the transformations based on aryldiazonium salts.

He Meng was born in September 1991 in Shijiazhuang of Hebei province. He is expecting B.S.degree in chemistry of Peking University. Now he is a senior student in Prof. Jianbo Wang's group.

Liang Jin was born in April 1991 in Shanghai of China. He has received his B.S. degree (in Prof. Jianbo Wang's group) from Peking University in 2013. Now he is a graduate student in Prof. Huwei Liu's group in chemistry department of Peking University.

Shengbo Tang was born in September 1992 in Jilin province of China. He is expecting B.S. degree in chemistry of Peking University. Now he is a junior student in Prof. Jianbo Wang's group. His hobbies are playing basketball and playing guitar.

Organic Syntheses

Shuai Wang was born in April 1993 in Tianjin of China. He is expecting B.S. degree in chemistry of Peking University. Now he is a junior student in Prof. Jianbo Wang's group. His favorite hobby is fitness.

Fanyang Mo was born in Liaoning Province of China. He received his B.Sc. and M.Sc. degrees from Beijing Institute of Technology (P. R. of China) in 2004 and 2006 under the supervision of Professor Zhiming Zhou. He then obtained his Ph.D. from Peking University, under the supervision of Prof. Jianbo Wang in 2010. He is currently a postdoctoral fellow in Prof. Guangbin Dong's group at University of Texas at Austin.

Yan Zhang was born in Shandong province of China. She received her B.S. and Ph. D. degrees from Lanzhou University. She continued her research as a postdoctoral fellow in Hong Kong University of Science and Technology (2002 - 2004), University of Innsbruck & Leibniz Institute of Surface Modification (IOM) e. V. (2004-2005), University of Missouri-St. Louis (2005-2006), and Auburn University (2006-2008). She has joined Peking University since 2008 as an associate professor. Her research focuses on the application of transition metal complexes of N-heterocyclic carbenes and synthesis of small biological compounds.

Jianbo Wang was born in Zhejiang province of China. He received his B.S. degree from Nanjing University of Science and Technology in 1983, and his Ph.D. from Hokkaido University in 1990. He was a postdoctoral associate at the University of Geneva from 1990 to 1993, and University of Wisconsin-Madison from 1993 to 1995. He began his independent academic career at Peking University in 1995. His research interests include the catalytic metal carbene transformations.

Austin G. Smith was born in Hartford, CT in 1984. He graduated in 2006 with a B.A. in chemistry from the College of the Holy Cross in Worcester, MA. He obtained his Ph.D from the University of North Carolina at Chapel Hill in 2011, where he studied stereoselective ring-opening reactions with donor-acceptor cyclopropanes under the direction of Professor Jeffrey Johnson. From 2011-2012 he was a postdoctoral research associate under the direction of Professor Huw M. L. Davies at Emory University. Austin joined the Chemical Process Research & Development group at Amgen in Thousand Oaks, CA in 2012.

rganic
yntheses

Preparation of Alkanesulfonyl Chlorides from *S*-Alkyl Isothiourea Salts *via N*-Chlorosuccinimide Mediated Oxidative Chlorosulfonation

Zhanhui Yang and Jiaxi Xu[1*]

State Key Laboratory of Chemical Resource Engineering, Department of Organic Chemistry, Faculty of Science, Beijing University of Chemical Technology, Beijing 100029, People's Republic of China

Checked by Miika Karjomaa, Yoichi Hirano, and Keisuke Suzuki

A. Bn—Cl + $\underset{H_2N}{\overset{S}{\underset{}{\parallel}}}NH_2$ $\xrightarrow[\text{1 h}]{\text{EtOH, reflux}}$ $\underset{Bn-S}{\overset{H_2N}{=}}NH_2^+ \; Cl^-$

1

B. $\underset{Bn-S}{\overset{H_2N}{=}}NH_2^+ \; Cl^-$ $\xrightarrow[\text{10 - 30 °C}]{\text{NCS, MeCN - 2 M HCl}}$ $BnSO_2Cl$

1   **2**

Procedure

During the process, there may be chlorine gas generated, so the experiments should be performed in a well ventilated hood.

A. *S-Benzyl Isothiouronium Chloride (**1**).* A 500-mL one-necked pear-shaped flask equipped with a 3-cm oval magnetic stir bar is sequentially charged with thiourea (19.1 g, 250 mmol, 1 equiv) (Note 1), benzyl chloride (31.7 g, 250 mmol, 1 equiv) (Note 2) and ethanol (250 mL) (Note 3). Then the flask is equipped with a condenser and the reaction mixture is heated to reflux with a 96 °C oil bath ($\Phi = 18$ cm, $h = 11$ cm) and stirred for 1 h. After the reaction mixture is cooled to ambient temperature, the ethanol is removed on a rotary evaporator (25 mmHg, 55 °C), and 57.9–65.7 g of a white solid is obtained as the title product **1** in quantitative yield (Note 4). Without further purification, the solid is ground with a ceramic mortar and pestle and used directly in Procedure B.

B. *Phenylmethanesulfonyl Chloride* (**2**). A 1-L three-necked round-bottomed flask is sequentially charged with *N*-chlorosuccinimide (133.5 g, 1000 mmol, 4 equiv) (Note 5) and MeCN (400 mL) (Note 6). The flask is immersed into a 10 °C water bath (Φ = 24.5 cm, h = 16 cm) and equipped with an overhead mechanical stirrer (paddle size: w = 8 cm, h = 1.5 cm), a thermometer fitted with a thermometer adapter and an addition funnel. After addition of 2 M HCl (70 mL), the addition funnel was exchanged with a solid-addition funnel. After the mixture is cooled to about 10 °C, the ground *S*-benzyl isothiouronium chloride solid (50.7 g, 250 mmol, 1 equiv) is slowly added through the solid addition funnel over the course of 30 min. In the initial 7–10 min of the addition, the reaction mixture turns into a clear yellow solution. During the addition, the temperature of the reaction mixture is kept below 30 °C. When the temperature reaches 25 °C, the addition is paused to allow the temperature to decrease to around 15 °C. Residual solid on the spatula, the solid-addition funnel and the addition neck of the flask is washed into the reaction mixture with a portion of the reaction mixture (15 mL) and then MeCN (10 mL). The 500-mL pear-shaped flask used in Step A is also rinsed with a portion of the reaction mixture (20 mL) and then with MeCN (20 mL). The resultant yellow solutions are transferred into the reaction system. The reaction mixture is stirred for an additional 15 min to completely consume the starting material (Note 7). Tap water (300 mL) is added to the flask using an addition funnel and a white precipitate forms. Half of the heterogeneous solution is transferred into a 1-L flask and MeCN is removed on a rotary evaporator (1 mmHg, 15 °C). The remaining half is transferred to the same 1-L flask and again the MeCN is removed on a rotary evaporator (1 mmHg, 15 °C). A large amount of white solid appears in the flask. The solid is collected by filtration through a Büchner funnel (Φ = 10 cm) and washed with tap water (100 mL) (Note 8). The white solid is transferred into a 1-L one-necked flask and dissolved in MeCN (300 mL), followed by an addition of tap water (150 mL). The resultant mixture is evaporated on a rotary evaporator (1 mmHg, 15 °C) to remove the MeCN and again a large amount of white solid appears. The solid is collected by filtration through a Büchner funnel (Φ = 10 cm) and washed with tap water (100 mL). The white solid is transferred into a 1-L Erlenmeyer flask, dissolved into CH_2Cl_2 (300 mL) (Note 3) and dried over

Na$_2$SO$_4$ (40 g) (Note 3). The dried solution is filtered into a 1-L one-necked pear-shaped flask through a sintered funnel and washed with CH$_2$Cl$_2$ (150 mL). The solvent is removed on a rotary evaporator (25 mmHg, 55 °C) and the resulting white solid is dried for 2 h at 0.5 mmHg in a vacuum drying oven (40 °C) (Note 9). The desired product **2** is obtained as a white solid (39.4–40.9 g, 83–86%) (Note 10).

Notes

1. Checkers purchased thiourea (≥99%) from Tokyo Chemical Industry Co., Ltd. and used it as received. Submitters purchased thiourea (≥99%) from Beijing Gaili Fine Chemical Co., Ltd. and used it as received.
2. Checkers purchased benzyl chloride (≥99%) from Tokyo Chemical Industry Co., Ltd. and used it as received. Submitters purchased benzyl chloride (≥99%) from Beijing Xingjin Chemical Factory and used it as received.
3. Checkers purchased ethanol (≥99.7%) from Wako Pure Chemical Industries, dichloromethane (≥99.5%) from Asahi Glass Co., Ltd. and anhydrous sodium sulfate (≥99%) from Shikoku Chemical Co. Submitters purchased ethanol (≥99.7%), dichloromethane (≥99.5%) and anhydrous sodium sulfate (≥99%) from Beijing Chemical Works and both checkers and submitters used the above mentioned chemicals as received.
4. *S*-Benzyl isothiouronium chloride has the following physical and spectroscopic properties: The crude solid started melting at 100 °C, but decomposed (turned brown) at 167 °C. The remaining solid turned liquid at 182 °C. ^{1}H NMR (600 MHz, D$_2$O) δ: 4.41 (s, 2 H), 7.40–7.55 (m, 5 H); ^{13}C NMR (600 MHz, D$_2$O) δ: 35.1, 128.5, 128.9, 129.1, 134.2, 170.5. Anal. calcd. of the crude material for C$_8$H$_{11}$ClN$_2$S: C, 47.40; H, 5.47; N, 13.82; S, 15.82. Found: C, 46.39; H, 5.45; N, 14.66; S, 16.80.
5. Checkers purchased *N*-chlorosuccinimide (≥98%) from Kanto Chemical Co., Ltd. Submitters purchased *N*-chlorosuccinimide (≥98%) from Beijing Ouhe Technology Co., Ltd. and both checkers and submitters used the chemicals as received.
6. Checkers purchased MeCN (≥99%) from Nacalai Tesque Inc. Submitters purchased MeCN (≥99%) from Tianjin Fuchen Chemical Reagents

Factory and both checkers and submitters used the chemicals as received.

7. When the submitters performed the procedure, the temperature of the tap water was about 10 °C, so they directly used the tap water in the water bath.

8. The reaction process is monitored by TLC on silica gel 60 F_{254} plate. For phenylmethanesulfonyl chloride, R_f = 0.67 (hexanes:EtOAc = 5:1 v/v); for S-benzyl isothiouronium chloride, R_f = 0.28 (CHCl$_3$:MeOH = 9:1 v/v). Plates were visualized by UV and an acidic phosphomolybdic acid stain [preparation: phosphomolybdic acid (11 g), sulfuric acid (23 mL), phosphoric acid (6.8 mL) and water (455 mL)].

9. Submitters reported that upon drying under an infrared lamp, the product was contaminated with trace amount of succinimide as indicated by ^{1}H NMR. However, when the product is synthesized on 50-mmol scale, this problem can be avoided. Checkers observed from ^{1}H NMR that the product was contaminated with < 1% of succinimide.

10. Phenylmethanesulfonyl chloride has the following physical and spectroscopic properties: mp 91–93 °C; ^{1}H NMR (600 MHz, CDCl$_3$) δ: 4.86 (s, 2 H), 7.43–7.51 (m, 5 H); ^{13}C NMR (600 MHz, CDCl$_3$) δ: 70.9, 126.1, 129.2 (2C), 130.3, 131.4 (2C); Anal. calcd. for C$_7$H$_7$ClO$_2$S: C, 44.10; H, 3.70; S, 16.82. Found: C, 44.05; H, 3.71; S, 17.01. IR (ATR): 2991, 2920, 1363, 1157, 1136, 772, 692 cm^{-1}. When the checkers dried the product, a small amount of product sublimed and formed colorless needles on the aluminum foil covering the beaker. The crystals were separated, identified by ^{1}H NMR as product, and combined with the bulk material for characterization.

Handling and Disposal of Hazardous Chemicals

The procedures in this article are intended for use only by persons with prior training in experimental organic chemistry. All hazardous materials should be handled using the standard procedures for work with chemicals described in references such as "Prudent Practices in the Laboratory" (The National Academies Press, Washington, D.C., 2011 www.nap.edu). All chemical waste should be disposed of in accordance with local regulations. For general guidelines for the management of chemical waste, see Chapter 8 of Prudent Practices.

These procedures must be conducted at one's own risk. *Organic Syntheses, Inc.*, its Editors, and its Board of Directors do not warrant or guarantee the safety of individuals using these procedures and hereby disclaim any liability for any injuries or damages claimed to have resulted from or related in any way to the procedures herein.

Discussion

The alkanesulfonyl chlorides represent an important class of compounds that have been widely applied as intermediates in synthetic organic chemistry and as building blocks in medicinal chemistry. In general, these compounds can be accessed (1) by chlorination of the corresponding sulfonic acids or their sodium salts with chlorinating reagents,[2] or (2) by oxidative chlorosulfonation of sulfur-containing compounds such as thiols, disulfides, thioacetates and thiocarbamates by oxidizing reagents. The first method suffers from drawbacks such as harsh conditions, long reaction time, low functional group tolerance, required use of an excess of chlorinating reagents, and the formation of acidic and/or toxic byproducts. The second method always involves the use of hazardous and toxic chlorine gas, which is not conveniently handled in laboratory operation. Therefore, a series of mild oxidizing reagents such as H_2O_2–$SOCl_2$,[3a-d] O_3–$SOCl_2$,[3e] SO_2Cl_2–KNO_3,[3f] Br_2–$POCl_3$,[3g] NCS–HCl,[3h-k] and TMSCl–KNO_3[3l] have been developed to overcome the drawbacks of chlorine gas. However, most of the substrates employed in these reactions have strong and repulsive odors, and these substrates have the potential to do great harm to environments and human health.

The present procedure describes a convenient preparation of alkanesulfonyl chlorides via *N*-chlorosuccinimide-mediated oxidative chlorosulfonation of *S*-alkyl isothiourea salts, of which the oxidative chlorosulfonation mediated by chlorine gas was firstly reported by Sprague and coworkers.[4] In this preparation, the drawbacks of both the pollutant thiol derivatives and hazardous and toxic chlorine gas are successfully overcome. The procedure is a convenient, environment- and worker-friendly method to synthesize alkanesulfonyl chlorides.

Org. Synth. **2014**, *91*, 116-124

DOI: 10.15227/orgsyn.091.0116

Table 1. Synthesis of Alkanesulfonyl Chlorides from *S*-Alkyl Isothiourea Salts by *N*-Chlorosuccinimide Mediated Oxidative Chlorosulfonation.[a]

$$RX + \underset{H_2N}{\overset{S}{\bigwedge}}NH_2 \xrightarrow[\text{1 h}]{\text{EtOH, reflux}} RS\underset{}{\overset{\overset{+}{N}H_2 \quad X^-}{\bigwedge}}NH_2 \xrightarrow{\text{NCS, MeCN - 2 M HCl}} RSO_2Cl$$

Entry	RX	Product	Yield (%)[b]
1	(benzyl–Cl)	(benzyl–SO₂Cl)	88
2	Me–(benzyl)–Cl	Me–(benzyl)–SO₂Cl	86
3	Cl–(benzyl)–Cl	Cl–(benzyl)–SO₂Cl	96
4	⟨C14⟩–Br	⟨C14⟩–SO₂Cl	92
5[c,d]	Br–⟨⟩–Br	ClO₂S–⟨⟩–SO₂Cl	99[c,d]

[a]Reaction in Entry 1 was conducted in 250 mmol scale, and reactions in Entries 2–6 in 50 mmol scale, based on the starting materials RX. [b]Isolated yield for two steps. [c]Reaction conducted at 30-40 °C. [d]Two equivalents of thiourea, 2 M HCl, and NCS were used.

References

1. State Key Laboratory of Chemical Resource Engineering, Department of Organic Chemistry, Faculty of Science, Beijing University of Chemical Technology, Beijing 100029, People's Republic of China, Fax/Tel: +86 10 64435565; E-mail: jxxu@mail.buct.edu.cn. The project was supported by The National Basic Research Program of China (No. 2013CB328905) and National Natural Science Foundation of China (Nos. 21372025 and 21172017).

2. (a) Bossard, H. H.; Mory, R.; Schmid, M.; Zollinger, H. *Helv. Chim. Acta* **1959**, *42*, 1653–1658. (b) Albright, J. D.; Benz, E.; Lanzilotti, A. E.; Goldman, L. *Chem. Commun.* **1965**, 413–414. (c) Fujita, S. *Synthesis* **1982**,

423–424. (d) Barco, A.; Benetti, S.; Pollini, G. P.; Taddia, R. *Synthesis* **1974**, 877–878. (e) Johary, N. S.; Owen, L. N. *J. Chem. Soc.* **1955**, 1307–1311. (f) Blotny, G. *Tetrahedron Lett.* **2003**, *44*, 1499–1501. (g) Brouwer, A. J.; Monnee, M. C. F.; Liskamp, R. M. J. *Synthesis* **2000**, 1579–1584. (h) He, F. D.; Meng, F. H.; Song, X. Q.; Hu, W. X.; Xu, J. X. *Org. Lett.* **2009**, *11*, 3922–3925.

3. (a) Monnee, M. C. F.; Marijne, M. F.; Brouwer, A. J.; Liskamp, R. M. J. *Tetrahedron Lett.* **2000**, *41*, 7991–7995. (b) Piatek, A.; Chapuis, C.; Jurczak, J. *Helv. Chim. Acta* **2002**, *85*, 1973–1988. (c) Humljan, J.; Gobec, S. *Tetrahedron Lett.* **2005**, *46*, 4069–4072. (d) Bahrami, K.; Khodaei, M. M.; Soheilizad, M. *J. Org. Chem.* **2009**, *74*, 9287–9291. (e) Kværnø, L.; Werder, M.; Hauser H.; Carreira, E. M. *Org. Lett.* **2005**, *7*, 1145–1148. (f) Park, Y. J.; Shin, H. H.; Kim, Y. II. *Chem. Lett.* **1992**, 1483–1486. (g) Meinzer, A.; Breckel, A.; Thaher, B. A.; Manicone, N.; Otto, H.-H. *Helv. Chim. Acta* **2004**, *87*, 90–105. (h) Kim, D. W.; Ko, Y. K.; Kim, S. H. *Synthesis* **1992**, 1203–1204. (i) Nishiguchi, A.; Maeda, K.; Miki, S. *Synthesis* **2006**, 4131–4134. (j) Liu, J.; Hou, S. L.; Xu, J. X. *Phosphorus Sulfur Silicon Relat. Elem.* **2011**, *186*, 2377–2391. (k) Meng, F. H.; Chen, N.; Xu, J. X. *Sci China: Chem.* **2012**, *55*, 2548–2553. (l) Prakash, G. K. S.; Mathew, T.; Panja, C.; Olah, G. A. *J. Org. Chem.* **2007**, *72*, 5847–5850. (m) Yang, Z. H.; Xu, J. X., *Synthesis* **2013**, *45*, 1675–1682.

4. (a) Johnson, T. B.; Sprague, J. M. *J. Am. Chem. Soc.* **1936**, *58*, 1348–1352. (b) Sprague, J. M.; Johnson, T. B. *J. Am. Chem. Soc.* **1937**, *59*, 1837–1840.

Appendix
Chemical Abstracts Nomenclature (Registry Number)

Thiourea; (62-56-6)
Benzyl chloride; (100-44-7)
N-Chlorosuccinimide; (128-09-6)
S-Benzyl isothiouronium chloride; (1334419-16-7)
Phenylmethanesulfonyl chloride; (1939-99-7)

Jiaxi Xu received his B. S. and Ph. D. degrees in chemistry at Peking University, followed by postdoctoral studies at Beijing Medical University. He also worked as a visiting scholar at the Chinese University of Hong Kong, and as postdoctoral fellow at Colorado State University and Vanderbilt University. He was nominated professor at Peking University in 2004. He moved to Beijing University of Chemical Technology in 2007. His scientific interests focus on the organic synthesis, asymmetric catalysis, and peptide chemistry.

Zhanhui Yang was born in 1987 and raised in Xuchang County, He'nan Province, China. In 2010, he received his B. S. degree from Beijing University of Chemical Technology. He is currently pursuing his doctoral studies under the supervision of Prof. Jiaxi Xu at the same university. His research now focuses on the simple and green synthesis of important sulfonic acid derivatives, the formation mechanism of β-sultams via reactions of sulfonyl chlorides and imines, and the carbene (carbenoid) reactions of diazosulfonyl compounds.

Miika Karjomaa was born in 1990 in Finland and he was accepted to Aalto University School of Chemical Technology in 2010. He completed his Bachelor's thesis on "Industrial applications of the Baeyer-Villiger reaction" in 2013. Currently, he is continuing his studies at the Suzuki-Ohmori laboratory in Tokyo Institute of Technology.

Organic
Syntheses

Yoichi Hirano was born in 1989 in Tokyo, Japan. He received his BS degree from Tokyo Institute of Technology in 2011, after which he continued his graduate studies in Professor Keisuke Suzuki's group. His study focuses on the synthesis of natural products

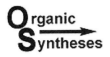
DABCO-*bis*(sulfur dioxide), DABSO, as an easy-to-handle source of SO₂: Sulfonamide preparation

Edward J. Emmett and Michael C. Willis[1]*

University of Oxford, Chemistry Research Laboratory, 12 Mansfield Road, Oxford, OX1 3TA, United Kingdom

Checked by Changming Qin and Huw M. L. Davies

Procedure

*A. 3-Methoxy-N-morpholinobenzenesulfonamide (**1**).* An oven-dried (Note 1), two-necked 250 mL round-bottomed flask containing a dried magnetic stirrer bar (26 × 12 mm oval) is fitted with rubber septa and allowed to cool to room temperature under vacuum (0.3 mmHg). 1,4-Diazabicyclo[2.2.2]octane bis(sulfur dioxide) adduct (DABSO) (3.60 g, 15.0 mmol, 0.6 equiv) (Note 2), 1,4-diazabicyclo[2.2.2]octane (DABCO)

(1.40 g, 12.5 mmol, 0.5 equiv), palladium(II) acetate (280 mg, 1.25 mmol, 5 mol%) and tri-*tert*-butylphosphonium tetrafluoroborate (510 mg, 1.75 mmol, 7 mol%) are weighed in air into a vial (Notes 3 and 4). The vacuum is switched to a positive argon pressure and the solids loaded into the flask. The flask is evacuated (0.3 mmHg) and filled with argon three times, being careful to avoid excessive dispersion of the solids. 3-Iodoanisole (3.0 mL, 5.85 g, 25.0 mmol, 1.0 equiv), 4-aminomorpholine (2.9 mL, 3.06 g, 30.0 mmol, 1.2 equiv) and 1,4-dioxane (140 mL) (Note 5) are added in this order *via* syringe through the septum. The flask, connected to the argon line, is clamped and placed in a pre-heated oil bath and stirred for 16 h at 70 °C (Note 6).

The reaction mixture, now a yellow solution with brown precipitate, is removed from the oil bath, allowed to cool to room temperature, diluted with DCM (50 mL) (Note 7), stirred for 5 min and filtered through a Celite® pad under low vacuum (Note 8). The reaction flask is rinsed clean (with sonication if required) with DCM (2 × 50 mL) and the Celite® pad washed with further DCM (100 mL). The filtrate is concentrated to dryness in a 500 mL round-bottomed flask on a rotary evaporator (40 °C, 40 mmHg) to give a partially solidified orange oil. The crude mixture is dissolved in DCM (approximately 100 mL) and silica gel (20 g) added. The DCM is removed on a rotary evaporator (40 °C, 40 mmHg) and the free-flowing silica with crude reaction mixture absorbed is dry-loaded onto a column of silica gel (Note 9). The column is run using a stepwise elution gradient from 50 – 100% Et$_2$O in hexane (Notes 7). The solvent is removed on a rotary evaporator (40 °C) and the solid dried under high-vacuum (room temperature, 0.1 mmHg, 4.0 h) to yield 3-methoxy-*N*-morpholinobenzenesulfonamide (**1**) as a white crystalline solid (5.14–5.45 g, 76–80 %) (Note 12).

B. 1-((3-Methoxyphenyl)sulfonyl)piperidine (**2**). An oven-dried (Note 1) three-necked 500 mL round-bottomed flask containing a dried magnetic stirrer bar (34 × 14 mm oval) is fitted with rubber septa and a thermometer and allowed to cool to room temperature under vacuum (0.3 mmHg). The vacuum is switched to a positive argon pressure and DABSO (8.4 g, 35.0 mmol, 1.25 equiv) (Note 2), weighed in air, is added to the flask. The flask is evacuated (0.3 mmHg) and filled with argon three times, being careful to avoid excessive dispersion of the solids. THF (160 mL) (Note 13) is added *via* syringe through the septum and the flask cooled to –45 °C (internal temperature, Note 14). 3-Methoxyphenylmagnesium bromide solution (1M in THF, 28 mL, 28.0 mmol, 1.0 equiv) (Note 15) is added dropwise to the stirred white suspension *via* syringe pump (Note 16) and the reaction left stirring at the same temperature for 1.0 h. Sulfuryl chloride (2.3 mL, 3.78 g, 28.0 mmol, 1.0 equiv) (Note 15) is added dropwise *via* syringe pump (Note 17) and the resultant white suspension left stirring at the same temperature for 5 min. The cooling bath is removed, the suspension allowed to warm to room temperature (approximately 1 h) and left stirring for a further 30 min (Note 17). Piperidine (11.0 mL, 9.5 g, 112 mmol, 4.0 equiv) (Note 15) is added dropwise *via* syringe pump (Note 18) at room temperature and the reaction mixture left stirring overnight (16 h).

Water (75 mL) (Note 19) and EtOAc (200 mL) (Note 7) are added in one portion to the flask at room temperature and the resultant mixture stirred for 5 min. The reaction mixture is transferred to a 1.0 L separating funnel and water (200 mL) added. The funnel is gently shaken, the layers separated and the aqueous layer extracted with further EtOAc (3 × 100 mL). The combined organic layers are washed with saturated brine (200 mL), dried (MgSO$_4$, 15.0 g), filtered through a medium porosity sinter funnel under low vacuum and concentrated on a rotary evaporator (40 °C, 40 mmHg) to give an off-white solid.

The solid is dissolved in DCM (approximately 100 mL) (Note 7) and transferred to a 250 mL round-bottomed flask and silica gel (16 g) added. The DCM is removed on a rotary evaporator (40 °C, 40 mmHg) and the free-flowing silica with crude product absorbed is dry-loaded onto a column of silica gel (Note 20). The column is run using a stepwise elution gradient from 10 – 45% Et$_2$O in hexane (Notes 7 and 21) The solvent is removed on a rotary evaporator (40 °C, 40 mmHg) and dried under high-vacuum (room temperature, 0.1 mmHg, 4.0 h) to yield 1-((3-methoxyphenyl)sulfonyl)piperidine (**2**) as a crystalline white solid (6.28–6.42 g, 88–90%) (Note 23).

Notes

1. "Oven-dried" (and "dried" thereafter) refers to equipment kept for at least 24 h in an >200 °C oven.
2. DABSO was prepared by the *Organic Syntheses* procedure "Preparation of DABSO from Karl-Fischer Reagent" *Org. Synth.* **2013**, *90*, 301. The

submitter purchased DABSO from Sigma-Aldrich (≥95%). The lumps were broken-up and the material weighed out as a fine powder.

3. DABCO was bought from Alfa-Aesar (98%) and sublimed (50 °C, 0.3 mmHg) prior to use. Following sublimation it was ground to give a fine white powder. Due to DABCO's hygroscopicity, care should be taken, where possible, to minimize contact with the air. Purified DABCO was stored under a nitrogen atmosphere.

4. Palladium(II) acetate (Pd 45.9-48.2%) and tri-*tert*-butylphosphonium tetrafluoroborate (97%) were purchased from Alfa Aesar and used as received.

5. 3-Iodoanisole was purchased from Sigma Aldrich (99%) and 4-aminomorpholine from Alfa Aesar (98%) and both reagents were used as received. 1,4-Dioxane (anhydrous, 99.8%) was bought from Sigma Aldrich and was degassed with dry argon bubbling for 30 min prior to use. The submitter reported that 4-aminomorpholine decomposes over time to morpholine, therefore purity was confirmed to be 95% by ^{1}H NMR spectroscopy before use.

6. The initial orange-beige suspension changed to a dark brown solution in ca. 30 min. A precipitate begins to form after ca. 1 h. The submitter reported that the reaction can be monitored *via* TLC - product $R_f = 0.2$, 3-iodoanisole $R_f = 0.75$ (eluent 75% Et_2O in petroleum ether) - although starting material is not fully consumed before catalyst deactivation occurs.

7. Dichloromethane (DCM; HPLC grade) was purchased from Sigma-Aldrich; Hexanes (Certified ACS), Et_2O (Certified ACS) and EtOAc (Certified ACS) were purchased from Fisher Chemical and used without further purification.

8. A Celite® 545 pad (30 g) was packed with DCM in a 7 cm diameter medium porosity sintered glass funnel under low vacuum.

9. Flash column chromatography was performed with Silicycle Silica Flash® Si P60 silica gel (40 – 63 μm) (125 g slurried in hexane) in an 8.0 cm diameter column with a thin band of sand protecting the top of the silica under a slightly positive pressure.

10. Stepwise elution gradient as follows: 700 mL 50%, 500 mL 60%, 500 mL 70%, 500 mL 80%, 100% Et_2O in hexanes until no further product was detected by TLC.

11. The desired product has an $R_f = 0.13$ when a TLC is run in 75% EtOAc in hexanes as the eluent (3-iodoaniosle $R_f = 0.84$). The product is

visualized using a UV lamp and staining in phosphomolybdic acid (giving a dark-blue spot).

12. Melting point was measured by the dry sample, mp 116–117 °C; ^{1}H NMR (400 MHz, CDCl$_3$, 7.27 ppm) δ: 2.64 (t, J = 4.8 Hz, 4 H (NCH$_2$CH$_2$O)$_2$), 3.62 (t, J = 4.8 Hz, 4 H, (NCH$_2$CH$_2$O)$_2$), 3.87 (s, 3 H, OMe), 5.60 (s, 1 H, NH), 7.14 (dd, J = 8.0, 2.0 Hz, 1 H, Ar-H), 7.43 (t, J = 8.0 Hz, 1 H, Ar-H), 7.49 (t, J = 2.0 Hz, 1 H, Ar-H), 7.57 (d, J = 8.0 Hz, 1 H, Ar-H); ^{13}C NMR (100 MHz, CDCl$_3$, 77.0 ppm) δ: 55.6, 56.4, 66.5, 112.5, 119.4, 120.1, 129.8, 139.7, 159.6; IR ν_{max} (film)/cm^{-1} 3212, 2963, 2856, 1597, 1478, 1433, 1361, 1317, 1287, 1242, 1157, 1109, 1034, 864, 685 (principal peaks); HRMS (FTMS+ p-NSI) found m/z 295.0721 [M+Na]$^+$, C$_{11}$H$_{16}$N$_2$O$_4$SNa requires m/z 295.0723. Reverse phase HPLC analysis reveals purity >99% (run on an Agilent Zorbax SB-C18, 5 μm, 4.6 × 150 mm column (23 °C) at a flow rate of 1.5 mL/min of 70:30 MeCN:H$_2$O observed at 210 nm giving a retention time of 1.32 min, 1.0 mg/mL in CH$_3$CN).

13. THF (HPLC) was used as dried by an in-house solvent purification system (SG Water USA LLC) having passed through anhydrous alumina columns. It was degassed by bubbling argon for 30 min before use. The submitter determined the water content in THF was <10 ppm by regular Karl Fischer titrations.

14. Cooling was achieved using a dry-ice-acetone bath in a Dewar flask and maintained within ±5 °C of the specified temperature by addition of dry ice as necessary.

15. 3-Methoxyphenylmagnesium bromide solution and piperidine (99%) were purchased from Sigma Aldrich and used as received. Sulfuryl chloride (97%) was purchased from Sigma Aldrich and fractionally distilled (with a Vigreux column) under argon to give a colorless fraction (oil bath temperature, 85 °C, 1 atm) on the day of use. Piperidine was used in significant excess because it was used as both a base and a nucleophile. In addition, residual SO$_2$ (from DABSO and the oxidative chlorination) can complex to piperidine, thereby removing it from availability for sulfonylation. A reduction in loading leads to a reduction in sulfonamide yield. The submitter determined the molarity of Grignard reagent to be at least 1M by titration with salicylaldehyde phenylhydrazone.[17]

16. Added at a rate of 0.4 mL/min. The internal temperature was maintained between –40 and –50 °C. The white suspension thinned as

the Grignard reagent was added before re-thickening to form another white suspension.

17. Added at a rate of 0.15 mL/min. The internal temperature was maintained between –40 and –50 °C. Upon warming to room temperature the white suspension turned to a yellow suspension.

18. Added at a rate of 0.3 mL/min. Although a slight exotherm was observed, the internal temperature did not exceed 40 °C at this addition rate. The yellow suspension became white by the end of the addition.

19. ASTM type II water, produced by RO/DI (HARLECO) was used as received.

20. Flash column chromatography was performed with Silicycle Silica Flash® Si P60 silica gel (40 – 63 μm) (200 g slurried in hexane) in an 8.0 cm diameter column with a thin band of sand protecting the top of the silica under slightly positive pressure.

21. Stepwise elution gradient as follows: 500 mL 10%, 500 mL 15%, 500 mL 20%, 1.5 L 25%, 1.5 L 35%, 45% Et$_2$O in hexanes until no product was detected by TLC.

22. The product has an R$_f$ = 0.44 when a TLC is run in 75 % EtOAc in hexane as the eluent. The product is visualized using a UV lamp.

23. Melting point was measured by the dry sample, mp 115–116 °C; ^{1}H NMR (400 MHz, DMSO-d$_6$, 2.50 ppm) δ: 1.30–1.35 (m, 2 H, N(CH$_2$CH$_2$)$_2$CH$_2$), 1.47–1.52 (m, 4 H, N(CH$_2$CH$_2$)$_2$CH$_2$), 2.85 (t, J = 5.2 Hz, 4 H, N(CH$_2$CH$_2$)$_2$CH$_2$), 3.83 (s, 3 H, OMe), 7.16 (t, J = 2.1 Hz, 1 H, Ar-H), 7.25–7.30 (m, 2 H, Ar-H), 7.55 (t, J = 8.0 Hz, 1 H, Ar-H); ^{13}C NMR (100 MHz, DMSO-d$_6$, 39.5 ppm) δ: 22.8, 24.7, 46.6, 55.6, 112.3, 118.7, 119.5, 130.5, 136.7, 159.5; IR ν$_{max}$ (film)/cm^{-1} 2940, 2851, 1597, 1478, 1359, 1340, 1318, 1287, 1241, 1167, 1098, 1040, 931, 856, 724, 688; (principal peaks); HRMS (FTMS+p-NSF) found m/z 256.1002 [M+H]$^+$, C$_{12}$H$_{18}$NO$_3$S requires m/z 256.1002. Reverse phase HPLC analysis reveals purity >99% (run on an Agilent Zorbax SB-C18, 5 μm, 4.6 × 150 mm column (23 °C) at a flow rate of 1.5 mL/min of 75:25 MeCN:H$_2$O observed at 210 nm giving a retention time of 1.95 min, 1.0 mg/mL in MeCN).

Handling and Disposal of Hazardous Chemicals

The procedures in this article are intended for use only by persons with prior training in experimental organic chemistry. All hazardous materials should be handled using the standard procedures for work with chemicals described in references such as "Prudent Practices in the Laboratory" (The National Academies Press, Washington, D.C., 2011 www.nap.edu). All chemical waste should be disposed of in accordance with local regulations. For general guidelines for the management of chemical waste, see Chapter 8 of Prudent Practices.

These procedures must be conducted at one's own risk. *Organic Syntheses, Inc.*, its Editors, and its Board of Directors do not warrant or guarantee the safety of individuals using these procedures and hereby disclaim any liability for any injuries or damages claimed to have resulted from or related in any way to the procedures herein.

Discussion

Despite its established utility in organic synthesis,[2] sulfur dioxide remains an underexploited reagent. The reasons for this are arguably the issues associated with both its toxicity and the specialist equipment required to handle gaseous reagents safely. A bench-stable, easy-to-use solid equivalent of sulfur dioxide would therefore be of considerable value to practicing synthetic chemists. We proposed that amine-sulfur dioxide complexes, many of which are solid, could provide such an equivalent.

Amine-sulfur dioxide charge transfer complexes of the form $R_3N{\rightarrow}SO_2$ (R = C or H), have been reported in the literature as early as 1900.[3] Although they have been studied widely, both theoretically[4] and spectroscopically,[5] to investigate the nature of the dative bonding, they appear to have found only a few practical uses. Olah used trimethyl(ethyl)amine complexes as reducing agents,[6] for example deoxygenating *N*-oxides to amines, nitro groups to nitriles and sulfoxides to sulfides; whilst Rudkevich developed a metalloporphyrin assay for sulfur dioxide concentration based on this Lewis acid-base chemistry.[7] However, they occur most extensively in industrial settings where amine solutions have been patented as sulfur dioxide gas effluent scrubbers.[8] We found that the complex formed from sulfur dioxide and 1,4-diazabicyclo[2.2.2]octane (DABCO), DABCO-

bis(sulfur dioxide), which we have abbreviated to DABSO, proved to be an excellent choice as it was an air-stable microcrystalline white powder which underwent a range of desired transformations functioning as a sulfur dioxide equivalent. This complex was first synthesized by Santos and Mello for spectroscopic investigation, but the compound, to the best of our knowledge, has not been reported on since.[9]

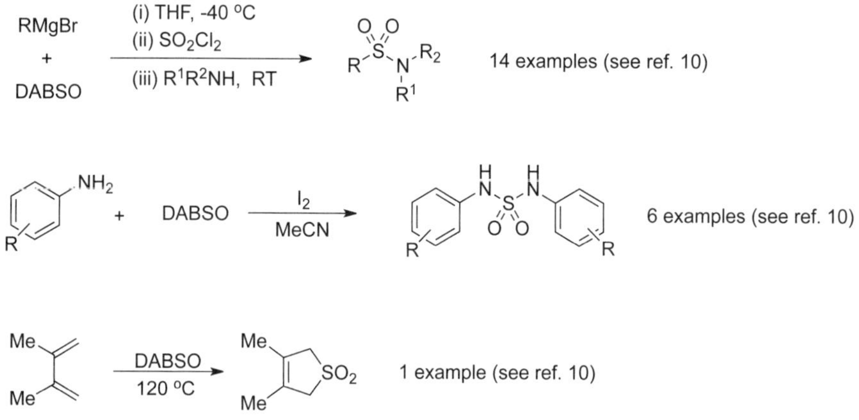

Scheme 1. Replacement of SO₂ gas with DABSO in known transformations

We successfully used DABSO in place of sulfur dioxide gas in a range of known transformations (Scheme 1);[10] sulfonamides, sulfamides and sulfolenes were synthesized by adapting procedures originally developed for the gaseous reagent.

In addition to this, we have developed the first palladium-catalyzed aryl halide cross coupling to incorporate sulfur dioxide.[11] Inspired by the prevalence of palladium-catalyzed carbonylations,[12] we noted several similarities between CO and SO₂. These included frontier molecular orbital symmetry,[13] their behaviors as ligands for transition metals[14] and their ability to insert into metal-alkyl bonds.[15] By using aryl, heterocyclic or alkenyl iodides in tandem with *N,N*-disubstituted hydrazines and DABSO with a Pd(OAc)₂/ⁱBu₃P catalytic system, *N*-aminosulfonamides were delivered in good to excellent yields (Scheme 2).[11] This amino*sulfonylation* is a sulfonylative analogue of the well precedented amino*carbonylation*, where aryl halide, carbon monoxide and amine combine under palladium catalysis to furnish amides.[12] It is also noted that the controlled loading of sulfur dioxide in only slight excess that DABSO allows, is key to this reactivity. Saturation with SO$_2$ gas in place of DABSO leads only to a very poor yield.

X = I or Br
Z = N, O or S

43 examples
(see ref. 11)

Scheme 2. Our novel palladium-catalyzed aminosulfonylation

In this *Organic Syntheses* publication, we demonstrate the utility of DABSO in two methods of sulfonamide formation; one *via* our novel palladium-catalyzed aminosulfonylation and the other using a Grignard reagent (Scheme 1). The sulfonamide functional group is particularly relevant to the pharmaceutical industry, present in 10% of the top 50 grossing pharmaceutical products of 2009.[16] Our methods of sulfonamide formation allow the –SO$_2$N– moiety to be introduced into a wide variety of positions in the molecule, demonstrating the potential for a diverse library synthesis. Current methods of sulfonamide formation (such as traditional electrophilic aromatic sulfonylation chemistry) would not provide such diversity without resorting to sulfur dioxide gas – a problem which is now eliminated by the use of DABSO.

References

1. University of Oxford, Chemistry Research Laboratory, 12 Mansfield Road, Oxford, OX1 3TA, United Kingdom. E-mail: michael.willis@chem.ox.ac.uk
2. For a review see: Florjanczyk, Z.; Raducha, D. *Pol. J. Chem.* **1995**, *69*, 481–508. For recent advancements see: Vogel, P.; Turks, M.; Bouchez, L.; Markovic, D.; Varela-Alvarez, A.; Sordo, J. A. *Acc. Chem. Res.* **2007**, *40*, 931–942.
3. Divers, E.; Ogawa, M. *J. Chem. Soc., Trans.* **1900**, *77*, 327–335.
4. Wong, M. W.; Wiberg, K. B. *J. Am. Chem. Soc.* **1992**, *114*, 7527–7535.

5. For selected examples see: (a) Faria, D. L. A.; Santos, P. S. *J. Raman Spectrosc.* **1988**, *19*, 471–478; (b) Hata, T.; Kinumaki, S. *Nature* **1964**, *203*, 1378–1379; (c) Oh, J. J.; LaBarge, M. S.; Matos, J.; Kampf, J. W.; Hillig, K. W.; Kuczkowski, R. L. *J. Am. Chem. Soc.* **1991**, *113*, 4732–4738; (d) Byrd, W. E. *Inorg. Chem.* **1962**, *1*, 762–768.

6. (a) Olah, G. A.; Arvanaghi, M.; Vankar, Y. D. *Synthesis* **1980**, 660–661; (b) Olah, G. A.; Vankar, Y. D.; Gupta, B. G. B. *Synthesis* **1979**, 36–37; (c) Olah, G. A.; Vankar, Y. D.; Arvanaghi, M. *Synthesis* **1979**, 984–985.

7. Leontiev, A. V.; Rudkevich, D. M. *J. Am. Chem. Soc.* **2005**, *127*, 14126–14127.

8. Klass, D. L.; Conrad J. R. Removal of Sulfur Dioxide from Waste Gases. U.S. Patent 4,208,387, Jun 17, 1980.

9. Santos, P. S.; Mello, M. T. S. *J. Mol. Struct.* **1988**, *178*, 121–133.

10. Woolven, H.; Gonzalez-Rodriguez, C.; Marco, I.; Thompson, A. L.; Willis, M. C. *Org. Lett.* **2011**, *13*, 4876–4878.

11. (a) Nguyen, B.; Emmett, E. J.; Willis, M. C. *J. Am. Chem. Soc.* **2010**, *132*, 16372–16373; (b) Emmett, E. J.; Richards-Taylor, C. S.; Nguyen, B.; Garcia-Rubia, A.; Hayter, B. R.; Willis, M. C. *Org. Biomol. Chem.* **2012**, *10*, 4007–4014.

12. For a review see: Brennfuhrer, A.; Neumann, H.; Beller, M. *Angew. Chem.Int. Edit.* **2009**, *48*, 4114–4133.

13. Lloyd, D. R.; Roberts, P. *J. Mol. Phys.* **1973**, *26*, 225–230.

14. For selected examples see: (a) Schenk, W. A. *Agnew. Chem. Int. Edit.* **1987**, *26*, 98–109; (b) D. L. Lichtenberger, A. Rai-Chaudhuri, and R. H. Hogan, in *Inorganometallic Chemistry*, Ed. T. P. Fehlner, Plenum Press, New York-London, 1992, Ch. 5, 223; (c) Kubas, G. J. *Acc. Chem. Res.* **1994**, *27*, 183–190.

15. For a selected examples see: (a) Wojcicki, A. *Acc. Chem. Res.* **1971**, *4*, 344–352; (b) Lefort, L.; Lachicotte, R. J.; Jones, W. D. *Organometallics* **1998**, *17*, 1420–1425.

16. Njardarson group:
    http://cbc.arizona.edu/njardarson/group/sites/default/files/Top200 PharmaceuticalProductsByWorldwideSalesin2009.pdf

17. Love, B. E.; Jones, E. G. *J. Org. Chem.* **1999**, *64*, 3755-3756.

Appendix
Chemical Abstracts Nomenclature (Registry Number)

DABCO: 1,4-Diazabicyclo[2.2.2]octane; (280-57-9)
SO$_2$: Sulfur dioxide; (7446-09-5)
DABSO, DABCO-*bis*(sulfur dioxide): 1,4-Diazoniabicyclo[2.2.2]octane, 1,4-disulfino-, bis(inner salt); (119752-83-9)
Palladium(II) acetate: Acetic acid, palladium(2+) salt (2:1); (3375-31-3)
Tri-*tert*-butylphosphonium tetrafluoroborate: Phosphine, tris(1,1-dimethylethyl)-, tetrafluoroborate(1-) (1:1); (131274-22-1)
3-Iodoanisole: Benzene, 1-iodo-3-methoxy-; (766-85-8)
4-Aminomorpholine: 4-Morpholinamine; (4319-49-7)
3-Methoxy-*N*-morpholinobenzenesulfonamide: Benzenesulfonamide, 3-methoxy-*N*-4-morpholinyl-; (1255365-27-5)
3-Methoxyphenylmagneisum bromide: Magnesium, bromo(3-methoxyphenyl)-; (36282-40-3)
1-((3-Methoxyphenyl)sulfonyl)piperidine: Piperidine, 1-[(3-methoxyphenyl)sulfonyl]-; (173681-65-7)

Michael Willis received his undergraduate education at Imperial College London, and his PhD from the University of Cambridge working with Prof. Steven V. Ley. After a postdoctoral stay with Prof. David A. Evans at Harvard University, as a NATO/Royal Society Research Fellow, he was appointed to a lectureship at the University of Bath in November 1997. In January 2007 he moved to the University of Oxford, where he is now a Professor of Chemistry and a Fellow of Lincoln College. His group's research interests are based on the development and application of new catalytic processes for organic synthesis.

Edward Emmett is originally from London and received his MChem from the University of Oxford in 2010. He joined Prof. Michael Willis's research group as an undergraduate student and remained to study for his DPhil in Organic Chemistry. His research focuses on the development of amine-sulfur dioxide complexes as SO_2 surrogates in both catalytic and non-catalytic methodologies.

Changming Qin was born in Shandong, China, in 1982. He did his undergraduate work at Ludong University on preparation polymer nanocomposites under the guidance of Prof. Yucai Hu. He obtained his Masters degree in organic chemistry in 2008 at Wenzhou University under the supervision of Prof. Huayue Wu while working on palladium-catalyzed transformations of aryl boronic acids. After graduation, he went to the University of Hong Kong to work with Prof. Chi-Ming Che in 2008-2009, and then joined Prof. Huw Davies' group at Emory University in 2010. His current research is focused on design and synthesis of chiral dirhodium catalysts and their application in novel asymmetric carbenoid transformations.

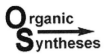
One-pot Preparation of (S)-N-[(S)-1-Hydroxy-4-methyl-1,1-diphenylpentan-2-yl]pyrrolidine-2-carboxamide from L-Proline

Wacharee Harnying, Nongnaphat Duangdee, and Albrecht Berkessel*[1]

Department of Chemistry (Organic Chemistry), University of Cologne, Greinstr. 4, 50939 Cologne, Germany

Checked by Takahiro Sakai and Keisuke Suzuki

A.

(S)-1 → PhMgBr → (S)-2

B.

(S)-3 + (S)-2

1. Boc₂O (1.1 equiv), NEt₃
2. ClCOOEt (1.0 equiv)
3.

→ (S,S)-4

4. conc. HCl
5. 6ɴ NaOH

→ (S,S)-5

Procedure

A. *(S)-2-Amino-4-methyl-1,1-diphenylpentan-1-ol (2)*. (Note 1) A 1-L three-necked round-bottomed flask (Note 2), equipped with a 4.5-cm oval Teflon-coated magnetic stirring bar, is charged with L-leucine methyl ester hydrochloride (1, 9.10 g, 50 mmol, 1.0 equiv) (Notes 3 and 4). The flask is equipped with a reflux condenser, fitted at the top with an argon balloon. Another neck is capped with a rubber septum and Et₂O (50 mL) (Note 5) is added to the flask via syringe. The remaining neck is equipped with a 100 mL dropping funnel. The reaction flask is cooled to 0–5 °C in an ice-water bath. By means of nitrogen pressure, the 3.0 M solution of phenylmagnesium bromide in Et₂O (83 mL, 250 mmol, 5.0 equiv) is transferred via an 18 gauge stainless cannula into the dropping funnel; this solution is added to the cold suspension of **1** over 10 min (Note 6). Upon completion of the addition, the reaction mixture is stirred for an additional 0.5 h in an ice-water bath and then overnight at room temperature (Note 7). The reaction mixture becomes a light yellow liquid mixed with white–brown precipitates. This heterogeneous mixture is cooled to 0–5 °C in an ice-water bath and cautiously quenched, with vigorous stirring, by slow addition of 10 wt% aqueous NH₄Cl (150 mL) (Note 8). Ethyl acetate (100 mL) (Note 9) is added and the ice-water bath is removed. The pH of the aqueous (lower) phase is adjusted to 7–8, as determined by pH paper, through addition of 4 M HCl (45-50 mL) in one portion. The resulting slurry is vigorously stirred until all solid dissolves (approximately 0.5 h) to give a clear two-phase mixture (Note 10). The mixture is then transferred to a 1-L separatory funnel using EtOAc (50 mL). The aqueous phase is separated and extracted with EtOAc (2 x 100 mL). The combined organic layers are washed with saturated NaHCO₃ solution (50 mL), water (50 mL) and brine (50 mL). The organic layer is dried over anhydrous MgSO₄ and filtered. The filtrate is concentrated by rotary evaporation (40 °C, 170 mmHg to 35 mmHg) (Note 11) and then further dried under vacuum (ca. 0.3 mmHg) at room temperature for 2–3 h. The title compound **2** is obtained as a pale yellow solid (12.2–12.9 g, 91–96%) (Notes 12 and 13) and used for the next step without further purification.

B.   *(S)-N-[(S)-1-Hydroxy-4-methyl-1,1-diphenylpentan-2-yl]pyrrolidine-2-carboxamide (5).* A 500-mL, oven-dried three-necked round-bottomed flask, equipped with a 4.5-cm oval Teflon-coated magnetic stir bar, is charged with L-proline (6.34 g, 55 mmol, 1.1 equiv) (Note 14). One neck of the flask is fitted with an argon balloon, and a second neck is equipped with a thermometer for measuring the internal temperature. The flask is charged with dichloromethane (120 mL) and the third neck capped with a rubber septum (Note 14). Di-*tert*-butyl dicarbonate (12.5 mL, 54.4 mmol, 1.1 equiv) is added by means of a syringe in one portion (Note 15). The resulting suspension is cooled to 0–5 °C using an ice-water bath. Triethylamine (27.0 mL, 194 mmol, 3.9 equiv) (Note 16) is added via a syringe over approximately 5 min. After completion of the addition, the ice-water bath is removed and the suspension is stirred at room temperature for 1 h. The reaction mixture becomes homogeneous as the proline reacts with Boc$_2$O to afford the intermediate N-Boc protected proline. The reaction mixture is then cooled to 0 °C in an ice-water bath. Ethyl chloroformate (5.0 mL, 53 mmol, 1.1 equiv) is added dropwise over 5 min by means of a syringe. During the addition, a white precipitate of triethylammonium hydrochloride is formed. Upon completion of the addition, the resulting slurry is stirred at 0 °C for an additional 0.5 h. The crude solid **2** is then added in one portion via a solid addition funnel. Dichloromethane (20 mL) (Note 14) is used to rinse the residue on the addition funnel into the reaction flask. After the addition is complete, the resulting milky suspension is stirred vigorously at 0 °C for 1 h and then allowed to warm to room temperature overnight (Note 17). The reaction mixture is then concentrated by rotary evaporation (40 °C, 100 mmHg) to dryness, providing the crude N-Boc proline amide **4** as a pale yellow solid. Dichloromethane (40 mL) and methanol (80 mL) are then added to the crude **4**, followed by dropwise addition of 12 N HCl (20 mL, ca. 240 mmol) (Note 18). After the addition, the flask is placed in an oil bath (50–55 °C) and the resulting slurry is vigorously stirred for 5 h. During this time, the slurry becomes homogeneous after 1.5 h as **4** is converted to the **5•HCl** salt. After the reaction is complete (Note 19), the resulting solution is concentrated by rotary evaporation (45 °C, 70 mmHg) until most of MeOH is removed, and the residue solidifies. The resulting solid is suspended in H$_2$O (100 mL) and EtOAc (200 mL) at 0–5 °C in an ice-water bath. The aqueous (lower) phase is adjusted to pH 8–9 with 6 M NaOH (35 mL). The ice-water bath is removed and the slurry is vigorously stirred at room temperature for 0.5 h. The resulting slurry of **5** (Note 20) is transferred to a 1-L separatory funnel using EtOAc (150 mL) to aid the transfer. After separation of the organic layer, the aqueous phase is extracted with EtOAc (2 x 100 mL). The combined organic layers are washed with 1 M NaOH (50 mL), H$_2$O (2 x 50 mL) and brine (50 mL), dried over MgSO$_4$, and filtered to give a pale-yellow solution (Note 21). The filtrate is concentrated by rotary evaporation (40 °C, 150 mmHg to 40 mmHg) (Note 22) and then further dried under vacuum (0.3 mmHg) at room temperature for 1–2 h, to give the crude product **5** as a white–pale yellow solid (14.9–18.0 g) (Note 23). The crude product is dissolved in EtOAc (11 mL per gram of crude **5**) at reflux (Note 24), and the product is allowed to crystallize at room temperature overnight. The resulting crystals are collected by suction filtration on a Büchner funnel, washed with 50 mL of ice-cold 5% EtOAc in hexane, and then air-dried to constant weight to provide the title compound **5** as a white crystalline solid (7.6–8.4 g, 42–46% based on **1**) (Notes 25 and 26).

Notes

1. The submitters recorded the following: To insure complete The checkers used phenylmagnesium bromide purchased from Sigma-Aldrich, 3.0 M in Et₂O solution. The submitters reported that the Grignard reagent was prepared using the following procedure, which was not checked: A 500-mL, oven-dried three-necked round-bottomed flask (Note 2), is equipped with a 4-cm oval Teflon-coated magnetic stir bar, a 125 mL pressure-equalizing addition funnel placed on the middle neck of the reaction flask, and a reflux condenser fitted at the top with an in-line oil bubbler connected to an argon line. The third neck is fitted with a glass stopper. The flask is purged with argon and an atmosphere of argon is maintained during the reaction. The flask is charged with dry magnesium turnings (6.1 g, 250 mmol, 5.0 equiv), and stirring is initiated to activate the magnesium surface. After stirring for 15 min, anhydrous Et₂O (50 mL) (Note 5) is added via the addition funnel. A solution of bromobenzene (39.3 g, 250 mmol, 5.0 equiv) (Note 3) in anhydrous Et₂O (100 mL) was transferred into the addition funnel and added dropwise. The clear reaction mixture becomes cloudy within 5 min, and soon thereafter begins to reflux and turns brown. The dropwise addition is continued for 45 min, at a rate to sustain a gentle reflux. After completion of the addition, the residue is rinsed from the addition funnel into the reaction flask with dry Et₂O (10 mL). The reaction mixture is then heated for 0.5 h using a warm-water bath of 45–50 °C. At this time, only a trace of magnesium metal is visible, and the solution has a dark brown, cloudy appearance. The flask is then removed from the warm-water bath, and the reaction mixture is stirred at room temperature for an additional 0.5–1 h.
2. Glassware was oven-dried (110 °C) before use.
3. The checkers used the following reagents in step A as received: L-leucine methyl ester hydrochloride (Tokyo Chemical Industry Co., Ltd., >98.0%). The submitters used the following reagents in step A: L leucine methyl ester hydrochloride (Alfa Aesar-Johnson Matthey Co., 99%).

4. The submitters recorded the following: To insure complete and clean conversion, L-leucine methyl ester hydrochloride (**1**) must be milled and/or delumped to a fine powder using a pestle and mortar and dried under vacuum (<0.1 mmHg, 2 h, 50 °C) prior to use. The checkers used commercial L-leucine methyl ester hydrochloride without pretreatment.

5. The checkers used anhydrous diethyl ether (Kanto Chemical Co., Inc., dehydrated, >99.5%), which was purified under argon by a solvent purifying unit (Wako Pure Chemical Industries, Ltd.). The submitters used anhydrous diethyl ether (Acros, 99%) that was freshly distilled under argon from Na/benzophenone.

6. Initially, the reaction is highly exothermic due to the Grignard reagent reacting with HCl resulting in the ether boiling. However, by slowly adding the Grignard reagent to the cooled suspension of **1**, this exotherm is easily controlled. The checkers added the Grignard reagent dropwise during 10 min. When half amount of Grignard reagent was added, the temperature had increased to 25 °C from 5 °C. Subsequently, the temperature gradually decreased to 17 °C.

7. The reaction is monitored by TLC; the checkers used the following analytical conditions. An aliquot (ca. 0.02 mL) is quenched by addition to saturated NH_4Cl (0.5 mL) and EtOAc (0.5 mL). The EtOAc layer was analyzed by TLC on silica gel with 1:9 $MeOH/CH_2Cl_2$ as eluent and visualization with phosphomolybdic acid. The starting material and product have the following R_f values: **1** as a free amine (0.62); **2** (0.55) on Merck TLC Plate 60 F_{254}.

8. The checkers noted that the addition of NH_4Cl solution is initially violent. When the first 50 mL of NH_4Cl solution was added carefully, the mixture solidified and the internal temperature rose to 35 °C from 6 °C. After the mixture was cooled in an ice bath to stop the boiling of diethyl ether, 100 mL of NH_4Cl was added dropwise over 5 min and the mixture became a yellow slurry.

9. For workup in steps A and B and for recrystallization, the checkers used ethyl acetate (Kanto Chemical Co., Inc., >99.5%). The submitters used ethyl acetate (tech. grade, BCD), which was distilled prior to use.

10. The submitters recorded the following: the addition of aqueous HCl can result in the formation of the **2•HCl** salt, which is not soluble in the two-phase mixture. After vigorous stirring at ca. pH 8, **2•HCl** is converted to **2**, resulting in dissolution of the solid, and giving a clear two-phase mixture. If the aqueous phase is not sufficiently basic, 4 M NaOH was added to adjust the pH.

11. The pressure should be lowered gradually in order to prevent bumping, since a precipitate of **2** is formed during the concentration. *CAUTION: Benzene, formed during the quench of the excess phenylmagnesium bromide, is evaporated during the concentration.*

12. Pure (*S*)-2-amino-4-methyl-1,1-diphenylpentan-1-ol (**2**) can be obtained by recrystallization from EtOAc and has the following physical and spectroscopic properties: colorless crystals; mp 134–137 °C; [a]$_D^{20}$ –91.6 (*c* 1.00, CHCl$_3$) lit.[9]: [a]$_D^{25}$ –96.6 (c 1.00, CHCl$_3$)]; ^{1}H NMR (CDCl$_3$, 600 MHz) δ: 0.87 (d, *J* = 6.4 Hz, 3 H), 0.89 (d, *J* = 6.4 Hz, 3 H), 1.06–1.11 (m, 1 H), 1.27 (ddd, *J* = 14.3, 10.4, 3.8 Hz, 1 H), 1.54–1.59 (m, 1 H), 2.31 (brs, 3 H), 3.98 (d, *J* = 10.4 Hz, 1 H), 7.14–7.19 (m, 2 H), 7.27 (t, *J* = 7.4 Hz, 2 H), 7.31 (t, *J* = 7.7 Hz, 2 H), 7.48 (d, *J* = 6.8 Hz, 2 H), 7.61 (d, *J* = 7.6 Hz, 2 H); ^{13}C NMR (CDCl$_3$, 150 MHz) d: 21.3, 24.0, 25.3, 39.4, 54.5, 79.1, 125.5, 125.8, 126.3, 126.6, 128.0, 128.4, 144.5, 147.1; IR (ATR): 3337, 3264, 2953, 2866, 1586, 1490, 1469, 1448, 1384, 1181, 1057, 1005, 903, 743 (s), 695 (s), 639 cm^{-1}; Anal. Calcd. for C$_{18}$H$_{23}$NO: C, 80.26; H, 8.61; N, 5.20. Found: C, 80.36; H, 8.82; N, 5.15.

13. The submitters report that the amount of **2** present in the crude product can be determined by ^{1}H NMR (400 MHz) analysis using methyl cinnamate as an external standard and using the CH resonances of methyl cinnamate (6.44 ppm) and **2** (3.98 ppm). Sample preparation: a mixture of the crude **2** (15 mg) and methyl cinnamate (8 mg) is dissolved in CDCl$_3$ (0.6 mL). Biphenyl (3-6%) is typically observed as the by-product (GC-MS). This does not affect the reaction in the next step.

14. The checkers used the following reagents and solvents in step B: L-proline (Tokyo Chemical Industry Co., Ltd., >99.0%), di-*tert*-butyl dicarbonate (Boc$_2$O, Wako Pure Chemical Industries, Ltd., >97%), ethyl chloroformate (Wako Pure Chemical Industries, Ltd., >95%), anhydrous CH$_2$Cl$_2$ (Kanto Chemical Co., Inc., dehydrated, >99.5%), which was purified under argon by solvent purifying unit (Wako Pure Chemical Industries, Ltd.), CH$_2$Cl$_2$ (Nacalai Tesque, Inc., 99.5%), MeOH (Nacalai Tesque, Inc., 99.8%), 35% HCl (Koso Chem.). The submitters used the following reagents and solvents in step B: L-proline (BioChemica AppliChem, 99%), di-*tert*-butyl dicarbonate (Boc$_2$O, Acros, 97%), ethyl chloroformate (Acros, 99%, AcroSeal®), anhydrous CH$_2$Cl$_2$ (Acros, 99.8%, extra dry over molecular sieves, stabilized, AcroSeal®), CH$_2$Cl$_2$ (Acros, 99.8%), MeOH (VWR BDH Prolabo, 99.9%), 37% HCl (BASF).

15. Prior to the transfer by syringe, Boc$_2$O is melted by warming the container in a warm-water bath (35-40 °C).

16. Triethylamine (Sigma-Aldrich, 99.5%) was freshly distilled under an argon atmosphere from CaH$_2$ prior to use.

17. The checkers monitored the progress of the reaction with the following analytical conditions: TLC on silica gel with 1:9 MeOH/CH$_2$Cl$_2$ as eluent and visualization with phosphomolybdic acid. The starting material and product have the following R$_f$ values: **2** (0.60); **4** (0.70) on Merck TLC Plate 60 F$_{254}$.

18. The addition of 12M HCl by pipette is slightly exothermic. The flask turns warm and fumes are evolved.

19. The reaction progress can be monitored by TLC; the checkers used the following analytical conditions: An aliquot (ca. 0.02 mL) was quenched by addition to 1 M NaOH (0.5 mL) and EtOAc (0.5 mL). The EtOAc layer was analyzed by TLC on silica gel with 1:9 MeOH/CH$_2$Cl$_2$ as eluent and visualization with phosphomolybdic acid. The starting material and product have the following R$_f$ values: **4** (0.63); **5** (0.19) on Merck TLC Plate 60 F$_{254}$.

20. Since **5** is sparingly soluble in EtOAc, the conversion of **5•HCl** to **5** proceeded via a slurry-to-slurry transformation. After the extraction process, **5** has a solubility in EtOAc of ca. 1 g/40 mL.

21. The submitters noted a yellow to yellow–brown solution might be obtained if a lower grade of MeOH and CH$_2$Cl$_2$ were used for the deprotection step. In that case, it is advisable to filter the solution through a pad of silica (5 g) in a sintered glass funnel using EtOAc (200 mL) to rinse the filter cake. This serves to remove the base-line impurities prior to recrystallization in order to obtain an optimal yield of the pure product.

22. The pressure should be lowered gradually in order to prevent bumping as a precipitate of **5** is formed during the concentration.

23. The submitters recorded an alternative procedure: The filtered EtOAc solution is concentrated by rotary evaporation (40 °C, 150 mmHg) to a volume of ca. 250 mL, to give an off-white slurry of the crude product. The slurry is then heated at reflux for dissolution of the solid and left at room temperature for crystallization.

24. The submitters observed the formation of side products upon long reflux of **5** in EtOAc. However, the recrystallization of **5** from EtOAc provided the best crystalline **5** compared to recrystallization from other solvents (e.g. MeOH, EtOH, CH$_2$Cl$_2$, MTBE).

25. (S)-N-[(S)-1-Hydroxy-4-methyl-1,1-diphenylpentan-2-yl]pyrrolidine-2-carboxamide (**5**) is obtained as a single isomer (by ^{1}H and ^{13}C NMR) and has the following physical and spectroscopic properties: mp 197–199 °C; $[a]_D^{20}$ –51.6 (c 1.00, CHCl$_3$) lit.^{3a}: $[a]_D^{25}$ –46 (c 1.2, CHCl$_3$)]; ^{1}H NMR (CDCl$_3$, 600 MHz) δ: 0.86 (d, J = 6.7 Hz, 3 H), 0.91 (d, J = 6.5 Hz, 3 H), 1.19–1.25 (m, 2 H), 1.42–1.49 (m, 2 H), 1.54–1.59 (m, 3 H), 1.86–1.92 (m, 2 H), 2.56 (dt, J = 10.0, 6.1 Hz, 1 H), 2.82 (dt, J = 9.9, 6.4 Hz, 1 H), 3.48 (dd, J = 9.3, 4.2 Hz, 1 H), 4.58 (t, J = 9.8 Hz, 1 H), 5.45 (br. s, 1 H), 7.12 (t, J = 7.3 Hz, 1 H), 7.18 (t, J = 7.4 Hz, 1 H), 7.24 (t, J = 7.5 Hz, 2 H), 7.30 (t, J = 7.6 Hz, 2 H), 7.55 (t, J = 7.6 Hz, 4 H), 7.94 (d, J = 8.0 Hz, 1 H); ^{13}C NMR (CDCl$_3$, 150 MHz) d: 21.5, 23.8, 25.3, 25.7, 30.4, 37.5, 46.9, 56.5, 60.2, 80.8, 125.6, 125.7, 126.3, 126.5, 127.8, 128.1, 145.1, 146.5, 175.9; IR (ATR): 3466, 3275, 3069, 2955, 2868, 1634, 1513, 1494, 1446, 1100, 1060, 885, 744 (s), 700 (s), 640 cm^{-1}; Anal. Calcd for C$_{23}$H$_{30}$N$_2$O$_2$: C, 75.37; H, 8.25; N, 7.64. Found: C, 75.55; H, 7.98; N, 7.74. The checkers established the enantiomeric purity of **5** thus obtained by conversion into (+)- and (–)-MTPA amides.^{10} Several sets of diagnostic peaks are observed without cross-over, proving the enantiomeric purity of the final product **5**. [MTPA amide derived from (+)-MTPACl: ^{1}H NMR (CDCl$_3$, 600 MHz) d: 0.90 (d, J = 6.9 Hz, 3 H, (CH$_3$)$_2$CH-)), 1.07 (d, J = 6.4 Hz, 3 H, (CH$_3$)$_2$CH-)), 3.60 (s, 3 H, CH$_3$O-); MTPA amide derived from (–)-MTPACl: ^{1}H NMR (CDCl$_3$, 600 MHz) d: 0.89 (d, J = 6.6 Hz, 3 H, (CH$_3$)$_2$CH-)), 0.98 (d, J = 6.4 Hz, 3 H, (CH$_3$)$_2$CH-)), 3.70 (s, 3H, CH$_3$O-).

(scheme image — reaction of (S,S)-5 with F$_3$C/OMe/Cl reagent, pyridine, THF, rt, 4 h to give MTPA amide product)

(S,S)-**5**

26. The submitters reported that concentration of the mother liquors and recrystallization from EtOAc (100 mL) afforded an additional 2 g (11%) of **5** of slightly lower purity (^{1}H NMR).

Handling and Disposal of Hazardous Chemicals

The procedures in this article are intended for use only by persons with prior training in experimental organic chemistry. All hazardous materials should be handled using the standard procedures for work with chemicals described in references such as "Prudent Practices in the Laboratory" (The National Academics Press, Washington, D.C., 2011 www.nap.edu). All chemical waste should be disposed of in accordance with local regulations. For general guidelines for the management of chemical waste, see Chapter 8 of Prudent Practices.

These procedures must be conducted at one's own risk. *Organic Syntheses, Inc.*, its Editors, and its Board of Directors do not warrant or guarantee the safety of individuals using these procedures and hereby disclaim any liability for any injuries or damages claimed to have resulted from or related in any way to the procedures herein.

Discussion

Proline-based chiral organocatalysts have been developed extensively and applied successfully in several reactions.[2] Singh and co-workers have designed proline amide catalysts of type **5**, bearing *gem*-diaryl substituents at the *β*-carbon, which play a key role in providing high enantioselectivity in the direct aldol reaction of ketones with aldehydes.[3] Catalyst **5** has proven to be one of the most effective organocatalysts for asymmetric direct aldol reactions of water-miscible ketones such as acetone with various aldehydes and reactive ketones, especially in aqueous reaction media or under solvent-free conditions.[3,4,5]

In the original method reported by Singh *et al.*,[3] a series of catalysts of type **5** were prepared from Boc-proline and the corresponding diphenylamino alcohols **2** in a sequential two-step method, i.e. condensation and Boc-deprotection, with isolation of the intermediate from the condensation reaction, i.e. **4**. As the first step, Boc-L-proline was condensed with the L-diphenylamino alcohol **2** via a mixed anhydride method using ethyl chloroformate[3] or *iso*-butyl chloroformate,[6] affording **4**. The final Boc-deprotection step was conducted by treatment of **4** with formic acid[3] or TFA,[6] furnishing **5** in 82–83% yield after recrystallization. In our hands, the removal of the Boc group using formic acid or TFA resulted in the formation of significant amounts of side products (e.g. the *N*-formyl pyrrolidinyl derivative). As a consequence, the purification of the resulting crude product by recrystallization led to a significantly lower yield of the desired **5**. Furthermore, this method starts from Boc-proline and pure diphenylamino alcohol **2** which are commercially available but rather expensive compared to their precursors proline and leucine methyl ester hydrochloride (**1**), respectively.

In summary, we have developed an economical and practical one-pot multistep synthesis in which the reaction product from the initial step is used directly without purification for the next step, thus affording **5** directly from proline.[7] Moreover, the chiral amino alcohol **2**, as the coupling partner, can be prepared and used directly without purification from the corresponding amino acid ester **1**. For the direct coupling process of proline with **2**, the Boc-protected proline is first generated *in situ* from proline and Boc$_2$O (1 equiv).[8] Successive treatment with ethyl chloroformate and the amino alcohol **2** yields the *N*-Boc proline amide **4**. Subsequent removal of the Boc-group using concentrated HCl in MeOH-CH$_2$Cl$_2$ as a clean, reliable and convenient deprotection method is another key feature of the procedure presented herein.

References

1. Department of Chemistry (Organic Chemistry), University of Cologne, Greinstr. 4, 50939 Cologne, Germany; Fax: (+49) 221-4705102, E-mail: berkessel@uni-koeln.de. This work was financially supported by the Deutsche Forschungsgemeinschaft (priority program "Organo-catalysis") and by the Fonds der Chemischen Industrie.
2. For reviews, see: (a) Berkessel, A.; Gröger, H. *Asymmetric Organocatalysis: From Biomimetic Concepts to Applications in Asymmetric Synthesis*; Wiley-VCH: Weinheim, 2005. (b) Dalko, P. I., Ed. *Enantioselective Organocatalysis*; Wiley-VCH: Weinheim, 2007. (c) Melchiorre, P.; Marigo, M.; Carlone, A.; Bartoli, G. *Angew. Chem. Int. Ed.* **2008**, *47*, 6138–6171.
3. (a) Raj, M.; Vishnumaya; Ginotra, S. K.; Singh, V. K. *Org. Lett.* **2006**, *8*, 4097–4099. (b) Maya, V.; Raj, M.; Singh, V. K. *Org. Lett.* **2007**, *9*, 2593–2595. (c) Raj, M.; Vishnumaya; Singh, V. K. *J. Org. Chem.* **2009**, *74*, 4289–4297. (d) Review: Raj, M.; Singh, V. K. *Chem. Commun.* **2009**, 6687–6703.
4. (a) Baer, K.; Kraußer, M.; Burda, E.; Hummel, W.; Berkessel, A.; Gröger, H. *Angew. Chem., Int. Ed.* **2009**, *48*, 9355–9358. (b) Rulli, G.; Duangdee, N.; Baer, K.; Hummel, W.; Berkessel, A.; Gröger, H. *Angew. Chem., Int. Ed.* **2011**, *50*, 7944–7947. (c) Duangdee, N.; Harnying, W.; Rulli, G.;

Neudörfl, J.-M.; Gröger, H.; Berkessel, A. *J. Am. Chem. Soc.* **2012**, *134*, 11196–11205.

5. (a) Song, L.; Chen, X.; Zhang, S.; Zhang, H.; Li, P.; Luo, G.; Liu, W.; Duan, W.; Wang, W. *Org. Lett.* **2008**, *10*, 5489–5492. (b) Zhang, H.; Zhang, S.; Liu, L.; Luo, G.; Duan, W.; Wang, W. *J. Org. Chem.* **2010**, *75*, 368–374.

6. Yang, R.-Y.; Da, C.-S.; Yi, L.; Wu, F.-C.; Li, H. *Lett. Org. Chem.* **2009**, *6*, 44–49.

7. Berkessel, A.; Harnying, W.; Duangdee, N.; Neudörfl, J.-M.; Gröger, H. *Org. Process Res. Dev.* **2012**, *16*, 123–128.

8. Huy, P.; Schmalz, H.-G. *Synthesis* **2011**, 954–960.

9. Zhou, Z.; Guo, Y. *Synth. Commun.* **2008**, *38*, 684–696.

10. (a) Dale, J. A.; Mosher, H. S. *J. Am. Chem. Soc.* **1973**, *95*, 512–519. (b) Clayden, J.; Lemiègre, L.; Pickworth, M. *Tetrahedron: Asymmetry* **2008**, *19*, 2218–2221.

Appendix
Chemical Abstracts Nomenclature (Registry Number)

Magnesium turning; (7439-95-4)
Bromobenzene; (108-86-1)
L-Leucine methyl ester hydrochloride; (7517-19-3)
L-Proline; (147-85-3)
Di-*tert*-butyl dicarbonate; (24424-99-5)
Triethylamine; (121-44-8)
Ethyl chloroformate; (541-41-3)

Albrecht Berkessel obtained his Ph.D. with Professor Waldemar Adam at the University of Würzburg in 1985. Later on, he joined the research group of Professor Ronald Breslow at Columbia University, New York. His habilitation at the University of Frankfurt/Main (associated to Professor Gerhard Quinkert) was completed in 1990. In 1992, he became Associate Professor at the University of Heidelberg. Since 1997, he is a Full Professor of Organic Chemistry at the University of Cologne. His research interests center around various aspects of catalysis, such as mechanism and method development, both in metal-based catalysis and organocatalysis, biomimetic and medicinal chemistry.

Wacharee Harnying (born 1975 in Thailand) completed her undergraduate education in Chemistry in 1996 at Khon Kaen University. She then moved to Mahidol University and completed her M.Sc. (Organic Chemistry) in 2000 under supervision of Professor Manat Pohmakotr. From 2000 to 2007, she served as a lecturer at Khon Kaen University. In parallel, she pursued a PhD (2001-2004) at RWTH-Aachen, Germany, under guidance of Professor Dieter Enders. In 2008-2009, she was a postdoctoral fellow of the Alexander von Humboldt-Foundation at the University of Cologne (Germany) with Professor Albrecht Berkessel, where she is now holding the position of a senior researcher.

Nongnaphat Duangdee was born in 1974 in Thailand. She received her Bachelor's Degree in Chemistry from Khon Kaen University in 1996. Since 1997, she joined the Department of Science Service (DSS, Thailand) where she worked as analytical chemist until 2009. In the meantime, she obtained her M.Sc. (Organic Chemistry) under the supervision of Professor Manat Pohmakotr at Mahidol University in 2002. In 2009, she began her doctoral studies in the group of Professor Albrecht Berkessel at Cologne University (Germany), investigating organocatalytic asymmetric aldol reactions with ketone acceptors, and she obtained her Ph.D. in 2013.

Takahiro Sakai was born in Niigata, Japan. He received his B. Sc. degree in 2010, and M. Sc. in 2013 at Kanazawa University under the supervision of Prof. Yutaka Ukaji. In the same year, he joined the research group of Prof. Keisuke Suzuki at the Tokyo Institute of Technology to pursue his Ph. D.

Enantioselective Rhodium-Catalyzed [2+2+2] Cycloaddition of Pentenyl Isocyanate and 4-Ethynylanisole: Preparation and Use of Taddol-pyrrolidine Phosphoramidite

Kevin M. Oberg, Timothy J. Martin, Mark Emil Oinen, Derek M. Dalton, Rebecca Keller Friedman, Jamie M. Neely, and Tomislav Rovis[*1]

Department of Chemistry, Colorado State University, Fort Collins, CO 80523

Checked by Yiyang Liu and Brian M. Stoltz

Procedure

> *Caution! Alkyl acyl azides can rapidly decompose with heat to release large amounts of nitrogen. Care should be taken during handling: do not attempt to convert neat and avoid handling neat.*

A. *Taddol-pyrrolidine phosphoramidite.* A 250-mL single-necked, round-bottomed flask equipped with an egg-shaped magnetic stir bar (12.7 x 25.4 mm) is flame-dried under vacuum. After cooling to 23 °C, (*R,R*)-Taddol (2.0 g, 4.29 mmol) (Note 1) is added to the round-bottomed flask, a rubber septum is fitted, the reaction flask is put under an atmosphere of N_2, and tetrahydrofuran (75 mL) (Note 2) is added via syringe. To this clear, colorless solution, triethylamine (2.4 mL, 17.2 mmol, 4 equiv) (Note 3) is added via syringe resulting in a clear solution with a slight yellow color. The reaction mixture is cooled to 0 °C with an ice bath and phosphorous trichloride (0.39 mL, 4.5 mmol, 1.05 equiv) (Note 4) is added dropwise over 2 min via syringe resulting in a white suspension. The ice bath is removed, the reaction is allowed to warm to 23 °C and stirred for 1 h. The reaction mixture is cooled to 0 °C with an ice bath, and pyrrolidine (1.8 mL, 21.4 mmol, 5 equiv) (Note 5) is added via syringe. The ice bath is removed, the reaction is allowed to warm to 23 °C and stir for 1 h. Diethyl ether (50 mL) is then added, and the reaction mixture is filtered through a medium-fritted funnel into a 250-mL round-bottomed flask. The solid residue in the reaction flask is washed with diethyl ether (2 x 25 mL) and the filtrates are added into the round-bottomed flask. The ether is transferred to a 500-mL separatory funnel and washed with deionized water (50 mL) (Note 6). The organic layer is dried over $MgSO_4$, vacuum-filtered through a 100 mL course-fritted filter funnel into a 250-mL round-bottomed flask using a water aspirator, and the $MgSO_4$ is rinsed with diethyl ether (2 x 10 mL). The filtrate is concentrated *in vacuo* using a rotovap, and the off-white solid put under high vacuum for 30 min. An egg-shaped magnetic stir bar (12.7 x 25.4 mm) and EtOAc (5 mL) are added to the round-bottomed flask containing the solid and a reflux condenser is attached. Using an oil bath, the mixture is heated to reflux with stirring and EtOAc (~20 mL) is added dropwise until all of the solids dissolve. The stir bar is removed from the clear, slightly yellow solution. The round-bottomed flask is allowed to cool slowly in the oil bath to 23 °C, placed in a –10 °C fridge for 12 h, and then placed in a –24 °C freezer for 12 h. The solid is collected with a Büchner funnel (5 cm) with medium porosity filter paper to yield 1.78 g of Taddol-pyrrolidine phosphoramidite as white crystals (73% yield) (Notes 7 and 8).

B. *Pentenyl isocyanate.* A 500-mL single-necked round-bottomed flask equipped with an egg-shaped magnetic stir bar (12.7 x 25.4 mm) is flame-dried under vacuum. After cooling to 23 °C, a rubber septum is fitted to the round-bottomed flask and the flask is put under an atmosphere of N_2. Dichloromethane (50 mL) (Note 9) and 5-hexenoic acid (10.4 mL, 87.6 mmol) (Note 10) are added via syringe to the round-bottomed flask and the flask is cooled to 0 °C with an ice bath. 1,8-Diazabicyclo[5.4.0]undec-7-ene (14.2 mL, 94.6 mmol, 1.08 equiv) (Note 11) is added to the round-bottomed flask via syringe over 5 min and the clear solution is stirred for 20 min. Diphenyl phosphoryl azide (20.4 mL, 94.6 mmol, 1.08 equiv) (Note 12) is added over 5 min via syringe resulting in a clear, yellow solution. The reaction mixture is stirred for 3 h at 0 °C. The ice bath is removed, the septum is removed, and hexanes (200 mL) (Note 13) is added. The reaction is stirred for 5 min and transferred to a 500-mL separatory funnel. After the layers separate, the lower, yellow dichloromethane layer is collected in a 250 mL Erlenmeyer flask and the upper, cloudy hexane layer is transferred to a 1-L round-bottomed flask. The dichloromethane layer is returned to the separatory flask and the 250 mL Erlenmeyer flask is rinsed with hexanes (200 mL) and transferred to the separatory funnel. The dichloromethane layer is extracted with hexanes. The lower dichloromethane layer is collected in the 250 mL Erlenmeyer flask and the cloudy hexane layer is transferred to the 1-L round-bottomed flask, which (without a septum) is placed into a 23 °C oil bath that is heated to 50 °C for 3 h and then at 55 °C for 3 h. After cooling to ambient temperature, the solvent is removed *in vacuo* using a 23 °C bath. This yellow solution is transferred to a 25-mL round-bottomed flask, and the 1 L flask is rinsed with minimal hexanes, which are transferred to the 25-mL round-bottomed flask. The 25-mL flask is concentrated *in vacuo* using a 23 °C bath (40 mmHg). The resultant yellow oil is purified via vacuum distillation and the first clear fraction distilling at 63 °C (50 mmHg) is collected in a 25 mL round-bottomed flask cooled to 0 °C. Pentenyl isocyanate (4.34 g, 45%) is isolated as a clear liquid (Notes 14 and 15).

C. *(R)-5-(4-Methoxyphenyl)-2,3,8,8a-tetrahydroindolizin-7(1H)-one.* An oven-dried 250-mL round-bottomed flask equipped with a polygon magnetic stir bar (6.4 x 25.4 mm) and an oven-dried reflux condenser with septum attached are loaded into an inert atmosphere (N_2) glove box (Note

16). Chlorobis(ethylene)rhodium(I) dimer (58 mg, 0.15 mmol, 0.005 equiv) (Note 17) and Taddol-pyrrolidine phosphoramidite (170 mg, 0.3 mmol, 0.01 equiv) are added to the round-bottomed flask. The reflux condenser is attached, the apparatus is removed from the glove box, and toluene (110 mL) (Note 18) is added via syringe resulting in a clear, gold solution. Toluene (5 mL) is added to a vial containing pentenyl isocyanate (3.33 g, 30 mmol) and 4-ethynylanisole (6.0 g, 45 mmol, 1.5 equiv) (Note 19) and this solution is added to the reaction mixture via syringe. The vial is rinsed with toluene (5 mL), which is added to the reaction vessel. Additional toluene (50 mL) is added to the reaction mixture resulting in a crimson solution. The reaction mixture is heated to 110 °C in an oil bath for 36 h resulting in a dark brown solution. The reaction mixture is concentrated *in vacuo*, and the crude reaction mixture is purified via flash chromatography (Note 20) resulting in the isolation of 6.26 g (86%) of (*R*)-5-(4-methoxyphenyl)-2,3,8,8a-tetrahydroindolizin-7(1H)-one as a light brown solid (89% ee) (Notes 21, 22, and 23).

Notes

1. (*R,R*)-Taddol was purchased from AK Scientific, Inc. and used as received.
2. Tetrahydrofuran (inhibitor-free, Chromasolv®, for HPLC, 99.9%) was purchased from Sigma-Aldrich, degassed with Ar and passed through two columns of neutral alumina.
3. Triethylamine was purchased from Sigma-Aldrich and distilled over KOH before use.
4. Phosphorous trichloride was purchased from Sigma-Aldrich and distilled before use.
5. Pyrrolidine was purchased from Sigma-Aldrich and distilled over KOH before use.
6. Use of deionized water is necessary. If tap water or acidic water is used, degradation of the ligand can be observed by ^{31}P NMR.
7. Physical characteristics of Taddol-pyrrolidine phosphoramidite: $[\alpha]^{25}_{D}$ = –157.8 (conc = 0.0106 g/mL, CHCl$_3$). ^{1}H NMR (500 MHz, CDCl$_3$) δ: 0.29 (s, 3 H), 1.27 (s, 3 H), 1.68–1.94 (m, 4 H), 3.14–3.31 (m, 2 H), 3.32–3.49 (m, 2 H), 4.83 (d, *J* = 8.5 Hz, 1 H), 5.21 (dd, *J* = 8.4, 3.2 Hz, 1 H), 7.16–7.35 (m, 12 H), 7.41 (d, *J* = 7.1 Hz, 2 H), 7.48 (d, *J* = 7.1 Hz, 2 H),

7.60 (d, J = 7.1 Hz, 2 H), 7.75 (d, J = 7.1 Hz, 2 H). ^{13}C NMR (126 MHz, CDCl$_3$) δ: 25.3, 26.0, 26.0, 27.5, 44.9, 45.0, 81.1, 81.2, 81.8, 82.2, 82.4, 82.5, 82.6, 111.7, 127.0, 127.1, 127.1, 127.2, 127.2, 127.4, 127.4, 127.6, 128.1, 128.7, 128.8, 129.0, 142.0, 142.3, 146.6, 146.9. ^{31}P NMR (121 MHz, CDCl$_3$) δ: 137.1. IR (NaCl, Thin Film) 3058, 2967, 2869, 1493, 1446, 1035, 1003, 878, 737, 699 cm^{-1}. Mp = 214–216 °C (EtOAc). HRMS (ESI) m/z [C$_{35}$H$_{37}$NO$_4$P]$^+$ calcd 566.2455, found 566.2445. Anal. calcd for C$_{35}$H$_{36}$NO$_4$P: C, 74.32; H, 6.42; N, 2.48; O, 11.31; P, 5.48, found C, 74.33; H, 6.71; N, 2.54; O, 11.09; P, 5.41.

8. On half scale (2.15 mmol Taddol), the yield was 68%.
9. Dichloromethane (methylene chloride, not stabilized, HPLC grade) was purchased from Fisher Scientific, degassed with Ar and passed through two columns of neutral alumina.
10. 5-Hexenoic acid was purchased from TCI and used as received.
11. 1,8-Diazabicyclo[5.4.0]undec-7-ene was purchased from AK Scientific, Inc. and distilled over KOH before use.
12. Diphenylphosphoryl azide was purchased from AK Scientific, Inc. and used as received.
13. Hexanes were distilled at ambient pressure over boiling chips.
14. Physical characteristics of pentenyl isocyanate: ^{1}H NMR (500 MHz, CDCl$_3$) δ: 1.66–1.77 (m, 2 H), 2.12–2.21 (m, 2 H), 3.32 (t, J = 6.7 Hz, 2 H), 5.02 (ddt, J = 10.2, 1.9, 1.2 Hz, 1 H), 5.07 (dq, J = 17.1, 1.7 Hz, 1 H), 5.77 (ddt, J = 17.0, 10.2, 6.7 Hz, 1 H). ^{13}C NMR (126 MHz, CDCl$_3$) δ: 30.2, 30.5, 42.2, 115.9, 136.8. IR (NaCl, Thin Film) 3369, 2932, 2275, 1687, 1641, 1524, 1211, 911 cm^{-1}.
15. On half scale (43.8 mmol of 5-hexenoic acid), the yield of the product was 44%.
16. The use of a glove box is for simplicity of set up due to the air sensitive nature of chlorobis(ethylene)rhodium(I) dimer. Use of standard Schlenk techniques in place of a glove box should provide similar results if the chlorobis(ethylene)rhodium(I) dimer is of high quality. Chlorobis(cyclooctadiene)rhodium(I) dimer may be used as an air stable alternative, but the submitter has observed lower yields (15–25% lower) when this catalyst is used on smaller scales. The phosphoramidite ligand is air stable and can be stored outside the glovebox, but the ligand is stored in the glovebox for ease of reaction setup.

17. Chlorobis(ethylene)rhodium(I) dimer was purchased from Strem, Inc., stored cold in an inert atmosphere glove box (N$_2$), and used as received.

18. Toluene (Chromasolv®, for HPLC, 99.9%) was purchased from Sigma-Aldrich, degassed with Ar and passed through one column of neutral alumina and one column of Q5 reactant.

19. 4-Ethynylanisole was purchased from AK Scientific, Inc. and used as received.

20. Column diameter: 6 cm, silica: 140 g (Silicycle, Inc. silica 60 (230-400 mesh)), eluant: 2.5 L (20:1 EtOAc:MeOH), fraction size: 50 mL (25 x 150 mm test tubes), product typically found in fractions 16-49. The first three fractions usually contain a small amount of impurity.

21. Physical characteristics of (R)-5-(4-methoxyphenyl)-2,3,8,8a-tetrahydroindolizin-7(1H)-one: 89% ee by HPLC: Chiralcel ODH column, 90:10 Hex:iPrOH, 1 mL/min, 330nm, RT$_{major}$ = 35.48 min, RT$_{minor}$ = 41.75 min. [α]$^{25}_D$ = +637.3 (conc = 0.0084 g/mL CHCl$_3$). ^{1}H NMR (500 MHz, CDCl$_3$) δ: 1.71–2.06 (m, 3 H), 2.25– 2.51 (m, 3 H), 3.27 (dt, J = 11.0, 7.1 Hz, 1 H), 3.50–3.58 (m, 1 H), 3.83 (s, 3 H), 3.98–4.09 (m, 1 H), 5.07 (s, 1 H), 6.91 (d, J = 8.9 Hz, 2 H), 7.33 (d, J = 8.8 Hz, 2 H). ^{13}C NMR (126 MHz, CDCl$_3$) δ: 24.6, 31.7, 41.5, 49.5, 55.3, 58.7, 99.8, 113.8, 128.5, 129.2, 160.8, 162.6, 191.9. R$_f$ = 0.16 (20:1 EtOAc:MeOH). IR (NaCl, Thin Film) 2969, 2875, 1623, 1606, 1510, 1472, 1242, 1176, 1030 cm^{-1}. Mp = 129–132 °C. HRMS (ESI) m/z [C$_{15}$H$_{18}$NO$_2$]$^+$ calcd 244.1332, found 244.1327. Anal. calcd for C$_{15}$H$_{17}$NO$_2$: C, 74.05; H, 7.04; N, 5.76; O, 13.15, found C, 74.08; H, 7.48; N, 5.80; O, 13.55.

22. On half scale (15 mmol of pentenyl isocyanate), the product was obtained as a viscous dark brown oil (86% yield, 88% ee) that slowly solidified after storing in –10 °C fridge for a week.

23. (R)-5-(4-Methoxyphenyl)-2,3,8,8a-tetrahydroindolizin-7(1H)-one can be recrystallized from EtOAc to yield light yellow crystals with 41% (98% ee) recovery. mp = 133–135 °C (EtOAc).

Handling and Disposal of Hazardous Chemicals

The procedures in this article are intended for use only by persons with prior training in experimental organic chemistry. All hazardous materials should be handled using the standard procedures for work with chemicals described in references such as "Prudent Practices in the Laboratory" (The National Academies Press, Washington, D.C., 2011 www.nap.edu). All chemical waste should be disposed of in accordance with local regulations. For general guidelines for the management of chemical waste, see Chapter 8 of Prudent Practices.

These procedures must be conducted at one's own risk. *Organic Syntheses, Inc.*, its Editors, and its Board of Directors do not warrant or guarantee the safety of individuals using these procedures and hereby disclaim any liability for any injuries or damages claimed to have resulted from or related in any way to the procedures herein.

Discussion

The use of chiral phosphoramidites is prevalent in organic chemistry.[2] In our lab, we have found phosphoramidites to be excellent ligands for many of our rhodium-catalyzed syntheses of nitrogen-containing heterocycles.[3] This procedure describes an improved synthesis of Taddol-pyrrolidine phosphoramidite, a chromatography free synthesis of pentenyl isocyanate, and use of these compounds in the enantioselective rhodium-catalyzed [2+2+2] cycloaddition of pentenyl isocyanate and 4-ethynylanisole.

Our first synthesis of Taddol-pyrrolidine phosphoramidite used column chromatography for purification of the phosphoramidite.[4] We have observed that the phosphoramidite is oxidized under acidic conditions and purification via column chromatography can result in partially oxidized ligand. Recrystallization avoids use of acidic silica for purification and makes the overall isolation of the ligand more convenient. Additionally, if the phosphoramidite is partially oxidized, recrystallization allows for isolation of the pure ligand without contamination with the oxidized ligand.

We have traditionally synthesized alkenyl isocyanates in one of two ways: conversion of the acyl azide to the isocyanate under reduced pressure[4] and column chromatography of the acyl azide followed by neat conversion.[5] These methods work well for small scale, but on larger scale, these approaches can potentially be dangerous if proper precautions are not observed. In order to make the synthesis more accessible, we developed a method where the acyl azide is not purified via column chromatography or isolated neat. The choice of 1,8-diazabicyclo[5.4.0]undec-7-ene as the base allows for good conversion to the acyl azide and makes purification by distillation easier due to its high boiling point.

The enantioselective rhodium-catalyzed [2+2+2] cycloaddition of alkenyl isocyanates and alkynes has been extensively investigated in our lab.[6] This procedure demonstrates the scalability of the reaction and the ability to lower catalyst loadings to 1 mol %.

References

1. Department of Chemistry, Colorado State University, Fort Collins, CO 80523, rovis@lamar.colostate.edu. We thank NIGMS (GM80442) for support. We thank Johnson Matthey for a loan of rhodium salts. T. R. thanks Roche for an Excellence in Chemistry Award and Amgen for unrestricted support. D. M. D. thanks NSF-LSAMP Bridge to the Doctorate Program and NIH Ruth M. Kirchstein Fellowship for support.
2. Teichert, J. F.; Feringa, B. L. *Angew. Chem., Int. Ed.* **2010**, *49*, 2486–2528.
3. (a) Perreault, S.; Rovis, T. *Chem. Soc. Rev.* **2009**, *38*, 3149–3159. (b) Friedman, R. K.; Oberg, K. M.; Dalton, D. M.; Rovis, T. *Pure Appl. Chem.* **2010**, *82*, 1353–1364. (c) Yu, R. T.; Rovis, T. *J. Am. Chem. Soc.* **2008**, *130*, 3262–3263. (d) Yu, R. T.; Friedman, R. K.; Rovis, T. *J. Am. Chem. Soc.* **2009**, *131*, 13250–13251. (e) Oberg, K. M.; Rovis, T. *J. Am. Chem. Soc.* **2011**, *133*, 4785–4787.
4. Yu, R. T.; Rovis, T. *J. Am. Chem. Soc.* **2006**, *128*, 12370–12371.
5. Lee, E. E.; Rovis, T. *Org. Lett.* **2008**, *10*, 1231–1234.
6. For a discussion of mechanism, see; Dalton, D. M.; Oberg, K. M.; Yu, R. T.; Lee, E. E.; Perreault, S.; Oinen, M. E.; Pease, M. L.; Malik, G.; Rovis, T. *J. Am. Chem. Soc.* **2009**, *131*, 15717–15728.

Appendix
Chemical Abstracts Nomenclature (Registry Number)

(*R,R*)-Taddol: 1,3-Dioxolane-4,5-dimethanol, 2,2-dimethyl-a4,a4, a5,aα-tetraphenyl-, (4*R*,5*R*)-; (93379-48-7)

Triethylamine: Ethanamine, *N,N*-diethyl-; (121-44-8)

Phosphorous trichloride; (7719-12-2)

Taddol-pyrrolidine phosphoramidite: Pyrrolidine, 1-[(3a*R*,8a*R*)-tetrahydro-2,2-dimethyl-4,4,8,8-tetraphenyl-1,3-dioxolo[4,5-e][1,3,2]dioxaphosphepin-6-yl]-; (913706-72-6)

5-Hexenoic acid; (1577-22-6)

1-8-Diazabicyclo[5.4.0]undec-7-ene: Pyrimido[1,2-*a*]azepine, 2,3,4,6,7,8,9,10-octahydro-; (6674-22-2)

Diphenyl phosphoryl azide: Phosphorazidic acid, diphenyl ester; (26386-88-9)

4-Pentenyl isocyanate: 1-Pentene, 5-isocyanato-; (2487-98-1)

Chlorobis(ethylene)rhodium(I) dimer: Rhodium, di-μ-chlorotetrakis(η2-ethene)di-; (12081-16-2)

4-Ethynylanisole: Benzene, 1-ethynyl-4-methoxy-; (768-60-5)

(*R*)-5-(4-methoxyphenyl)-2,3,8,8a-tetrahydroindolizin-7(1H)-one: 7(1H)-Indolizinone, 2,3,8,8a-tetrahydro-5-(4-methoxyphenyl)-,(8a*R*)-; (913626-94-5)

Tomislav Rovis was born in Zagreb in the former Yugoslavia but was raised in Southern Ontario, Canada. Following his undergraduate studies at the University of Toronto, he earned his Ph. D. degree at the same institute in 1998 under the direction of Professor Mark Lautens. From 1998-2000, he was an NSERC postdoctoral fellow at Harvard University with Professor David A. Evans. In 2000, he began his independent career at Colorado State University and was promoted in 2005 to Associate Professor and in 2008 to Professor. He currently holds the John K. Stille Chair in Chemistry.

Kevin M. Oberg received his B. A. from Gustavus Adolphus College, where he worked under the supervision of Professor Brian A. O'Brien. He is now pursuing his graduate studies at Colorado State University under the guidance of Professor Tomislav Rovis. His graduate research focuses on the development of metal-catalyzed cycloadditions.

Timothy J. Martin received his B. Sc. in chemistry from University of Delaware. He received his Ph. D. in 2011 from University of North Carolina - Chapel Hill under the direction of Professor Michael Crimmins. His graduate studies were focused on the synthesis of amphidinol 3 and his post-doctoral studies focused on nitrogen heterocycle synthesis using rhodium catalysis.

Mark Emil Oinen graduated with a B.Sc. in chemistry from State University of New York college at Brockport 2005, under the supervision of Professor Margaret Logan. He received his M. Sc. In Chemistry in 2010 from the Colorado State University under the guidance of Professor Tomislav Rovis. Mark is currently an associate research scientist for Crestone Pharmaceuticals in Fort Collins, CO.

Derek M. Dalton was born in Aurora, Colorado in 1981. After earning a B. A. (Religion) at the Colorado College in 2004 and a B. Sc. (Chemistry) at the University of Colorado Denver in 2007, he entered into his current position as a Ph. D. candidate at Colorado State University where he investigates metal-catalyzed cycloadditions with Tomislav Rovis.

Rebecca Keller Friedman (born 1983) graduated with a B.A. in chemistry from Washington University in St. Louis in 2005. She received her Ph. D. in 2010 from Colorado State University under the guidance of Tomislav Rovis working on the development of rhodium-catalyzed [2+2+2] and [4+2+2] cyclizations. She then joined the labs of Xiang Wang (University of Colorado, Boulder) as a post-doctoral researcher, focusing on the method development of gold-catalyzed [3,3]-rearrangements of indole derivatives. She is currently working as a scientific analyst for Stratfor: Global Intelligence.

Jamie M. Neely received her B. Sc. in chemistry from the University of Missouri in Columbia, where she worked under the supervision of Professor Timothy Glass. She began her graduate studies in 2008 at Colorado State University under the advisement of Professor Tomislav Rovis. Her current research focuses on the synthesis of heterocycles via C-H activation.

Yiyang Liu received his B.S. degree in Chemistry from Peking University in 2010 under the direction of Prof. Jianbo Wang and Prof. Yan Zhang. He then moved to the California Institute of Technology and began his doctoral studies under the guidance of Prof. Brian Stoltz. His graduate research focuses on chemical synthesis using renewable resources.

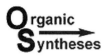

Organic
Syntheses

Synthesis and Use of a Trifluoromethylated Azomethine Ylide Precursor: Ethyl 1-Benzyl-*trans*-5-(trifluoromethyl)pyrrolidine-3-carboxylate

Daniel M. Allwood, Duncan L. Browne and Steven V. Ley[*1]

Department of Chemistry, University of Cambridge, Lensfield Road, Cambridge, UK, CB2 1EW

Checked by Hannah E. Peterlin, Neil Languille and Margaret Faul

Procedure

A. *N-Benzyl-1-(trimethylsilyl)methanamine.* A dry 2-L, three-necked, round-bottomed flask equipped with a 7.5-cm Teflon-coated magnetic stir bar, a rubber septum, an internal thermometer and a reflux condenser fitted with a nitrogen inlet (Note 1, Note 2) is charged with acetonitrile (500 mL). The stirring is started and benzylamine (43.7 mL, 42.9 g, 400 mmol, 2.0 equiv) is added *via* syringe, followed by (chloromethyl)trimethylsilane

(27.9 mL, 24.5 g, 200 mmol, 1.0 equiv) *via* syringe over 5 min (Note 3). The mixture is heated on a 2000-mL hemispherical Glas-Col heating mantle using a J-Kem temperature controller until an internal reaction temperature of 84 °C is achieved ($\Delta T = 95$ °C per hour, 40 min total time). After 20 min at this temperature, the mixture begins to precipitate a white crystalline solid (Note 4). The reaction is maintained at this temperature for 15.5 h and is monitored by TLC analysis on silica gel with 20% EtOAc/hexanes as eluent and visualization with acidic ammonium molybdate and ultraviolet radiation (254 nm) (Note 5). The benzylamine starting material has $R_f = 0.00$–0.21 (blue, UV active), while the product has $R_f = 0.08$–0.40 (weak to no stain, UV active).

The reaction is cooled to room temperature by allowing reaction mixture to cool slowly over 90 min, vacuum filtered through a one-half inch pad of Celite on a 60 mL medium porosity sintered filtration funnel and rinsed with 100 mL of hexane before being concentrated by rotary evaporation (35 °C, 75 mmHg) to a volume of *ca*. 100 mL (Note 6). The resulting liquid is suspended between hexane (250 mL) and water (250 mL) and transferred to a 1-L separatory funnel. The organic phase is removed and the aqueous phase is extracted with hexane (2 × 250 mL). The combined organic phases are washed with saturated aqueous sodium chloride (100 mL), dried over anhydrous magnesium sulfate (*ca* 20 g) and filtered through a one-half inch pad of Celite on a 60 mL medium porosity glass filtration funnel. The solvent is removed by rotary evaporation (35 °C, 15 mmHg) leaving a pale yellow liquid, which is transferred to a 50-mL one-neck round-bottomed flask and purified by fractional vacuum distillation (bp 84–94 °C/5 mmHg) over a still head apparatus featuring a 6-cm water-cooled condenser set at 4 °C, a ground glass joint thermometer and a cow collection adapter to give N-benzyl-1-(trimethylsilyl)methanamine (21.0–26.1 g, 54–67%) as a colorless liquid (Notes 7 and 8).

B.   *N-Benzyl-2,2,2-trifluoro-1-methoxy-N-((trimethylsilyl)methyl)ethan-1-amine.* A 250-mL, three-necked, round-bottomed flask equipped with a 3.5-cm Teflon-coated magnetic stirbar, a rubber septum, an internal thermometer and a reflux condenser fitted with a nitrogen inlet (Note 1) is charged with anhydrous magnesium sulfate (12.0 g, 100 mmol, 1.0 equiv) and *p*-toluenesulfonic acid (860 mg, 5.0 mmol, 0.05 equiv) *via* powder addition funnel followed by dichloromethane (45 mL) and trifluoroacetaldehyde methyl hemiacetal (41.9 mL, 52.0 g, 400 mmol, 4.0 equiv) *via* syringe (Note 9). The vessel is then heated on a 250-mL hemispherical Glas-Col heating mantle using a J-Kem temperature controller until an internal reaction temperature of 55 °C was achieved (ΔT over 15 min). N-Benzyl-1-(trimethylsilyl) methanamine (21.9 mL, 19.3 g, 100 mmol, 1.0 equiv) is added dropwise *via* syringe over 10 min at 45 °C internal temperature. The suspension is maintained at this temperature for 2 h and is monitored by ^{1}H NMR (400 MHz, CDCl$_3$), observing the disappearance of a peak at 2.07 ppm, corresponding to the amine starting material. The reaction is cooled to room temperature by removal of the heating mantle, vacuum filtered through a one-half inch pad of Celite on a 30 mL medium porosity glass filtration funnel and rinsed with 50 mL of dichloromethane before being concentrated by rotary evaporation (50 °C, 15 mmHg) (Note 10). The resulting pale yellow liquid is dissolved in 5% EtOAc/hexane (1 mL per g) (Note 5) and applied to 200 g of Merck neutral alumina 90 packed in a 5-cm diameter column with a 2 cm sand layer on top. The product is eluted with 1 L of 5% EtOAc/hexane (Note 5) collecting 50 mL fractions to afford 23.9-24.3 g, (78-80%) of a colorless liquid, consisting mainly (78-81%) of N-benzyl-2,2,2-trifluoro-1-methoxy-N-((trimethylsilyl)methyl)ethan-1-amine, which is sufficiently pure for further synthetic use (Note 11).

An analytically pure sample may be prepared *via* flash column chromatography by dissolving the crude liquid in 1% EtOAc/hexanes (1 mL per g) (Note 5) and applying it to Sigma-Aldrich Merck grade 9385 60 Å (230-400 mesh) silica gel (10 g per g of crude material) packed in a 10-cm diameter column with a 2 cm sand layer on top. The product is eluted with 1% EtOAc/hexanes (Note 5) to afford N-benzyl-2,2,2-trifluoro-1-methoxy-N-((trimethylsilyl)methyl) ethan-1-amine as a colorless liquid (bp 100–104 °C/5 mmHg) (Note 12).

C. *Ethyl 1-benzyl-trans-5-(trifluoromethyl)pyrrolidine-3-carboxylate.* A 250-mL three-necked round-bottomed flask equipped with a 3.5-cm Teflon-coated magnetic stirbar, a rubber septum, internal thermocouple and a nitrogen inlet (Note 1) is charged with dichloromethane (105 mL). N-Benzyl-2,2,2-trifluoro-1-methoxy-N-((trimethylsilyl)methyl)ethan-1-amine (12.8 g, 41.9 mmol, 1.2 equiv) and ethyl acrylate (3.8 mL, 3.5 g, 35.0 mmol, 1.0 equiv) are then added *via* syringe. Trimethylsilyl trifluoromethanesulfonate (1.3 mL, 1.6 g, 7.0 mmol, 0.2 equiv) is added dropwise *via* syringe (Note 13). The reaction is stirred at room temperature for 16 h and is monitored by TLC analysis on silica gel with 5% EtOAc/hexane as eluent and visualization with potassium permanganate (Note 4). The hemiaminal starting material has R_f = 0.75 (yellow), while the product has R_f = 0.31 (yellow). The reaction is quenched at room temperature by addition of saturated aqueous NaHCO₃ (100 mL) from a measuring cylinder and is transferred to a 500-mL separatory funnel (Note 14). The organic phase is removed and the aqueous phase is extracted with dichloromethane (2 × 100 mL). The combined organic phases are dried over anhydrous magnesium sulfate (*ca* 10 g) and filtered through a one-half inch pad of Celite on a 30 mL medium porosity glass filtration funnel (Note 10). The solvent is removed by rotary evaporation (35 °C, 15 mmHg) leaving a pale yellow liquid which is dissolved in 1% EtOAc/hexane (1 mL per g) (Note 5) and applied to 500 g of Sigma-Aldrich Merck grade 9385 60 Å (230-400 mesh) silica gel packed in a 10-cm diameter column. The product is eluted with 2 L of 1% EtOAc/hexanes followed by 5 L of 2% EtOAc/hexane (Note 5) to afford ethyl 1-benzyl-*trans*-5-(trifluoromethyl) pyrrolidine-3-carboxylate (3.54–4.93 g, 34–47%) as a colorless liquid (bp 104–108 °C/5 mmHg) (Notes 15, 16 and 17).

Notes

1. All glassware is oven-dried at 200 °C overnight and allowed to cool to ambient temperature under nitrogen, which is maintained throughout the course of the reaction.
2. A glass inlet is recommended. Use of a septum and needle resulted in corrosion of the needle and incorporation of red color in the reaction mixture, due to evolution of hydrochloric acid during the reaction.
3. Water content in acetonitrile (EMD, HPLC grade, 99.9%) was measured by Karl Fisher titration prior to use, found to contain 50-80 ppm water, and was used as received. Benzylamine (99.5+%) was obtained from Sigma-Aldrich and used as received. Chloromethyltrimethylsilane (98%) was obtained from Oakwood Products Inc. and used as received.
4. Agitation must be sufficient to prevent precipitated solids from sticking to internal thermocouple. Significant aggregation of solids was found to result in overall lower yield (64-70% crude weight-adjusted yield as measured by Quantitative-NMR versus benzyl benzoate internal standard).
5. Hexane (Sigma-Aldrich, HPLC >98.5%) and EtOAc (Sigma-Aldrich, anhydrous 99.8%) were used as received with no further purification.

6. Celite 545 filter agent (Sigma-Aldrich) wetted with *ca.* 20 mL hexane prior to filtration.

7. Distillation was performed by warming the reaction mixture using a bath composed of Aluminum beads at bath temperature of 130–170 °C; the collection flasks were cooled to 0 °C. Desired product was observed to distill at head temperature of 84–94 °C; residual solvent (hexane) was observed in early fractions and was discarded. The distillation was stopped after six hours, leaving some residual product behind (7-12% by Quantitative-NMR versus benzyl benzoate internal standard, observed in all experiments). Total isolated yield is variable, dependent on apparatus and total distillation time.

8. The compound displays the following physical and spectral properties: ^{1}H NMR (400 MHz, CDCl$_3$) δ: 0.06 (s, 9 H), 1.17 (br, 1 H), 2.07 (s, 2 H), 3.82 (s, 2 H), 7.24–7.28 (m, 1 H), 7.31–7.37 (m, 4 H); ^{13}C NMR (100 MHz, CDCl$_3$) δ: –2.6, 39.5, 58.1, 126.8, 128.1, 128.3, 140.6; FTIR (neat, v_{max} cm^{-1}) 3028 (w), 2954 (w), 2894 (w), 2780 (w), 1490 (w), 1450 (m), 1355 (w), 1250 (s), 1181 (w), 1102 (w), 1075 (w), 1030 (w), 850 (s), 836 (s), 765 (w), 734 (s), 694 (s), 670 (w); R_f 0.10–0.40 (20% EtOAc/hexane); HRMS (ESI+) calculated for C$_{11}$H$_{20}$NSi [M+H]$^+$ 194.1365, found 194.1364. Anal. Calcd. for C$_{11}$H$_{19}$NSi: C 68.33, H 9.90, N 7.24, Si 14.52; found C 68.29, H 9.87, N 7.17, Si 14.47.

9. Dichloromethane (Sigma-Aldrich, HPLC grade ≥99.9%) was measured by Karl Fisher titration prior to use, found to contain <50 ppm water, and was used as received. Trifluoroacetaldehyde methyl hemiacetal (technical grade, >90% by ^{1}H NMR) was purchased from Oakwood Products Inc. and was used as received. *p*-Toluenesulfonic acid monohydrate (>98.5%) was purchased from Sigma-Aldrich and was dried by Dean-Stark azeotropic distillation in toluene prior to use (after treatment, contains 1.7% water by Karl Fisher titration using Metrohm KF oven). Magnesium sulfate (laboratory reagent grade) was purchased from Fisher scientific and dried in an oven at 200 °C overnight prior to use (after treatment, contains <0.1% water by dry Karl Fisher titration using Metrohm KF oven).

10. Celite 545 filter agent (Sigma-Aldrich) wetted with *ca.* 5 mL dichloromethane prior to filtration.

11. Submitters isolated higher purity product (>90%) in overall lower yield (65-71%) indicating variability in retained versus discarded fractions and efficiency of chromatography. Crude weight-adjusted reaction

yield is 69-75% as measured by Quantitative-NMR analysis versus benzyl benzoate internal standard.

12. Purification on silica gel gives analytically pure product, but lower recovery of material (less than can be accounted for by simply removing the impurities), suggesting that the compound may be unstable to silica chromatography. The product proved intractable to distillation and co-distills with byproducts observed in the crude material. The compound displays the following physical and spectral properties: ^{1}H NMR (400 MHz, CDCl$_3$) δ: 0.08 (s, 9 H), 2.30 (d, J=15.1 Hz, 1 H), 2.38 (d, J = 15.1 Hz, 1 H), 3.48 (s, 3 H), 3.81 (d, J = 13.9 Hz, 1 H), 3.94 (d, J = 13.9 Hz, 1 H), 4.15 (qd, J = 5.7, 1.5 Hz, 1 H), 7.28–7.31 (m, 1 H), 7.33–7.37 (m, 4 H); ^{13}C NMR (100 MHz, CDCl$_3$) δ: –1.4, 40.0, 56.4, 57.6, 89.5 (q, J = 30.4 Hz), 124.1 (q, J = 290.0 Hz), 127.3, 128.5, 128.7, 138.7; ^{19}F NMR (376 MHz, CDCl$_3$) δ: –76.0 referenced externally to CF$_3$-toluene; the submitters report a chemical shift of –73.3 ppm; FTIR (neat, υ_{max} cm^{-1}) 2954 (w), 2830 (w), 1490 (w), 1455 (w), 1373 (w), 1268 (m), 1245 (m), 1150 (s), 1121 (s), 1089 (s), 1071 (s), 1024 (w), 990 (w), 968 (w), 837 (s), 763 (w), 739 (m), 721 (w), 697 (s); R_f 0.75 (5% EtOAc/hexanes); HRMS (ESI+) calculated for C$_{14}$H$_{23}$NOF$_3$Si [M+H]$^+$ 306.1501, found 306.1515. Anal. Calcd. for C$_{14}$H$_{22}$NOF$_3$Si: C 55.06, H 7.26, N 4.59, F 18.66, Si 9.20; found C 54.82, H 6.98, N 4.78, F 17.83, Si 9.14.

13. Dichloromethane (Sigma-Aldrich, HPLC grade ≥99.9%) was measured by Karl Fisher titration prior to use, found to contain <50 ppm water, and was used as received. Ethyl acrylate (99% with 10-20 ppm MEHQ as inhibitor) was obtained from Sigma-Aldrich and was used as received. Trimethylsilyl trifluoromethanesulfonate (99%) was obtained from Fluka (Sigma-Aldrich) and was used as received.

14. During quench with aqueous sodium bicarbonate a small exotherm was observed, reaching maximum temperature of 23 °C.

15. Weight-adjusted crude yield before chromatography is 5.17-6.67 g (49-64%) as measured by Quantitative-NMR versus benzyl benzoate internal standard. Submitters' isolated yield was 5.01-5.48 g (47-52%).

16. After using 5 L of 2% EtOAc/hexane, impure fractions containing product continued to elute from the chromatography column. Discrepancies in yield may be due to variability in discarded versus retained fractions, or due to the volatility of the desired product.

17. A small fraction of the desired product is recovered as a mixture with other diastereo-/regio-isomers, which may be re-purified to improve the isolated yield. The isomers form in a 30:5:5:1 ratio (2,4 anti : 2,4 syn :

2,3 anti : 2,3 syn) as previously reported.[3] The compound displays the following physical and spectral properties: ^{1}H NMR (400 MHz, CDCl$_3$) δ: 1.24 (t, J = 7.1 Hz, 3 H), 2.23 (ddd, J = 13.4, 7.9, 2.9 Hz, 1 H), 2.35 (m, 1 H), 2.59 (app t, J = 9.3 Hz, 1 H), 3.13 (m, 1 H), 3.20 (app t, J = 7.7 Hz, 1 H), 3.45 (dqd, J = 10.3, 7.0, 3.1 Hz, 1 H), 3.64 (d, J = 13.3 Hz, 1 H), 4.13 (app qd, J = 7.1, 2.1 Hz, 2 H), 4.20 (d, J = 13.2 Hz, 1 H), 7.28 (t, J = 4.2 Hz, 1 H), 7.33–7.34 (app d, J = 4.5 Hz, 4 H); ^{13}C NMR (100 MHz, CDCl$_3$) δ: 14.1, 29.4 (q, J = 2.0 Hz), 42.2, 56.2, 59.5, 60.9, 63.2 (q, J = 29.4 Hz), 126.8 (q, J = 280.3 Hz), 127.3, 128.4, 128.5, 138.4 (C4), 173.0; ^{19}F NMR (376 MHz, CDCl$_3$) δ: –78.8 (CF$_3$); FTIR (neat, v_{max} cm^{-1}) 2980 (w), 2814 (w), 1730 (s), 1455 (w), 1390 (w), 1373 (w), 1279 (m), 1139 (s), 1110 (m), 1028 (m), 925 (w), 860 (w), 742 (m), 699 (s); R$_f$ 0.47 (10% EtOAc/hexanes); HRMS (ESI+) calculated for C$_{15}$H$_{19}$NO$_2$F$_3$ [M+H]$^+$ 302.1368, found 302.1378. Anal. Calcd. for C$_{15}$H$_{18}$NO$_2$F$_3$: C 59.79, H 6.02, F 18.92, N 4.65, found C 59.53, H 5.95, F 19.80, N 4.79.

Handling and Disposal of Hazardous Chemicals

The procedures in this article are intended for use only by persons with prior training in experimental organic chemistry. All hazardous materials should be handled using the standard procedures for work with chemicals described in references such as "Prudent Practices in the Laboratory" (The National Academies Press, Washington, D.C., 2011 www.nap.edu). All chemical waste should be disposed of in accordance with local regulations. For general guidelines for the management of chemical waste, see Chapter 8 of Prudent Practices.

These procedures must be conducted at one's own risk. *Organic Syntheses, Inc.*, its Editors, and its Board of Directors do not warrant or guarantee the safety of individuals using these procedures and hereby disclaim any liability for any injuries or damages claimed to have resulted from or related in any way to the procedures herein.

Discussion

The introduction of fluorine into molecular scaffolds targeted by the pharmaceutical, agrochemical and materials chemistry sectors can confer superior property profiles compared to their non-fluorous congeners.[4] The unison of methods generating rapid complexity with simple fluorinated starting materials is a strategy that offers the potential to deliver interesting building blocks.[5] Furthermore, the presence of fluorine in such complexity-generating reactions provides an interesting means to control chemo-, regio-, diastereo- and even enantioselectivities through steric and electronic effects.

With regard to azomethine ylides[6] bearing a trifluoromethyl group, few approaches are known. Those that are present in the literature are generally not shown to be broad of scope and suffer from longevity of starting material preparation. Meffert reports trapping of a geminally disubstituted trifluoromethylated azomethine ylide derived from a parent azlactone with dimethylacetylenedicarboxylate.[7] Tanaka *et al* report a thermal ring opening/1,3 dipolar cycloaddition approach from acyl-trifluoromethylated aziridines. [8] While Viehe assessed 1,3 dipolar cycloadditions of ylides derived from trifluorothioacetamides;[9] a process which Domingo later studied computationally.[10]

The ylide precursor reported here is prepared in a straightforward fashion in two synthetic steps from bulk, readily available starting materials. Furthermore it has been shown to perform well in cycloaddition reactions with a set of dipolarophiles that are typically used in these types of processes, including mono-, di- and tri-substituted alkenes, alkynes and diimides. Seemingly, the CF$_3$ group helps impart useful levels of regio- and diastereo-control, with a clear preference for the least sterically encumbered transition state; leading to the 2,4 anti pyrrolidines and 2,4 substituted 3-pyrrolines as the major isomers which are generally isolable from the mixture by flash column chromatography.[2]

Table 1. Examples of the substrate scope of 1,3 dipolar cycloaddition reactions using the trifluoromethylated azomethine ylide

dipolarophile	major product	yields and selectivities
Et–N (maleimide)	Et–N ... N–Bn (CF₃)	83% yield 1 diastereomer
dimethyl maleate (CO₂Me, CO₂Me)	Bn–N pyrrolidine with CO₂Me, CO₂Me, F₃C	83% yield 1 diastereomer
diethyl fumarate (CO₂Et, EtO₂C)	Bn–N pyrrolidine CO₂Et, CO₂Et, F₃C	74% yield 6:1 dr
F₃C, CF₃, EtO₂C alkene	Bn–N pyrrolidine CF₃, CF₃, CO₂Et, F₃C	65% yield 1 regioisomer 22:1 dr
SO₂Tol alkyne	Bn–N pyrroline SO₂Tol, F₃C	74% yield 7:2 regioselectivity
CO₂Et alkene	Bn–N pyrrolidine CO₂Et, F₃C	58% yield 6:1 regioselectivity 30:5:5:1 dr

yield corresponds to clean samples of major isomers shown
dr reported as *2,4 anti: 2,4 syn: 2,3 anti: 2,3 syn*

The cycloadducts may be *N*-debenzylated without incident and the 3-pyrrolines oxidized to the corresponding pyrroles with manganese dioxide. The ester described herein is also synthetically versatile, undergoing hydrolysis, reduction and hydrogenolysis under standard conditions in high yields. The adducts may be further functionalized using standard procedures for amide coupling, tosylation/displacement, reductive amination and sulfonylation chemistry in high yield and without erosion of relative stereochemistry.[11]

Scheme 1. Synthetic uses of ethyl 1-benzyl-*trans*-5-(trifluoromethyl) pyrrolidine-3-carboxylate[11]

References

1. Department of Chemistry, University of Cambridge, Lensfield Road, Cambridge, UK, CB2 1EW. e-mail: dma33@cam.ac.uk. DMA acknowledges financial support from Pfizer.

2. This is an adaptation of a procedure reported by Carey, J. S. *J. Org. Chem.* **2001**, *66*, 2526–2529.

3. Tran, G.; Meier, R.; Harris, L.; Browne, D. L.; Ley, S. V. *J. Org. Chem.* **2012**, *77*, 11071–11078.

4. (a) Purser, S.; Moore, P. R.; Swallow, S.; Gouverneur, V. *Chem. Soc. Rev.* **2008**, *37*, 320–330. (b) Jeschke, P. *ChemBioChem* **2004**, *5*, 570–589. (c) Kirsch P.; Bremer, M. *Angew. Chem. Int. Ed.* **2000**, *39*, 4216–4235.

5. (a) Wilson, P. G.; Percy, J. M.; Redmond, J. M.; McCarter, A. W. *J. Org. Chem.* **2012**, *77*, 6384–6393. (b) Kyne, S. H.; Percy, J. M.; Pullin, R. D. C.; Redmond, J. M.; Wilson, P. G. *Org. Biomol. Chem.* **2011**, *9*, 8328–8339. (c)

Audouard, C.; Garayt, M. R.; Kérourédan, E.; Percy, J. M.; Yang, H. *J. Fluor. Chem.* **2005**, *126*, 609–621.

6. (a) Gothelf, K. V.; Jørgensen, K. A. *Chem. Rev.* **1998**, *98*, 863–910. (b) Coldham, I.; Hufton, R. *Chem. Rev.* **2005**, *105*, 2765–2810. (c) Pandey, G.; Banerjee, P.; Gadre, S. R. *Chem. Rev.* **2006**, *106*, 4484–4517.

7. Burger, K.; Meffert, A.; Bauer, S. *J. Fluor. Chem.* **1977**, *10*, 57–62.

8. Tanaka, K.; Nagatani, S.; Ohsuga, M.; Mitsuhashi, K. *Bull. Chem. Soc. Jpn.* **1994**, *67*, 589–591.

9. (a) Laduron, F.; Ates, C.; Viehe, H. G. *Tetrahedron Lett.* **1996**, *37*, 5515–5518. (b) Laduron, F.; Viehe, H. G. *Tetrahedron* **2002**, *58*, 3543–3551.

10. Domingo, L. R. *J. Org. Chem.* **1999**, *64*, 3922–3929.

11. Allwood, D.M.; Ley, S.V., Unpublished results.

Appendix
Chemical Abstracts Nomenclature (Registry Number)

(Chloromethyl)trimethylsilane; (2344-80-1)
Benzenemethanamine; (100-46-9)
Benzenemethanamine, *N*-[(trimethylsilyl)methyl]-; (53215-95-5)
Ethanol, 2,2,2-trifluoro-1-methoxy-; (431-46-9)
Benzenesulfonic acid, 4-methyl-; (104-15-4)
Sulfuric acid magnesium salt (1:1); (7487-88-9)
Benzenemethanamine, *N*-(2,2,2-trifluoro-1-methoxyethyl)-*N*-
[(trimethylsilyl)methyl]; (1415606-26-6)
2-Propenoic acid, ethyl ester; (140-88-5)
Methanesulfonic acid, 1,1,1-trifluoro-, trimethylsilyl ester; (27607-77-8)
3-Pyrrolidinecarboxylic acid, 1-(phenylmethyl)-5-(trifluoromethyl)-, ethyl
ester, (3*R*,5*S*)-*rel*-; (1415606-48-2)

Organic
Syntheses

Steven V. Ley received his PhD from Loughborough University in 1972, after which he carried out post-doctoral research with Professor Leo Paquette at Ohio State University, followed by Professor Derek Barton at Imperial College London. In 1975, he joined that Department as a lecturer and became Head of Department in 1989. In 1992, he moved to the 1702 BP Chair of Organic Chemistry at the University of Cambridge and became a Fellow of Trinity College. He was elected to the Royal Society in 1990 and was President of the Royal Society of Chemistry (RSC) 2000-02. Steve has been the recipient of many prizes and awards including the Yamada-Koga Prize, Nagoya Gold Medal, ACS Award for Creative Work in Synthetic Organic Chemistry and the Paul Karrer Medal.

Daniel M. Allwood was born in 1986 in Mansfield, England. He completed his undergraduate degree in chemistry at the University of Warwick in 2008. Following this, he joined Prof. Steven V. Ley's group at the University of Cambridge as a Ph.D. student, where he worked on the design, biological evaluation and asymmetric synthesis of novel non-peptidomimetic XIAP inhibitors. Currently, he is a postdoctoral research associate in the Ley group, investigating various areas of synthetic methodology.

Duncan L. Browne was born in 1983 in Dunstable, England. He graduated with a Masters degree in chemistry in 2006 from the University of Sheffield and with a Ph.D. in Organic Synthesis in 2009 from the same institution (with Joseph P. A. Harrity). Following this, he was awarded a one year Doctoral Prize Fellowship from the EPSRC. In 2010 he moved to the ITC and Whiffen Laboratories at the University of Cambridge as a Postdoctoral Research Associate with Steven V. Ley, where he has been developing new flow chemistry tools, techniques and synthetic methods.

Hannah E. Peterlin is currently a fourth year student at Northeastern University working toward a B.S. degree in chemistry. She has participated in the cooperative education program through Northeastern and has held various positions within the field of chemistry. Her latest internship was in the Chemical Process R&D group at Amgen Inc. Upon graduation from Northeastern University, she plans to pursue a career in the pharmaceutical industry.

A native of Massachusetts, Neil Langille attended Rochester Institute of Technology, receiving his B.S. degree in Chemistry. He then pursued his Ph. D. at Boston University under the guidance of James Panek; his graduate research focused on hydrometalations and transition metal mediated cross-coupling reactions applied toward natural product synthesis. Following completion of his Ph.D. in 2005, Neil performed post-doctoral studies in Timothy Jamison's laboratory at MIT, applying cascade reactions toward natural product targets. In 2007, Neil joined the Chemical Process R&D department at Amgen in Cambridge, MA. Since joining Amgen, Neil has had the privilege of working on small molecule projects in oncology and neuroscience therapeutic areas.

.

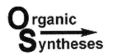
Enantioselective Organocatalytic α-Arylation of Aldehydes

Pernille H. Poulsen, Mette Overgaard, Kim L. Jensen and Karl Anker Jørgensen[*1]

Center for Catalysis, Department of Chemistry, Aarhus University, 8000 Aarhus C, Denmark

Checked by Yu-Peng He and Dawei Ma

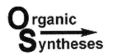

Procedure

(2S,3R)-3-Isopropyl-2,3-dihydronaphtho[1,2-b]furan-2,5-diyl diacetate. A solution of (S)-2-(diphenyl-(trimethylsilyloxy)methyl)pyrrolidine (0.50 g, 1.52 mmol, 0.10 equiv) in EtOH (5 mL) is transferred to a 50-mL round-

bottomed flask equipped with a 1.6-cm oval Teflon-coated magnetic stir bar (Notes 1 and 2). To the flask is added H_2O (1.5 mL, 1.5 g, 83 mmol, 5.5 equiv) using a 2 mL disposable syringe, benzoic acid (0.093 g, 0.76 mmol, 0.050 equiv) and 3-methylbutanal (8.2 mL, 6.6 g, 76 mmol, 5.0 equiv) via a 10 mL disposable syringe (Note 3). Ethanol (2 mL) is used to ensure that no reagents are left on the side of the flask. The mixture is stirred at room temperature for 15 min (Note 4) followed by addition of 1,4-naphthoquinone (2.408 g, 15.23 mmol, 1.0 equiv) and EtOH (8.2 mL) to ensure that no 1,4-naphthoquinone is left on the side of the flask (Note 5). The reaction mixture is stirred at room temperature until complete (20 h) (Note 6). The reaction is transferred to a 250 mL separatory funnel, diluted with EtOAc (60 mL) and washed with saturated aqueous $NaHSO_3$ (4 x 50 mL) and H_2O (2 x 50 mL) (Note 7). The combined aqueous layer is extracted with EtOAc (2 x 2 x 55 mL) (Note 8). The combined organic layer is dried over $MgSO_4$ (19 g) and filtered by suction through cotton wool using a glass funnel. Ethyl acetate (200 mL) is used to wash the $MgSO_4$ and the filtrate is concentrated by rotary evaporation (40 °C bath, 150-27 mmHg). The resulting oil, containing (2R,3R)-3-isopropyl-2,3-dihydronaphtho[1,2-b]furan-2,5-diol (Note 9), is transferred to a 250-mL round-bottomed flask using CH_2Cl_2 which is then evaporated (40 °C bath, 150-27mmHg). The flask is equipped with a 2-cm oval Teflon-coated magnetic stir bar followed by addition of DMAP (0.183 g, 1.52 mmol, 0.10 equiv). The flask is capped with a rubber septum and flushed with argon. The mixture is diluted with dry CH_2Cl_2 (76.2 mL) (Note 10), followed by addition of Et_3N (5.3 mL, 3.9 g, 38 mmol, 2.5 equiv) and acetic anhydride (3.6 mL, 3.9 g, 38 mmol, 2.5 equiv) using a 10 mL and 5 mL disposable syringe, respectively (Note 11). The reaction is stirred at room temperature for 30 min and then transferred to a 250–mL separatory funnel. The mixture is diluted with CH_2Cl_2 (50 mL) and washed with 1M aqueous HCl (2 x 100 mL), H_2O (1 x 100 mL), saturated aqueous $NaHCO_3$ (2 x 100 mL) and H_2O (1 x 100 mL). The combined aqueous layer is extracted with CH_2Cl_2 (3 x 100 mL) (Note 12) and the combined organic layer dried over $MgSO_4$ (14 g). The suspension is filtered by suction through cotton wool using a glass funnel into a 1-L round-bottomed flask using CH_2Cl_2 (200 mL) to transfer and wash the $MgSO_4$. To the filtrate is added Celite (17 g) and the solution is concentrated by rotary evaporation (40 °C bath, 550-27mmHg). The crude product is purified by flash chromatography using silica gel (Note 13) to furnish (2S,3R)-3-isopropyl-2,3-dihydronaphtho[1,2-b]furan-2,5-diyl diacetate (3.79 g, 76% yield, >20:1 dr (Note 14)) as a clear orange oil

(Note 15). Enantiomeric excess is determined to be 97% ee by chiral HPLC analysis (Notes 16 and 17).

Notes

1. The following reagents and solvents are used as received: Ethanol (VWR, Technosolv, 96% vol), pentane (Sigma-Aldrich, chromasolv, ≥99%), ethyl acetate (Sigma-Aldrich, chromasolv, ≥99.7%), dichloromethane (Sigma-Aldrich, chromasolv, ≥99.8%), (S)-(-)-α,α-diphenyl-2-pyrrolidinemethanol trimethylsilyl ether (Sigma-Aldrich, 95%), benzoic acid (FlukaChemica, >99%), isovaleraldehyde (Sigma-Aldrich, 97%), 1,4-naphthoquinone (Sigma-Aldrich, 97%), 4-(dimethylamino)pyridine (Sigma-Aldrich, 99%), triethyl amine (Sigma-Aldrich, ≥99%), acetic anhydride (Riedel-de Haën, ACS reagent, 99%), magnesium sulfate hydrate (Sigma-Aldrich), Celite® 545 (Sigma-Aldrich), silica gel (Sigma-Aldrich, high purity grade, pore size 60 Å, 230-400 mesh particle size), chloroform-d (Sigma-Aldrich, 99.8 atom% D), hexane (Sigma-Aldrich, chromasolv, ≥97.0%), 2-propanol (Sigma-Aldrich, chromasolv, ≥99.8%). Deionized water is used throughout. The following salts are used as saturated aqueous solutions made by dissolving the salt in H_2O until saturation is reached: $NaHSO_3$ (Sigma-Aldrich, ACS reagent, mix of $NaHSO_3$ and $Na_2S_2O_5$), $NaHCO_3$ (Sigma-Aldrich, -40 +140 mesh, Na_2CO_3 2-5%). 1M HCl was prepared by mixing 459 mL H_2O and 41 mL concentrated HCl (VWR, analaR NORMAPUR. 37).
2. (S)-2-(Diphenyl-(trimethylsilyloxy)methyl)pyrrolidine is weighed out in a glass vial and transferred using 5 mL EtOH.
3. The composition of the solvent (H_2O to EtOH ratio) and the ratio of quinone to aldehyde are important due to the oxidation-reduction chemistry of the quinone. Previous studies have shown that 5 equiv of aldehyde and 5 equiv of H_2O are required in order to minimize reduction of the quinone to the hydroquinone.[2]
4. A slight heating of the mixture is observed upon addition of the acid. Consequently, the reaction mixture is stirred for 15 min to ensure that the mixture is cooled to room temperature, prior to addition of the quinone.
5. The EtOH is added using a 20 mL disposable syringe and the total amount of added EtOH is 15.2 mL.

6. The conversion is monitored using ^{1}H NMR spectroscopy. 1,4-Naphthoquinone shows a multiplet at 7.77 ppm, corresponding to 2 hydrogen atoms. When this signal is no longer observable, the reaction is judged complete.

7. This step is introduced in order to reduce the large amount of aldehyde present on this scale. Aldehydes are known to form water soluble bisulfite adducts with $NaHSO_3$.

8. The combined aqueous layer has a total volume of approximately 400 mL, therefore 200 mL portions are extracted with EtOAc (55 mL) two times in a 250-mL separatory funnel.

9. This product is unstable on silica, and the checkers did not purify the crude reaction mixture. ^{1}H NMR and ^{13}C NMR spectra of the crude reaction mixture, as acquired by the checkers, are provided. The submitters report that the hemiacetal can be purified by flash chromatography using Iatrobeads (spherical silica gel). Purified (2R,3R)-3-isopropyl-2,3-dihydronaphtho[1,2-b]furan-2,5-diol (grey foam, >20:1 dr, 99% ee) has the following physical and spectroscopic data: R_f = 0.50 (50% EtOAc/50% pentane); $[\alpha]^{20}_D$ = –117.3 (c = 1.0, CH_2Cl_2); FT-IR (ATR): 3329, 2960, 1595, 1442, 1395, 1260, 1228, 1154, 1066, 1027, 1008 cm^{-1}; ^{1}H NMR (400 MHz, CDCl$_3$) δ: 0.85 (d, J = 6.8 Hz, 3H), 0.92 (d, J = 6.8 Hz, 3H), 1.93 (pd, J = 6.9, 5.2 Hz, 1H), 3.08 (dd, J = 5.2, 1.7 Hz, 1H), 3.72 (s, 1H), 5.53 (s, 1H), 5.85 (d, J = 1.8 Hz, 1H), 6.64 (s, 1H), 7.38–7.47 (m, 2H), 7.89 (dd, J = 7.3, 2.0 Hz, 1H), 8.08 (dd, J = 7.4, 2.0 Hz, 1H); ^{13}C NMR (100 MHz, CDCl$_3$) δ: 19.0, 19.4, 30.5, 57.6, 104.1, 106.4, 119.8, 120.8, 121.3, 122.0, 124.5, 125.1, 126.0, 145.7, 147.2; HRMS calculated for: [C$_{15}$H$_{16}$O$_3$+Na]$^+$ 267.0992; found 267.0994.

10. Dichloromethane is obtained from a solvent purification system and transferred using a 20 mL disposable syringe.

11. Triethylamine and acetic anhydride are used in excess because two hydroxyl groups are acetylated.

12. CO_2 gas evolves when the acid and base solutions are combined.

13. A glass column (5.5 x 28 cm) is wet-packed (10% EtOAc/90% pentane) with silica (200 g). The Celite, with the product adsorbed, is added and the column is topped with 0.5 cm sand. The product is eluted with 10%

EtOAc/90% pentane (3 L), collecting 14 fractions of 50 mL followed by 20 mL fractions. TLC (UV visualization) is used to follow the chromatography. Fractions 33-61 are concentrated by rotary evaporation (40 °C bath, 550-27mmHg), then held under vacuum (0.08 mmHg) at room temperature overnight.

14. Diastereomeric ratio is determined by ^{1}H NMR analysis of the purified product. Only one diastereoisomer is observed.

15. When the procedure was performed on half-scale, the reaction provided 1.85 g (75%) of the product. (2S,3R)-3-Isopropyl-2,3-dihydronaphtho[1,2-b]furan-2,5-diyl diacetate has the following physical and spectroscopic data: R_f = 0.34 (25% EtOAc/75% pentane); FT-IR (ATR): 2962, 1762, 1599, 1459, 1435, 1368, 1205, 1160, 1063 cm^{-1}; $[\square]^{20}_D$ = –183.1 (c = 2.0, CH$_2$Cl$_2$); ^{1}H NMR (400 MHz, CDCl$_3$) δ: 0.98 (d, J = 7.2 Hz, 6H), 2.09 (m, 4H), 2.46 (s, 3H), 3.33 (d, J = 5.0 Hz, 1H), 6.76 (s, 1H), 7.17 (s, 1H), 7.50 (m, 2H), 7.81 (m, 1H), 8.01 (m, 1H); ^{13}C NMR (100 MHz, CDCl$_3$) δ: 19.3, 19.3, 21.1, 21.3, 30.7, 55.4, 102.0, 115.4, 119.8, 120.7, 121.6, 122.1, 126.3, 126.6, 126.9, 141.0, 151.5, 169.9, 170.1; HRMS calculated for: [C$_{19}$H$_{20}$O$_5$+Na]$^+$ 351.1203; found 351.1201.

16. Enantiomeric excess is determined to be 96% ee by chiral HPLC using the following conditions: Chiralpak IB column (particle size: 5 μm; dimensions: 4.6 mmØ x 250 mmL) 95% hexanes/5% isopropanol, 0.5 mL/min. Retention times are: 11.3 min (minor), 12.2 min (major). A Photodiode Array Detector is used.

17. In order to determine the retention times for both enantiomers a racemic mixture of the enantiomers is prepared: In a vial (S)-2-(diphenyl-(trimethylsilyloxy)methyl)pyrrolidine (3.3 mg, 0.010 mmol, 0.050 equiv) and (R)-2-(diphenyl-(trimethylsilyloxy)methyl)pyrrolidine (3.3 mg, 0.010 mmol, 0.050 equiv) are dissolved in EtOH (0.2 mL). To the vial is added H$_2$O (20 μL, 20 mg, 1.1 mmol, 5.5 equiv), benzoic acid (1.4 mg, 0.010 mmol, 0.050 equiv) and 3-methylbutanal (107 μL, 86.1 mg, 1.00 mmol, 5.00 equiv). Finally, 1,4-naphthoquinone (31.6 mg, 0.20 mmol, 1.00 equiv) is added and the mixture stirred for 24 h at room temperature. The reaction mixture is diluted with EtOAc and washed with saturated aqueous NaHSO$_3$ and H$_2$O. The organic layer is dried over MgSO$_4$ and the filtrate concentrated by rotary evaporation. To a vial containing the crude mixture is added DMAP (2.4 mg, 0.020 mmol, 0.10 equiv), dry CH$_2$Cl$_2$ (1.0 mL), Et$_3$N (70 μL, 51 mg, 0.50 mmol, 2.5 equiv) and acetic anhydride (47 μL, 51 mg, 0.50 mmol, 2.5 equiv) under argon atmosphere. The reaction is stirred at room temperature

for 30 min and then diluted with CH_2Cl_2 and washed with 1M aqueous HCl, H_2O, saturated aqueous $NaHCO_3$ and H_2O. The organic layer is dried over $MgSO_4$. To the filtrate is added Celite and the suspension concentrated by rotary evaporation. The crude product is purified by flash chromatography using silica gel (10% EtOAc/90% pentane) to furnish the product *rac*-**5a**.

Handling and Disposal of Hazardous Chemicals

The procedures in this article are intended for use only by persons with prior training in experimental organic chemistry. All hazardous materials should be handled using the standard procedures for work with chemicals described in references such as "Prudent Practices in the Laboratory" (The National Academies Press, Washington, D.C., 2011 www.nap.edu). All chemical waste should be disposed of in accordance with local regulations. For general guidelines for the management of chemical waste, see Chapter 8 of Prudent Practices.

These procedures must be conducted at one's own risk. *Organic Syntheses, Inc.*, its Editors, and its Board of Directors do not warrant or guarantee the safety of individuals using these procedures and hereby disclaim any liability for any injuries or damages claimed to have resulted from or related in any way to the procedures herein.

Discussion

In organocatalysis, two different mechanistic approaches are commonly considered: covalent and non-covalent catalysis. For the covalent approach, chiral secondary amines are very powerful catalysts for the enantioselective functionalization of carbonyl compounds. Over the last decade a large number of α-, β- and γ-functionalizations, involving enamine, iminium-ion and dienamine intermediates, respectively, have been reported.[3] The development of enantioselective α-arylation reactions of carbonyl compounds is very important in contemporary organic synthesis as this reaction introduces a $C(sp^3)$-$C(sp^2)$ bond. This reaction has been a challenge and prior to the present study, this reaction has mostly relied on transition-metal catalyzed approaches.[4]

The present procedure describes an organocatalytic, enantioselective α-arylation of aldehydes, in which quinones are applied as the aromatic partner, to afford optically active α-arylated aldehydes. Following the addition/aromatization sequence, subsequent hemiacetal formation affords the 2,3-dihydrobenzofuran products.

In order to make the previously reported α-arylation protocol[2] suitable for a larger scale synthesis, some modifications have been developed. The present procedure is improved by lowering the catalyst loading from 20 mol% to 10 mol%, using 5 mol% benzoic acid. Furthermore, to simplify the purification process, an acetylation with acetic anhydride has been applied as it stabilizes the products without loss of enantiopurity.

Table 1. Scope of the organocatalytic, enantioselective α–arylation of aldehydes[a]

Entry	Quinone	R	T [°C]	Product	Yield[a] [%]	ee[a] [%]	dr
1	1a	i-Pr (2a)	rt	5a	74-77	97-98	>20:1
2	1a	n-Hex (2b)	−11	5b	57-58	97-99	>20:1
3	1a	Et (2c)	0	5c	50-56	93-94	>20:1
4	1b	i-Pr (2a)	0	5d	66-70	98-99	>20:1

[a]The presented intervals represent two runs.

With the improved reactions conditions, the generality of the procedure has been illustrated by varying both the quinone and aldehyde components (Table 1). Reactions with 1,4-naphthoquinone were performed using three different aliphatic aldehydes, all resulting in the desired products (**5a-c**) in good yields ranging from 50–77% over two steps and high enantio- and diastereoselectivities (93–99% and >20:1 dr). Furthermore, 2,6-dichlorobenzoquinone was reacted with 3-methylbutanal affording product **5d** in 66–70% yield, 98–99% ee and >20:1 dr.

The suggested mechanism for the direct α-arylation consists of two catalytic cycles. The stereogenic center is formed in the first cycle - the reaction of the enamine intermediate with the quinone. In the second cycle a series of proton-transfer reactions leads to the optically active α-arylated aldehyde (obtained as a hemiacetal). H_2O is suggested to be crucial for the proton-transfer reactions.[2]

References

1. Center for Catalysis, Department of Chemistry, Aarhus University, 8000 Aarhus C, Denmark. E-mail: kaj@chem.au.dk. This work was made possible by grants from Aarhus University, Carlsberg Foundation and FNU.
2. Alemán, J.; Cabrera, S.; Maerten, E.; Overgaard, J.; Jørgensen, K. A. *Angew. Chem., Int. Ed.* **2007**, *46*, 5520.
3. For reviews on organocatalysis, see: (a) Jensen, K. L.; Dickmeiss, G.; Jiang, H.; Albrecht, Ł.; Jørgensen, K. A. *Acc. Chem. Rev.* **2011**, *45*, 248. (b) Xu, L.-W.; Li, L.; Shi, Z.-H. *Adv. Synth. Catal.* **2010**, *352*, 243. (c) Mielgo, A.; Palomo, C. *Chem.-Asian J.* **2008**, *3*, 922. (d) Bertelsen, S.; Jørgensen, K. A. *Chem. Soc. Rev.* **2009**, *38*, 2178. (e) Nielsen, M.; Worgull, D.; Zweifel, T; Gschwend, B.; Bertelsen, S.; Jørgensen, K. A. *Chem. Commun.* **2011**, *47*, 632. (f) MacMillan, D. W. C. *Nature* **2008**, *455*, 304.
4. For a representative review, see: (a) Bellina, F.; Rossi, R. *Chem. Rev.* **2010**, *110*, 1082. For some examples, see (b) Martín, R.; Buchwald, S. L. *Angew. Chem., Int. Ed.* **2007**, *46*, 7236. (c) Vo, G. D.; Hartwig, J. F. *Angew. Chem., Int. Ed.* **2008**, *47*, 2127. (d) García-Fortanet, J.; Buchwald, S. L. *Angew. Chem., Int. Ed.* **2008**, *47*, 8108.

Appendix
Chemical Abstracts Nomenclature (Registry Number)

(S)-2-(Diphenyl-(trimethylsilyloxy)methyl)pyrrolidine; (848821-58-9)
Benzoic acid; (65-85-0)
3-Methylbutanal; (590-86-3)
1,4-Naphthoquinone; (130-15-4)
N,N-Dimethylaminopyridine; (1122-58-3)
Triethylamine; (121-44-8)
Acetic anhydride; (108-24-7)

Pernille Hartveit Poulsen was born in 1988 in Hjørring, Denmark. She graduated from Aarhus University with a B. Sc. degree in chemistry in 2011. She is currently pursuing her Ph. D. under guidance of Prof. Karl Anker Jørgensen at the Center for Catalysis, Aarhus University.

Mette Kiilerich Overgaard was born in 1987 in Aarhus, Denmark. She graduated from Aarhus University with a B. Sc. degree in chemistry in 2011. She is currently pursuing her M. Sc. under guidance of Prof. Karl Anker Jørgensen at the Center for Catalysis, Aarhus University.

Organic
Syntheses

Kim L. Jensen was born in Svendborg, Denmark, in 1985. In 2012, he earned his Ph. D. with Prof. Karl Anker Jørgensen at the Center for Catalysis, Aarhus University. He is currently pursuing postdoctoral research with Prof. Timothy F. Jamison at Massachusetts Institute of Technology (MIT).

Karl Anker Jørgensen received his Ph. D. from Aarhus University in 1984. He was a post-doc. with Prof. Roald Hoffmann, Cornell University, 1985. In 1985, he became an Assistant Professor at Aarhus University and in 1992 he moved up the ranks to Professor. His research interests are the development, understanding and application of asymmetric catalysis.

Yu-Peng received his Ph. D. from Max-Planck-Institute of Molecular Physiology in Germany with Prof. Herbert Waldmann in 2010. He became an Associate Professor at Liaoning Shihua University in 2011. Currently, he is as visiting scholar with Prof. Dawei Ma in SIOC.

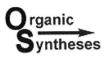

Synthesis of 4,5-Disubstituted 2-aminothiazoles from α,β-Unsaturated Ketones: Preparation of 5-Benzyl-4-methyl-2-aminothiazolium Hydrochloride Salt

Antonio Bermejo Gómez,[‡1] Nanna Ahlsten,[‡1] Ana E. Platero-Prats[2] and Belén Martín-Matute[*1]

Department of Organic Chemistry, Arrhenius Laboratory, Stockholm University, SE-106 91 Stockholm, Sweden

Checked by Michael T. Tudesco and John L. Wood

A.

NaBH$_4$, EtOH

4°C → rt, 65 min
H$_2$O, H$_3$O$^+$

1

B.

1

[Cp*IrCl$_2$]$_2$ (0.5 mol%)
THF-H$_2$O (1:2 v/v)

18h

2

C.

2

Thiourea

EtOH, 100°C, 17h
62.4% - 67% over 3 steps

3

Procedure

A. *4-Phenyl-3-buten-2-ol (1)*. A 500-mL three-necked, round-bottomed flask equipped with an octagonal Teflon-coated magnetic stir bar

(3.5 x 1.5 cm) is flame-dried and placed under an atmosphere of nitrogen. A thermometer adapter (with a thermometer inserted) is attached to the left neck. The flask is charged with (E)-4-phenyl-3-buten-2-one (10.00 g, 68.4 mmol, 1.0 equiv) and absolute ethanol (100 mL) (Note 1), and stirred at room temperature for 15 min until the solid dissolves. The flask is then immersed in an ice-water bath and cooled to 4 °C. Sodium borohydride (2.59 g, 68.5 mmol, 1.0 equiv) (Note 2) is added in one portion (internal temperature during addition: 4 °C), and the reaction mixture is maintained at this temperature for 15 min. The ice-water bath is removed and the reaction is stirred for 50 min, during which time the solution warms to room temperature (Note 3). The flask is cooled in an ice-water bath to 4 °C and a 125-mL pressure-equalizing addition funnel is attached to the center neck of the flask. Deionized-water (30 mL) is added dropwise over 6 min through the addition funnel (internal temperature during addition: 4 °C – 6 °C) (Note 4). A solution of 5 mL of concentrated hydrochloric acid (Note 5) in 40 mL of deionized-water is added dropwise through the addition funnel over 15 min (internal temperature during addition: 6 °C – 20 °C) (Note 6). The ice-water bath is then removed and the reaction is stirred for 30 min, during which time the solution warmed to room temperature. The reaction mixture is transferred to a 500-mL single-necked, round-bottomed flask using 25 mL of ethyl acetate (Note 7), and then the volatile components are removed by rotary evaporation (bath temperature: 40 °C; ≈ 40 mmHg). The crude material (Note 8) is extracted in a 250-mL separation funnel with ethyl acetate (1 × 50 mL). The organic phase is washed with a saturated solution of $NaHCO_3$ (1 x 20 mL) and then a saturated solution of NaCl (1 × 20 mL). The aqueous phases are extracted with ethyl acetate (2 × 25 mL), and all the combined organics are dried in a 250-mL Erlenmeyer flask with $MgSO_4$ (2 g) (Note 9) and then filtered through a funnel with a cotton plug into a 500-mL single-necked, round-bottomed flask. The Erlenmeyer and the funnel are rinsed with ethyl acetate (20 mL). The solvent is removed using a rotary evaporator (bath temperature 40 °C; ≈ 40 mmHg), and the residue is dried under vacuum for 2 h (room temperature, 0.2 mmHg). The colorless oil (10 g) is used in the next step without further purifications (Notes 10, 11, and 12).

B. *3-Chloro-4-phenylbutan-2-one* (**2**). A 500-mL three-necked, round-bottomed flask, equipped with an octagonal Teflon coated magnetic stir bar (3.5 x 1.5 cm) and a thermometer adapter (with a thermometer inserted) in the left neck, is placed under an atmosphere of air. The flask is charged with the crude *4-phenyl-3-buten-2-ol* (**1**) (10 g, 67.5 mmol, 1.0 equiv) from step A, using 14 mL of THF (Note 13) to transfer the oil into the flask. Deionized-water (28 mL) is then added to the flask to keep the solvents at a 1:2 v/v relationship. A solution of [Cp*IrCl$_2$]$_2$ (273 mg, 0.342 mmol, 0.5 mol%) (Notes 14 and 15) in a mixture of THF (76 mL) and water (152 mL) (1:2 v/v) is added (Note 16), followed by the addition of N-chlorosuccinimide (11.2 g, 82.1 mmol, 1.2 equiv) in one portion (internal temperature during additions: 23 °C) (Note 17). The reaction mixture is stirred vigorously at room temperature for 18 h (Note 18), after which time it is transferred to a 500-mL single-necked, round-bottomed flask using 25 mL of ethyl acetate. The volatiles are then removed by rotary evaporation (bath temperature: 40 °C; ≈ 40 mmHg). To this residue (Note 19) a saturated solution of NaCl (20 mL) is added, and the mixture is extracted in a 500-mL separatory funnel with ethyl acetate (4 × 40 mL). The combined organic phases are washed with a saturated solution of NaCl (3 × 25 mL), dried in a 250-mL Erlenmeyer flask over MgSO$_4$ (3 g), and then filtered through a funnel with a cotton plug into a 500-mL single-necked, round-bottomed flask. The Erlenmeyer and the funnel are rinsed with ethyl acetate (20 mL). The solvent is removed using a rotary evaporator (bath temperature 40 °C; ≈ 40 mmHg) and the residue is dried under vacuum for 1 h (room temperature, 0.200 mmHg). The obtained red-brown oil (14.2 g) is used in the next step without further purification (Notes 20 and 21).

C. *5-Benzyl-4-methyl-2-aminothiazolium hydrochloride (3).* A 300-mL three-necked, round-bottomed flask equipped with an octagonal Teflon coated magnetic stir bar (3.5 × 1.5 cm), a cooling condenser in the center neck, and a thermometer adapter (with a thermometer inserted) in the left neck is flame-dried and placed under an atmosphere of nitrogen. The flask is charged with the crude of *3-chloro-4-phenylbutan-2-one (2)* (14.2 g) from step B and absolute ethanol (50 mL). Thiourea (5.21 g, 68.5 mmol, 1.0 equiv) (Note 22) is then added (internal temperature during addition: 23 °C) and the reaction mixture is heated at 100 °C in an oil bath for 17 h (internal temperature: 78 °C) (Note 23). The reaction is cooled down, with stirring, to room temperature and then immersed into an ice-water bath (internal temperature: 5 °C) (Note 24). The cooling condenser is removed and replaced with a 125-mL pressure-equalizing addition funnel, through which diethyl ether (60 mL) (Note 25) is added dropwise over 10 min (internal temperature: 5 °C). The reaction is then stirred for 15 min at 5 °C. The precipitated brown solid is filtered through a 150-mL fritted glass funnel (frit pore size M) and then washed with cooled (5 °C) diethyl ether (3 × 20 mL). The solid is dried under air for 30 min and vacuum-dried for

1 h (room temperature, 0.2 mmHg) to afford 5-benzyl-4-methyl-2-aminothiazolium hydrochloride (**3**) (10.29 g, 42.7 mmol, 62.4% over 3 steps) as a brown solid (Notes 26, 27, and 28).

Notes

1. (*E*)-4-Phenyl-3-buten-2-one (*Acros Organics*, 99%) and ethanol (*Fisher Scientific*, ACS Reagent Grade, absolute, anhydrous) were used as received. The submitters used (*E*)-4-phenyl-3-buten-2-one purchased from *Sigma-Aldrich Co.* (99%) and ethanol purchased from *VWR* (GPR Rectapur, 99.5%) as received.
2. Sodium borohydride (*Sigma-Aldrich Co.*, 98%, powder) was used as received. The submitters used sodium borohydride purchased from *Fisher Scientific* (98%, granules) as received.
3. The reaction can be monitored by TLC analysis on glass-backed extra hard layer TLC plates (*Silicycle*, 60 Å, 250 µm thickness, containing F-254 indicator) using a 10:90 solution of EtOAc:pentane as eluent, and visualized with $KMnO_4$ stain. The ketone starting material has $R_f = 0.29$ (yellow spot) and the allylic alcohol product has $R_f = 0.19$ (yellow spot).
4. The stirring must be vigorous because a white solid is formed with the addition of water (boron salts).
5. Hydrochloric acid (*EMD Chemicals*, ACS Reagent Grade, 37%) was used as received. The submitters used hydrochloric acid purchased from *VWR* (AnalaR Normapur, 37%) as received.
6. It is important to add the water before the acidic solution to avoid side-reactions of the allylic alcohol (**1**) in the reaction media.
7. Ethyl acetate (*Fisher Scientific*, Analytical reagent grade) was used as received.
8. After evaporation of the solvent, the crude is a biphasic system.
9. Magnesium sulfate anhydrous (*Alfa Aesar*, 99.5%, anhydrous powder) was used as received.
10. The crude reaction mixture, which may include a small quantity of ethyl acetate, contains allylic alcohol **1**, which has the following spectroscopic properties that correspond with the data described in literature:[3] ^{1}H NMR (500 MHz, $CDCl_3$) δ: 1.38 (d, *J* = 6.4 Hz, 3 H), 1.57 (bs, 1 H), 4.50 (dq, *J* = 6.4 Hz, *J* = 1.2 Hz, 1 H), 6.27 (dd, *J* = 16 Hz, *J* = 6.4 Hz, 1 H), 6.58 (d, *J* = 16 Hz, 1 H), 7.22–7.26 (m, 1 H), 7.30–7.33 (m,

2 H), 7.38–7.39 (m, 2 H); ^{13}C NMR (125 MHz, CDCl$_3$) δ: 23.4, 68.9, 126.4, 127.6, 128.6, 129.4, 133.5, 136.7.

11. The crude should be immediately used in the next step to avoid decomposition.

12. Pure 4-phenyl-3-buten-2-ol (**1**) was prepared by carrying out the reaction on half scale, and then purifying the crude product mixture by column chromatography. The crude oil is loaded onto 35.3 g silica gel (SilicaFlash® F60 (40-63 μm/230-500 mesh) purchased from Silicycle) that had been dry-packed in a 3 cm diameter chromatography column and wetted with a 5:95 solution of EtOAc:pentane. The product is eluted with 300 mL of a 5:95 solution of EtOAc:pentane, followed by 200 mL of a 10:90 solution of EtOAc:pentane, followed by 200 mL of a 20:80 solution of EtOAc:pentane, followed by 200 mL of a 30:70 solution of EtOAc:pentane, and the eluent is collected in 15 mL fractions in 16x125mm test tubes. Fractions 10-39 (R$_f$ = 0.19 (yellow spot), visualized with KMnO$_4$, 10:90 solution of EtOAc:pentane as eluent) are combined in a 1-L round-bottomed flask and concentrated under reduced pressure (bath temperature 23 °C; ≈ 20 mmHg) to give pure 4-phenyl-3-buten-2-ol (**1**) as a colorless oil. Compound **1** has the following spectroscopic properties: ^{1}H NMR (500 MHz, CDCl$_3$) δ: 1.38 (d, J = 6.4 Hz, 3 H), 1.81 (bs, 1 H), 4.49 (q, J = 6.4 Hz, 1 H), 6.27 (dd, J = 16, 6.4 Hz, 1 H), 6.57 (d, J = 16 Hz, 1 H), 7.23–7.26 (m, 1 H), 7.30–7.33 (m, 2 H), 7.38–7.39 (m, 2 H). ^{13}C NMR (125 MHz, CDCl$_3$) δ: 23.4, 68.9, 126.4, 127.6, 128.5, 129.3, 133.5, 136.7. IR (neat) cm^{-1}: 3346, 3026, 2972, 1494, 1449. HRMS (ESI) Exact mass calcd for C$_{10}$H$_{11}$ [M–OH]$^+$: 131.0855, found: 131.0856.

13. Tetrahydrofuran (non-stabilized THF purchased from Fisher Scientific and passed through a column of activated alumina.) was used. The submitters used tetrahydrofuran purchased from *Sigma-Aldrich Co.*, (ACS Reagent Grade, containing 250 ppm BHT as inhibitor) as received.

14. Dichloro(pentamethylcyclopentadienyl)iridium (III) dimer (*Strem Chemicals*, 98%) was used as received. The submitters note that the iridium dimer can be synthesized by the procedure described in the literature[4] as well, obtaining the same results.

15. A mixture containing [Cp*IrCl$_2$]$_2$ (273 mg) in a THF-water mixture (1:2 v/v) (228 mL) is stirred with an octagonal Teflon coated magnetic stir bar (3.7 x 0.7 cm) in a 500-mL Erlenmeyer flask for 2 h, followed by sonication for 30 minutes. Alternatively, the solution can be prepared at the same time as step A is carried out and let stir for about 4 h with an

octagonal Teflon coated magnetic stir bar (3.7 x 0.7 cm). Both methods worked equally well.

16. When the solution of [Cp*IrCl$_2$]$_2$ in the THF-water (1:2 v/v) mixture was added to the crude mixture containing the allylic alcohol **1**, the system becomes biphasic.

17. N-Chlorosuccinimide (*Sigma-Aldrich*, 98%) was used as received.

18. The reaction can be monitored by ^{1}H NMR spectroscopy: a drop of the organic phase (top layer) is dissolved in CDCl$_3$. Allylic alcohol consumption is confirmed by a decreased intensity of the ^{1}H NMR resonances at 6.58 (d, J = 16 Hz, 1 H) and at 6.27 (dd, J = 16 Hz, J = 6.4 Hz, 1 H). Simultaneously, formation of the α-chloroketone product (**2**), can be followed by monitoring the peak at 4.41 (dd, J = 8, 6.2 Hz, 1 H).

19. After evaporation, the crude mixture is biphasic.

20. The ^{1}H NMR spectrum of α–chloroketone (**2**, crude mixture) corresponds with the NMR data previously described in the literature:[5b] ^{1}H NMR (500 MHz, CDCl$_3$) δ: 2.29 (s, 3 H), 3.08 (dd, J = 14.3 Hz, J = 8 Hz, 1 H), 3.34 (dd, J = 14.3 Hz, J = 6.2 Hz, 1 H), 4.41 (dd, J = 8 Hz, J = 6.2 Hz, 1 H), 7.18–7.34 (m, 5 H); ^{13}C NMR (125 MHz, CDCl$_3$) δ: 26.8, 39.8, 63.8, 127.2, 128.6, 129.3, 136.1, 202.6.

21. Pure 3-chloro-4-phenylbutan-2-one (**2**) was prepared by carrying out the reaction on half scale using pure 4-phenyl-3-buten-2-ol (**1**), and then purifying the crude reaction mixture by column chromatography. The crude oil is loaded onto 65.3 g silica gel that had been dry-packed in a 5 cm diameter chromatography column and wetted with *n*-pentane. The product is eluted with 300 mL of *n*-pentane, followed by 200 mL of a 1:99 solution of EtOAc:pentane, followed by 1-L of a 2:98 solution of EtOAc:pentane, and the eluent is collected in 60 mL fractions in 2.5 x 20 cm test tubes. Fractions 10-17 (R$_f$ = 0.58 (yellow spot), visualized with KMnO$_4$, 10:90 solution of EtOAc:pentane as eluent) are combined in a 1-L round-bottomed flask and concentrated under reduced pressure (bath temperature 23 °C; ≈ 20 mmHg) to give pure 3-chloro-4-phenylbutan-2-one (**2**) as a yellow oil. 3-chloro-4-phenylbutan-2-one (**2**) has the following spectroscopic properties: ^{1}H NMR (500 MHz, CDCl$_3$) δ: 2.29 (s, 3 H), 3.08 (dd, J = 14.3, 8.1 Hz, 1 H), 3.34 (dd, J = 14.3, 6.2 Hz, 1 H), 4.41 (dd, J = 8, 6.2 Hz, 1 H), 7.21–7.33 (m, 5 H). ^{13}C NMR (125 MHz, CDCl$_3$) δ: 26.8, 39.8, 63.8, 127.2, 128.6, 129.3, 136.1, 202.6. IR (neat) cm^{-1}: 3030, 2928, 1715, 1357, 1157. HRMS (ESI) Exact mass calcd for C$_{10}$H$_{11}$ClONa [M+Na]$^+$: 205.0391, found: 205.0394.

22. Thiourea (*Sigma-Aldrich*, 99%) was used as received.

23. The reaction can be monitored by ^{1}H NMR spectroscopy: a drop of the hot mixture is dissolved in DMSO-d_6. The reaction is finished when the peak at 4.41 (dd, J = 8, 6.2 Hz, 1 H) from α-chloroketone **2** has disappeared.

24. A brown solid starts to precipitate when the reaction is cooled to room temperature and/or 5 °C. Vigorous stirring must be maintained to avoid agglomeration of the solid.

25. Diethyl ether (*Fisher Scientific*, anhydrous, BHT Stabilized, ACS Reagent Grade) was used as received. The submitters used diethyl ether purchased from *VWR* (GPR Rectapur, >99) as received.

26. 5-Benzyl-4-methyl-2-aminothiazolium hydrochloride (**3**) has the following physical and spectroscopic properties: mp = 219–221 °C (decomp.). ^{1}H NMR (500 MHz, DMSO-d_6) δ: 2.20 (s, 3 H), 3.44 (bs) (water), 3.93 (s, 2 H), 7.23–7.26 (m, 3 H), 7.33 (t, J = 7.4 Hz, 2 H), 9.24 (s, 2 H),[6] 13.37 (bs, 1 H).[6] ^{13}C NMR (125 MHz, DMSO-d_6) δ: 11.5, 30.4, 116.8, 126.8, 128.3, 128.7, 131.0, 138.8, 167.8. IR (neat) cm^{-1}: 3242, 3192, 3058, 2919, 2652, 1623, 1573, 1453, 1075, 830, 760, 698. HRMS (ESI) Exact mass calcd for $C_{11}H_{13}N_2S$ [M+H]$^+$: 205.0794, found: 205.0798; Anal. Calcd for $C_{11}H_{13}N_2SCl$: C, 54.88; H, 5.44; N, 11.64; S, 13.32; Cl, 14.73; Found: C, 54.72; H, 5.63; N, 11.57; S, 13.28; Cl, 14.58 (these elemental analysis values were obtained from a sample that had been prepared by running the reaction with pure 3-chloro-4-phenylbutan-2-one (**2**); when a sample that had been prepared by running the reaction with crude 3-chloro-4-phenylbutan-2-one (**2**) was used, the CHN results were not in agreement with the calculated values despite identical ^{1}H, ^{13}C, IR, and HRMS data for both samples).

27. When the three-step sequence was run on half scale, 5.50 g (22.9 mmol) of 5-benzyl-4-methyl-2-aminothiazolium hydrochloride (**3**) was isolated as a brown solid (67% yield over three steps).

28. The submitters report a 76% yield over the three steps. The submitters also report a crystal structure of **3** that was solved from single crystal X-Ray diffraction data, confirming the substitution pattern as well as the protonation of the imidazolic nitrogen (this data was not attempted to be reproduced by the checkers).[7]

Handling and Disposal of Hazardous Chemicals

The procedures in this article are intended for use only by persons with prior training in experimental organic chemistry. All hazardous materials should be handled using the standard procedures for work with chemicals described in references such as "Prudent Practices in the Laboratory" (The National Academies Press, Washington, D.C., 2011 www.nap.edu). All chemical waste should be disposed of in accordance with local regulations. For general guidelines for the management of chemical waste, see Chapter 8 of Prudent Practices.

These procedures must be conducted at one's own risk. *Organic Syntheses, Inc.*, its Editors, and its Board of Directors do not warrant or guarantee the safety of individuals using these procedures and hereby disclaim any liability for any injuries or damages claimed to have resulted from or related in any way to the procedures herein.

Discussion

2-Aminothiazoles are privileged structures that are found in numerous biologically active compounds, with applications as antibiotics, anti-inflammatory and psychotropic agents, among others.[8] These heterocycles can be synthesized in a straightforward manner *via* condensation of α-chlorocarbonyls with thiourea. A challenge usually encountered is, however, the selective synthesis of 4,5-disubstituted 2-aminothiazoles. This is due to the unavailability and/or tedious methods to prepare the α-chlorocarbonyls precursors in high yields and with complete selectivity.

Recently, we reported a method to synthesize selectively α-halogenated ketones from allylic alcohols. The transformation is catalyzed by [Cp*IrCl$_2$]$_2$, which in the presence of a fluorinating (Selectfluor®)[9] or a chlorinating (*N*-chlorosuccinimide, **Table 1**)[5] reagent affords α-halocarbonyls (halogen = F, Cl) as single constitutional isomers in good to excellent yields.

Table 1. Examples of iridium-catalyzed tandem isomerization/C–Cl
bond formation of allylic alcohols.

Allylic alcohol	α-Chloroketone	Isolated Yield (%)
		88
		89
		91
		89
		80
		91

The chlorination reaction (**Table 1**) was used to synthesize a variety of
4,5-disubstituted 2-aminothiazoles from allylic alcohols. Thus, condensation
of selected α–chlorocarbonyls with thiourea followed by neutralization of
the thiazolium salt with sodium bicarbonate afforded the corresponding 2-
aminothiazole in excellent yield after two steps (**Table 2**).[5]

Table 2. Synthesis of 4,5-disubtituted 2-aminothiazoles from allylic alcohols.

Allylic alcohol	α-Chloroketone	2-Amino thiazole	Isolated Yield (%) (two steps)
			86
			90
			82
			95

Here, we describe the large-scale synthesis of 5-benzyl-4-methyl-2-aminothiazolium hydrochloride (**3**) from enone (*E*)-4-phenyl-3-buten-2-one in 3 steps without purifications by column chromatography. In the first step, allylic alcohol **1** was synthesized from (*E*)-4-phenyl-3-buten-2-one by reduction with NaBH$_4$. The crude of **1** was directly used in the iridium-catalyzed tandem isomerization/C–Cl bond formation, yielding α-chloroketone **2**. In the last step, the crude reaction mixture containing **2** was treated with thiourea affording 2-aminothiazolium hydrochloride salt **3** (**Scheme 1**). The final product (**3**) was isolated by precipitation from the reaction mixture.

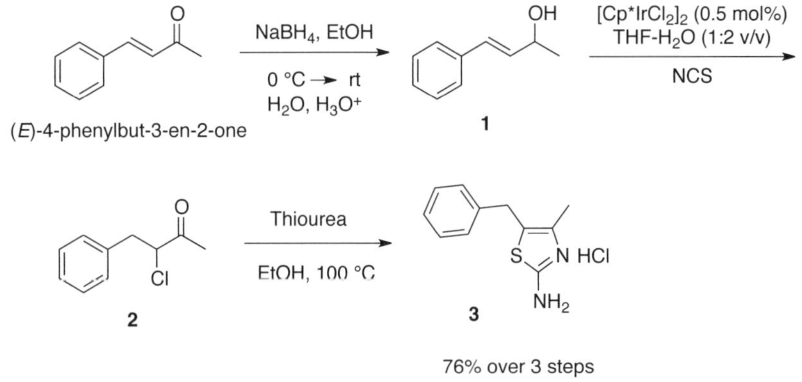

Scheme 1. Synthesis of 5-benzyl-4-methyl-2-aminothiazolium hydrochloride
(3) from (*E*)-4-phenyl-3-buten-2-one.

The structure of **3** was unambiguously confirmed by NMR spectroscopy,
HRMS, elemental analysis and single crystal X-Ray diffraction. Single
crystals were obtained by recrystallization from hot ethanol. The analysis
also indicates that protonation occurred on the nitrogen of the thiazole ring.

Compound **3** crystallizes in the monoclinic crystal system ($P2_1/c$ space
group). The cell parameters determined for this structure are: $a = 11.7869(5)$
Å, b = 7.5852(3) Å, c = 13.5158(5) Å, β = 92.805(4) °, V = 1206.95(8) Å^3.[7] The
ORTEP representation of its asymmetric unit is shown in **Figure 1**. The
imidazolic and amino H atoms were located in a difference Fourier map
and refined without any constraint. The remaining Ar-H and CH$_2$-group
hydrogen atoms were positioned geometrically and were constrained to
ride on their parent atoms, with C—H = 0.95 Å (for Ar-H) and C—H = 0.97
Å (for CH$_2$-) and Uiso(H) = 1.2Ueq(C). The methyl group (C4) is disordered.
The corresponding methyl hydrogen atoms were positioned geometrically
and were constrained to ride on their parent atom, with C—H = 0.93 Å; and
Uiso(H) = 1.5Ueq(C). The refined model describes a 30-70% disorder of
methyl groups.

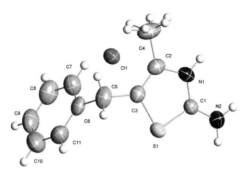

Figure 1. ORTEP representation of the asymmetric unit of 2-aminothiazole **3**. Ellipsoids are displayed at the 50% probability level. Hydrogen labels were omitted for clarity.

The supramolecular interactions in compound **3** are governed by weak hydrogen bonding-type interactions among chlorides and the protonated imidazolic and amino nitrogen atoms (**Figure 2**). Three different weak hydrogen bonds are determined, the strongest being the one corresponding to the protonated imidazole group. The distances and angles of these weak hydrogen bonds are presented in **Table 3**.

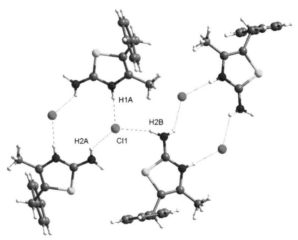

Figure 2. Hydrogen bonding-type interactions described in **3**. As it is depicted in the figure, each molecule interacts with other three neighbors through the described supramolecular interactions.

Table 3. Distances and angles of hydrogen bonds found in compound 3.

D-H···A[1]	D-H[2] (Å)	H···A[3] (Å)	D···A[4] (Å)	<D-H···A[5] (°)
N(1)i-H(1A)i..Cl(1)[1]	0.86(2)	2.19(3)	3.052(2)	175(2)
N(2)a-H(2A)a..Cl(1)[2]	0.84(2)	2.50(2)	3.182(2)	139(2)
N(2)a-H(2B)a..Cl(1)[3]	0.89(2)	2.34(2)	3.187(2)	1.60(2)

	symmetry operator codes	
[1] x, y+1, z	[2] -x+1, -y+1, -z+1	[3] x, -y+3/2, z+1/2

D: donor atom, A: acceptor atom. [1] Names of donor, hydrogen and acceptor atoms involved in the hydrogen bond. [2] Distance D – A. [3]Distances H – A. [4]Distance D – A. [5] Angle D – H – A. Ni: nitrogen atom coming from the imidazole group. Na: nitrogen atom coming from the amino group.

We have presented here a straightforward and easy procedure for the synthesis of 4,5-disbstitued 2-aminothiazolium hydrochloride salts. The method is exemplified in the synthesis of 2-aminothiazole **3** from a readily available enone. High yields of **3** are obtained in 3 steps, avoiding difficult or expensive purifications. The key step of this process is the tandem allylic alcohol isomerization/C–Cl bond formation catalyzed by iridium, which is a highly efficient method to synthesize α-(mono)chloroketones selectively.

References

1. Department of Organic Chemistry, Arrhenius Laboratory, Stockholm University, SE-106 91 Stockholm, Sweden. E-mail: belen@organ.su.se. ‡These authors (A.B.G. and N.A.) contributed equally. Financial support from the Swedish Research Council (Vetenskapsrådet), the Knut and Alice Wallenberg Foundation, and the Berzelii Center EXSELENT is gratefully acknowledged.
2. Department of Materials and Environmental Chemistry, Arrhenius Laboratory, Stockholm University, SE-106 91 Stockholm, Sweden. Financial support from the Knut and Alice Wallenberg Foundation is gratefully acknowledged.

3. Akai, S.; Hanada, R.; Fujiwara, N.; Kita, Y.; Egi, M. *Org. Lett.* **2011**, *12*, 4900.

4. White, C.; Yates, A.; Maitlis, P. M.; Heinekey, D. M. *Inorg. Synth.* **1992**, *29*, 228.

5. (a) Martín-Matute, B.; Bermejo, A.; Ahlsten, N. Patent Application P-76116-USP. (b) Ahlsten, N.; Bermejo, A.; Martín-Matute, B. *Angew. Chem. Int. Ed.* **2013**, *52*, 6273-6276. (c) For a related bromination reaction, see: Bermejo Gómez, A.; Erbing, E.; Batuecas, M.; Vázquez-Romero, A.; Martín-Matute, B. *Chem. Eur. J.* **2014**, *20*, 10703-10709.

6. These two signals could appear at different shifts, and depend on the concentration of the sample.

7. The crystallographic data of **3** can be obtained free of charge from the Cambridge Crystallographic Data Center. For atomic coordinates, equivalent isotropic displacements parameters, bond length, angles and anisotropic displacement parameters check the CIF file (CCDC 935780).

8. (a) Tsuji, K.; Ishikawa, H. *Bioorg. Med. Chem. Lett.* **1994**, *4*, 1601. (b) Haviv, F.; Ratajczyk, J. D.; DeNet, R. W.; Kerdesky, F. A.; Walters, R. L.; Schmidt, S. P.; Holms, J. H.; Young, P. R.; Carter, G. W. *J. Med. Chem.* **1988**, *31*, 1719. (c) Jaen, J. C.; Wise, L. D.; Caprathe, B. W.; Tecle, H.; Bergmeier, S.; Humblet, C. C.; Heffner, T. G.; Meltzner, L. T.; Pugsley, T. A. *J. Med. Chem.* **1990**, *33*, 311. (d) Bell, F. W.; Cantrell, A. S.; Hoberg, M.; Jaskunas, S. R.;Johansson, N. G.; Jordon, C. L.; Kinnick, M. D.; Lind, P.; Morin, J. M.; Noreen, R.; Oberg, B.; Palkowitz, J. A.; Parrish, C. A.; Pranc, J.; Zhang, H.; Zhou, X.-X. *J. Med. Chem.* **1995**, *38*, 4929.

9. (a) Ahlsten, N.; Martín-Matute, B. *Chem. Commun.* **2011**, *47*, 8331. (b) Ahlsten, N.; Bartoszewicz, A.; Agrawal, S.; Martín-Matute, B. *Synthesis* **2011**, *16*, 2600.

Appendix
Chemical Abstracts Nomenclature (Registry Number)

(*E*)-4-Phenyl-3-buten-2-one; (1896-62-4)
Sodium borohydride; Sodium tetrahydridoborate; (16940-66-2)
4-Phenylbut-3-en-2-ol; (17488-65-2)
Dichloro(pentamethylcyclopentadienyl)iridium (III) dimer; (12354-84-6)
N-Chlorosuccinimide; (128-09-6)
3-Chloro-4-phenylbutan-2-one; (20849-77-8)
Thiourea; Sulfourea; Thiocarbamide; (62-56-6)
5-Benzyl-4-methyl-1,3-thiazol-2-amine hydrochloride; (95767-21-8)

Belén Martín-Matute was born in Madrid (Spain) and obtained her Ph.D. at the Universidad Autónoma de Madrid (UAM 2002) with Prof. A. M. Echavarren. After a postdoctoral stay at Stockholm University (Sweden) with Prof. J.-E. Bäckvall working on dynamic kinetic resolutions, she joined the UAM as an Assistant Professor (2005–2007). She returned to Stockholm in 2007 and in 2014 became Full Professor. She received the Sigma Aldrich Young Chemist Award from the Spanish Royal Society of Chemistry in 2007, and the Lindbomska Award by the Swedish Royal Academy of Sciences in 2013. Her research is focused on homogeneous and heterogeneous transition metal catalysis, as well as enzymatic transformations and mechanistic investigations.

Antonio Bermejo Gómez was born in Sevilla (Spain) in 1982. He obtained his Ph.D at the Universidad de Sevilla (US 2011) with Prof. R. Fernández and Prof. J. M. Lassaletta in the field of asymmetric catalysis using chiral hydrazones as ligands. He started his postdoctoral research work at Stockholm University (Sweden) with Prof. B. Martín-Matute in 2011. His ongoing research is focused on different transformations of allylic alcohols and synthesis of biological active compounds (collaborations with the companies *Cambrex KA* and *AstraZeneca*).

Nanna Ahlsten obtained her Ph.D in 2013 with Prof. Belén Martín-Matute at the Department of Organic Chemistry at Stockholm University. Her Ph.D studies were focused on the development of transition metal catalyzed isomerization and tandem isomerization/ halogenation reactions of allylic alcohols. She is currently a postdoctoral research assistant in the group of Dr Igor Larrosa at Queen Mary University of London (UK).

Ana E. Platero Prats was born in Barcelona (Spain) in 1984. She obtained her Ph.D at the Instituto de Ciencia de Materiales de Madrid (ICMM-CSIC) in 2011 with Prof. E. Gutiérrez Puebla and Dr N. Snejko in the field of green metal-organic frameworks for catalytic and adsorption applications. She started her postdoctoral research work at Stockholm University (Sweden) within a collaboration between Prof. X. Zou and Prof. B. Martín-Matute in 2012. Her ongoing research is focused on the functionalization and characterization of metal-organic frameworks, as well as the application of modern crystallographic tools to study these complex materials.

Michael T. Tudesco obtained his B.S. degree in Chemistry at the University of North Carolina at Chapel Hill in 2012, performing undergraduate research under the supervision of Professor Michel R. Gagné. He is currently pursuing his Ph.D. at Baylor University under the guidance of Professor John L. Wood.

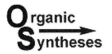
Sodium Methoxide-Catalyzed Direct Amidation of Esters

Submitted by Kazushi Agura,[1] Takashi Ohshima,*[1] Yukiko Hayashi,[2] and Kazushi Mashima*[2]

[1]Graduate School of Pharmaceutical Sciences, Kyushu University, Fukuoka 812-8582, Japan; [2]Department of Chemistry, Graduate School of Engineering Science, Osaka University, Osaka 560-8531, Japan

Checked by Mengyang Fan and Dawei Ma

A.

$$\text{NaOMe (1 mol\%)} \atop \text{MS 3Å}$$
toluene, 55 °C

B.

$$\text{NaOMe (10 mol\%)} \atop \text{4-CF}_3\text{-C}_6\text{H}_4\text{-OH (30 mol\%)} \atop \text{MS 3Å}$$
toluene, 70 °C

Procedure

A. *N-Hexylbenzamide.* A 50-mL, one-necked, round-bottomed Schlenk flask equipped with a 2-cm ellipsoidal magnetic stir bar and a septum is connected to Schlenk line and dried with a heat gun under reduced pressure (1.5 mmHg) (Note 1). After the flask cooled to room temperature, it is evacuated and backfilled with argon for three times. The septum is removed and sodium methoxide (17.7 mg, 0.328 mmol, 1 mol%) (Note 2) and an activated powder of 3Å molecular sieve (808 mg) (Note 3) are added quickly to the flask under an argon atmosphere (Note 4). The septum is replaced and the flask is filled with argon and toluene (8.0 mL) (Note 5), *n-*

HexNH$_2$ (5.5 mL, 41.6 mmol, 1.3 equiv) (Note 6), and methyl benzoate (4.0 mL, 32.1 mmol, 1.0 equiv) (Note 7) are added via syringe. The reaction mixture is stirred and heated at 55 °C (Note 8) with an oil bath for 20 h under an argon flow conditions. The progress of the reaction is monitored by TLC (R_f = 0.30, hexanes/EtOAc = 4/1). The resulting mixture is quenched by adding aqueous saturated NH$_4$Cl (30 mL), then EtOAc (30 mL). The biphasic suspension is filtered through a Celite pad (5 g) and the residue is washed with EtOAc (100 mL). The filtered solution is transferred to a 250-mL separatory funnel, and the organic layer is separated, washed with aqueous saturated NH$_4$Cl (30 mL), brine (30 mL), dried over Na$_2$SO$_4$ (10 g), filtered through a cotton plug, and concentrated using a rotary evaporator (30 °C, 4 mmHg). The residue is purified by flash column chromatography (diameter = 6 cm; 550 mL of silica gel; hexanes/EtOAc = 4/1 to 2/1) to give 5.42–5.54 g (82–84% yield) of *N*-hexylbenzamide (Note 9) as a white solid.

B. *(S)-tert-Butyl (1-(benzylamino)-4-methyl-1-oxopentan-2-yl)carbamate.* A 50-mL, one-necked, round-bottomed Schlenk flask equipped with a 2-cm ellipsoidal magnetic stir bar and a septum is connected to Schlenk line and dried with a heat gun under reduced pressure (1.5 mmHg) (Note 1). After the flask cooled to room temperature, it is evacuated and backfilled with argon for three times. The septum is removed and sodium methoxide (108 mg, 2.01 mmol, 10 mol%) (Note 2) and an activated powder 3 Å molecular sieve (530 mg) (Note 3) are added to the flask under an argon atmosphere quickly. The septum is replaced and the flask is filled with argon gas and a yellow solution of 4-CF$_3$-C$_6$H$_4$-OH (973 mg, 6.00 mmol, 30 mol%) (Note 10) in toluene (5.0 mL) (Note 5), BnNH$_2$ (2.8 mL, 26.1 mmol, 1.3 equiv) (Note 11), and (*S*)-Boc-Leu-OMe (4.95 g, 20.2 mmol, 1.0 equiv) (Note 12) are added. The reaction mixture is stirred and heated at 70 °C (Note 8) with an oil bath for 99 h under argon flow conditions. The progress of the reaction is monitored by TLC (R_f = 0.30, hexanes/EtOAc = 4/1). The resulting mixture is quenched with aqueous saturated NH$_4$Cl (30 mL). After adding EtOAc (30 mL), the biphasic suspension is filtered through a Celite pad (5 g) and the residue is washed with EtOAc (100 mL). The filtered solution is transferred to a 250-mL separatory funnel, and the organic layer is separated and washed with aqueous saturated NH$_4$Cl (30 mL), water (30 mL), and brine (30 mL), dried over Na$_2$SO$_4$ (10 g), filtered through a cotton plug, and concentrated using a rotary evaporator (30 °C, 4 mmHg). The residue is purified by flash column chromatography (diameter = 6 cm;

550 mL silica gel; hexane/EtOAc = 8/1 to 2/1) to give 5.31–5.47 g (82–85% yield and 98% ee) of (*S*)-Boc-Leu-NHBn (Note 13) as a white solid.

Notes

1. Maintaining anhydrous conditions is critically important to achieve a high turnover frequency and maintain good reproducibility, because contamination by water decomposes the NaOMe catalyst to NaOH, which further reacts with ester to generate inactive sodium carboxylate. The submitter used a 50-mL, one-necked, round-bottomed flask equipped with a three-way cock. To ensure the anhydrous conditions, the checker used a 50-mL, one-necked, round-bottomed Schlenk flask attached to Schlenk line.
2. From a freshly opened bottle, NaOMe (powder, 95%, purchased from Aldrich) is quickly transferred to a Schlenk tube under a flow of argon and stored under an argon atmosphere.
3. The powdered 3Å molecular sieves (powder < 50 μm, purchased from Acros) are activated by heating at 200 °C through use of an oil bath under reduced pressure (1.5 mmHg) for 5 h. The activated sieves are stored in a Schlenk tube under argon atmosphere.
4. The submitter used a two-leg glass adapter to transfer the NaOMe or the molecular sieve from the Schlenk tube to the flask. The checkers did not use such apparatus, but connected the flask to an argon line and quickly transferred the NaOMe or the molecular sieve under a flow of argon.
5. The toluene (purchased from Sinopharm Chemical Reagent Co., Ltd) is dried over CaH$_2$ overnight at room temperature and then distilled over CaH$_2$ under standard pressure and stored in a side-arm flask under argon atmosphere.
6. *n*-HexNH$_2$ (purchased from Aldrich, 99%) is dried over CaH$_2$ overnight at room temperature and then distilled under ordinary pressure and stored in a Schlenk tube under argon atmosphere.
7. Methyl benzoate (purchased from Aldrich, 99%) is dried over CaH$_2$ overnight at room temperature and then distilled at atmospheric pressure and stored in a Schlenk tube under argon atmosphere.
8. The reaction temperature significantly affects the rate of the NaOMe-catalyzed amidation. Although the 8-mmol scale reactions reported in

the original manuscript[3] are performed at 50 °C (oil bath), for procedures A and B (32.1 mmol scale), the oil bath temperature is increased to 55 °C to improve heat transfer efficiency.

9. The analytical data of N-hexylbenzamide are as follows: white solid; mp 41–43 °C; R_f = 0.30 (hexanes/EtOAc = 4/1); IR (KBr disk, v/cm^{-1}) 3342, 2965, 2956, 2921, 2857, 1632, 1577, 1529, 1489, 1481, 1466, 1376, 1350, 1313, 1275, 1078, 928, 859, 805, 718, 695, 634; ^{1}H NMR (500 MHz, CDCl$_3$) δ: 0.90 (t, J = 6.8 Hz, 3 H), 1.30-1.44 (m, 6 H), 1.62 (m, 2 H), 3.46 (dt, J = 6.6, 6.0 Hz, 2 H), 6.08 (br s, 1 H), 7.41–7.51 (m, 3 H), 7.76 (d, J = 7.6 Hz, 2 H); ^{13}C NMR (125 MHz, CDCl$_3$) δ: 14.0, 22.5, 26.6, 29.6, 31.5, 40.1, 126.8, 128.5, 131.2, 134.9, 167.5; MS (ESI+) m/z (relative intensity) 206.2 ([M+H$^+$], 100%), 228.2 ([M+Na$^+$], 22%); HRMS (ESI+) m/z calcd. for C$_{13}$H$_{20}$NO 206.1539, found 206.1539; Anal. calcd for C$_{13}$H$_{19}$NO: C, 76.06, H, 9.33, N, 6.82, found: C, 76.09, H, 9.45, N, 6.79.

10. p-Trifluoromethylphenol (purchased from Aldrich, 97%) is dissolved in toluene (Note 4) and stored in a Schlenk tube under argon atmosphere.

11. Benzylamine (purchased from Aldrich, 99%) is dried over CaH$_2$ overnight at room temperature and then distilled under reduced pressure and stored in a Schlenk tube under argon atmosphere.

12. N-(tert-Butoxycarbonyl)-L-leucine methyl ester ((S)-Boc-Leu-OMe) (97%) was purchased from Sigma-Aldrich and used directly without further purification.

13. The analytical data of (S)-Boc-Leu-NHBn are as follows: white solid; mp 77–79 °C; R_f = 0.30 (hexanes/EtOAc = 4/1); IR (KBr disk, v/cm^{-1}) 3294, 3089, 2961, 2870, 1682, 1655, 1534, 1454, 1392, 1366, 1321, 1274, 1247, 1171, 1046, 1027, 714, 695; ^{1}H NMR (500 MHz, CDCl$_3$) δ: 0.93 (d, J = 4.5 Hz, 3 H), 0.95 (d, J = 4.5 Hz, 3 H), 1.42 (s, 9 H), 1.46–1.54 (m, 1 H), 1.64–1.78 (m, 2 H), 4.11 (m, 1 H), 4.45 (m, 2 H), 4.84 (br s, 1 H), 6.42 (br s, 1 H), 7.24–7.35 (m, 5 H); ^{13}C NMR (125 MHz, CDCl$_3$) δ: 22.0, 22.9, 24.7, 28.2, 41.2, 43.4, 53.1, 80.0, 127.4, 127.6, 128.6, 138.1, 155.8, 172.5; MS (ESI+) m/z (relative intensity) 343.3 ([M+Na$^+$], 100%), 265.3 (56%); HRMS (ESI+) m/z calcd for C$_{18}$H$_{29}$N$_2$O$_3$ 321.2173, found 321.2166; Anal. calcd for C$_{18}$H$_{28}$N$_2$O$_3$: C, 67.47, H, 8.81, N, 8.74, found: C, 67.50, H, 8.78, N, 8.73; [α]$_{589}^{27}$ –24.8 (c 1.03 in CH$_2$Cl$_2$); The enantiomeric excess (%ee) was determined to be 98% by HPLC using CHIRALPAK OD-3 column (2% i-PrOH/hexane, 1.0 mL/min, 254 nm): t_R (minor, 13.3 min) t_R (major, 19.3 min). The racemic mixture was prepared through the condensation of rac-Boc-Leu-OH and benzylamine using HOBt and EDCI.

Handling and Disposal of Hazardous Chemicals

The procedures in this article are intended for use only by persons with prior training in experimental organic chemistry. All hazardous materials should be handled using the standard procedures for work with chemicals described in references such as "Prudent Practices in the Laboratory" (The National Academies Press, Washington, D.C., 2011 www.nap.edu). All chemical waste should be disposed of in accordance with local regulations. For general guidelines for the management of chemical waste, see Chapter 8 of Prudent Practices.

These procedures must be conducted at one's own risk. *Organic Syntheses, Inc.*, its Editors, and its Board of Directors do not warrant or guarantee the safety of individuals using these procedures and hereby disclaim any liability for any injuries or damages claimed to have resulted from or related in any way to the procedures herein.

Discussion

Amides are one of the most ubiquitous and important functional groups in natural and synthetic organic compounds, and the amide bond formation has been studied intensively in organic synthesis. The most common method of synthesizing amides is the coupling reaction of carboxylic acids and amines using stoichiometric amounts of condensation reagents. Amidation of esters with amines, which is a key transformation in biologic peptide synthesis on ribosomes, is another important synthetic method for amide bond formation due to the environmental benefits of this reaction and the operational benefits of esters, such as their handling ease and high stability as well as their high solubility in most organic solvents compared with carboxylic acid. In non-enzymatic amide formation, however, simple alkyl esters are viewed as an inert scaffold and rather harsh reaction conditions, such as high temperature, high pressure, or the use of more than stoichiometric amounts of strongly basic reagents, are required to promote amidation.

Ester-amide exchange reactions using alkali metal alkoxides such as NaOMe[5] and KO-*t*-Bu have been reported, but more than stoichiometric amounts or sub-stoichiometric amounts of these reagents are necessary in these systems. In contrast, this is the first report that maintenance of the anhydrous conditions successfully promotes the reactions with only a catalytic amount of NaOMe (1-10 mol%). Because this NaOMe-catalyzed amidation proceeds with high efficiency under mild conditions (as low as room temperature), a variety of functionalized aliphatic and aromatic

Table 1. NaOMe-catalyzed amidation with various esters and amine[a,b]

A reaction scheme is shown:

$$R^1\text{-}C(=O)\text{-}OMe + HNR^2R^3 \xrightarrow[\substack{\text{toluene (4.0 M)} \\ 50\,^\circ\text{C, time}}]{\text{NaOMe (5 mol\%)}} R^1\text{-}C(=O)\text{-}NR^2R^3$$

Products in table:

- NC-aryl amide: 92% yield (20 h)
- F₃C-aryl amide: 99% yield (20 h)
- MeO-aryl amide: 71% yield (44 h)
- cinnamyl amide: 85% yield (48 h)
- phenylpropanoyl amide: 94% yield (20 h)
- cyclohexanecarbonyl amide: 62% yield (20 h)
- Boc-glycine amide: 88% yield (20 h)
- Cbz-glycine amide: 97% yield (26 h)
- F₃C-aryl benzylamide: 90% yield (20 h)
- F₃C-aryl cyclohexylamide: 93% yield (20 h)
- F₃C-aryl piperidine amide: 94% yield (36 h)
- F₃C-aryl aminoethanol amide: 70% yield[c] (20 h)

[a] Reaction condition; A reaction mixture of ester (8.0 mmol), amine (10.4 mmol), NaOMe (0.4 mmol, 5 mol% based on ester) and toluene (2.0 mL) was heated at 50 °C with oil bath under an argon atmosphere. [b] Isolated yield. [c] 21% of amino ester was also obtained.

methyl esters as well as cyclic lactones are smoothly converted to the corresponding amides (Table 1). Furthermore, adding a desiccant such as MS3Å or Drierite can minimize catalyst loading to 1 mol% (*Procedure A*).

When chiral α-amino ester derivatives are used as the substrate, epimerization is a major problem due to the strongly basic conditions. This severe epimerization is successfully rectified by the addition of the rather acidic alcohol 4-trifluorophenol. Under the optimized conditions using 4-trifluorophenol, a catalytic ester-amide exchange reaction of various *N*-Boc protected chiral α-amino esters with benzylamine proceeds in high yield without epimerization (Table 2).

Table 2. NaOMe-catalyzed amidation with various chiral α-amino esters[a,b,c]

NaOMe (10 mol%)
4-CF$_3$-C$_6$H$_4$-OH (30 mol%)
MS 3Å
toluene (4.0 M)
temp., time

BocHN-CH(R^1)-C(O)-OMe + H$_2$NBn → BocHN-CH(R^1)-C(O)-NHBn

81% yield 97% ee (50 °C, 72 h)	80% yield 98% ee (50 °C, 72 h)	72% yield[d] 99% ee (70 °C, 99 h)
80% yield 99% ee (70 °C, 99 h)	80% yield 99% ee (50 °C, 99 h)	84% yield 99% ee (70 °C, 99 h)

[a] Reaction condition; A reaction mixture of ester (2.0 mmol), benzyl amine (2.6 mmol), NaOMe (0.2 mmol, 10 mol% based on ester), 4-trifluoromethylphenol (0.6 mmol, 30 mol% based on ester), MS 3Å (50 mg) and toluene (0.5 mL) was heated with oil bath under an argon atmosphere. [b] Isolated yield [c] ee was determined by chiral HPLC analysis. [d] 4.0 equiv of amine was used.

References

1. Graduate School of Pharmaceutical Sciences, Kyushu University, Fukuoka 812-8582, Japan.
2. Department of Chemistry, Graduate School of Engineering Science, Osaka University, Osaka 560-8531, Japan.
3. Ohshima, T.; Hayashi, Y.; Agura, K.; Fujii, Y.; Yoshiyama, A.; Mashima, K. *Chem. Commun.* **2012**, *48*, 5434.
4. Xu, Z.; DiCesare, J. C.; Baures, P. W. *J. Comb. Chem.* **2010**, *12*, 248.
5. (a) Bunnett, J.; Davis, G. *J. Am. Chem. Soc.* **1960**, *82*, 665. (b) De Feoand, R. J.; Strickler, P. D. *J. Org. Chem.* **1963**, *28*, 2915.
6. (a) Varma, R. S.; Naicker, K. P. *Tetrahedron Lett.* **1999**, *40*, 6177. (b) Perreux, L.; Loupy, A.; Delmotte, M. *Tetrahedron* **2003**, *59*, 2185.

Appendix
Chemical Abstracts Nomenclature (Registry Number)

Benzoic acid, methyl ester; (93-58-3)
Benzenemethanamine; (100-46-9)
Sodium methoxide; (124-41-4)
Benzamide, *N*-(phenylmethyl)-; (1485-70-7)
1-Hexanamine; (111-26-2)
Benzamide, *N*-hexyl-; (4773-75-5)
L-Leucine, N-[(1,1-dimethylethoxy)carbonyl]-, methyl ester; (63096-02-6)
Phenol, 4-(trifluoromethyl)-; (402-45-9)
Carbamic acid, *N*-[(1*S*)-3-methyl-1-[[(phenylmethyl)amino]carbonyl]butyl]-, 1,1-dimethylethyl ester; (101669-45-8)

Takashi Ohshima received his Ph.D. from The University of Tokyo in 1996 under the direction of Professor Masakatsu Shibasaki. He joined Otsuka Pharmaceutical Co., Ltd. for one year. After two years as a postdoctoral fellow at The Scripps Research Institute with Professor K. C. Nicolaou (1997-1999), he returned to Japan and joined Professor Shibasaki's group in The University of Tokyo as an Assistant Professor. He was appointed as Associate Professor of Osaka University in 2005. In 2010, he was promoted to Professor of Kyushu University. He has received the Fujisawa Award in Synthetic Organic Chemistry (2001), Pharmaceutical Society of Japan Award for Young scientists (2004), The Japanese Society for Process Chemistry Award for Excellence (2008), 9th Green Sustainable Chemistry Award with MEXT Award (2010), and Asian Core Program Lectureship Award (2012).

Kazushi Mashima received his Doctor degree (1986) from Osaka University under the supervision of Professor A. Nakamura. He became an Assistant Professor at Institute for Molecular Science, Okazaki National Institutes in 1983, Faculty of Engineering, Kyoto University in 1989, and then to Faculty of Science, Osaka University in 1991. He was appointed as an Associate Professor at Faculty of Engineering Science, Osaka University in 1994, and then a full Professor at Graduate School of Engineering Science, Osaka University in 2003. He worked with Professor M. A. Bennett, Australian National University in 1992 and Professor W. A. Herrmann, Technisch Universität München in 1993. He has received The Chemical Society of Japan Award for Creative Work for 2008, The 9th Green and Sustainable Chemistry Award, Awarded by the Ministry of Education, Culture, Sports, Science and Technology in 2010, and The Award of the Society of Polymer Science, Japan in 2010.

Kazushi Agura was born in 1986 in Kyoto, Japan. He obtained his Master's degree from Graduate School of Engineering Science, Osaka University, Osaka, Japan, in the laboratory of Professor Kazushi Mashima. He then joined the Ph.D. program at Graduate School of Pharmaceutical Sciences, Kyushu University, Fukuoka, Japan, in the laboratory of Professor Takashi Ohshima and is studying the development of new chiral bimetallic catalyst. He then received the Ph.D. from Kyushu University, Fukuoka, Japan, in 2014 under the direction of Professor Takashi Ohshima. He is currently working in Shionogi & Co., Ltd.

Yukiko Hayashi was born in 1986 in Kobe, Japan. After obtaining her B.Sc. degree from Osaka University in 2008, she received Ph.D. from Osaka University in 2013 under the supervision of Prof. Kazushi Mashima. She was Research Fellow of the Japan Society for the Promotion of Science in 2010-2013. She is currently working in Noritake Co., Ltd.

Mengyang Fan was born in 1989 in Xuzhou, China. He obtained his B.Sc. degree from the Department of Chemistry and Chemical Engineering, Southeast University in 2011. He then joined the Ph.D. program in the laboratory of Professor Dawei Ma at Shanghai Institution of Organic Chemistry and is studying transition metal catalyzed C-H bond activation.

Synthesis of Tetrasubstituted 1*H*-Pyrazoles by Copper-mediated Coupling of Enaminones with Nitriles

Mamta Suri and Frank Glorius[*1]

Organisch-Chemisches Institut der Westfälischen Wilhelms-Universität
Münster, Corrensstrasse 40, 48149 Münster, Germany

Checked by Lan-Ting Xu and Dawei Ma

A.

Ph–NH₂ (1.0 equiv) + [Me, O, CO₂Me] (1.0 equiv) → Zn(ClO₄)₂ (5 mol%), MgSO₄ (1 equiv), DCM (2.5 M), RT → [Me, Ph–N–H, CO₂Me]

B.

[Me, Ph–N–H, CO₂Me] (1.0 equiv) + [N≡C–CF₃ aryl] (3.0 equiv) → Cu(OAc)₂ (1.5 equiv), DMF (1 M), 110 °C, 26 h → [Me, Ph–N–N, CO₂Me, CF₃]

Procedure

A. *(Z)-Methyl 3-(phenylamino)but-2-enoate.*[2] A flame-dried round-bottomed, single-neck flask (capacity: 166 mL, diameter: 45 mm) is equipped with a 3 cm rod-shaped, Teflon-coated magnetic stir bar under an argon atmosphere (Note 1). Zinc perchlorate-hexahydrate (Zn(ClO₄)₂·6H₂O) (1.6 g, 4.4 mmol, 5 mol%) (Note 2) and magnesium sulfate (10.7 g, 88.5 mmol, 1 equiv) (Note 3) are weighed into the flask under air. Dichloromethane (35 mL, 2.5 M), methyl acetoacetate (9.5 mL, 88.5 mmol, 1.0 equiv) and aniline (8.1 mL, 88.5 mmol, 1.0 equiv) (Note 4) are added via cannula in the respective order into the flask resulting in a white suspension. The flask is tightly closed with a glass stopper and the suspension is stirred at room temperature for 24 h (Note 5). Upon completion of the reaction, dichloromethane (20 mL) is added and the resulting mixture is filtered through a short pad of sand (approx. 0.5 cm x 2.5 cm) and Celite (approx. 3 cm x 2.5 cm), prepacked with dichloromethane. The solid is washed thoroughly with dichloromethane (4 × 30 mL) and the combined filtrates are concentrated by rotary evaporation (40 °C water bath, 700-600 mmHg) to obtain a viscous paste (Notes 6 and 7). The crude product was purified by re-crystallization (Note 8) to furnish 13.5–13.6 g (80–81%) of (Z)-methyl 3-(phenylamino)but-2-enoate as colorless needle-shape crystalline solid (Note 9).

B. *Methyl 5-methyl-1-phenyl-3-(3-(trifluoromethyl)phenyl)-1H-pyrazole-4-carboxylate:* A flame-dried, single-necked, 250-mL round bottomed flask is equipped with a 3 cm rod-shaped, Teflon-coated magnetic stir bar under an argon atmosphere (Note 1). Copper(II) acetate (5.40 g, 29.7 mmol, 1.5 equiv) (Note 10) and (Z)-methyl 3-(phenylamino)but-2-enoate (3.80 g, 19.8 mmol, 1.0 equiv) are weighed into the flask under air. N,N-Dimethylformamide (19.8 mL, 1.0 M) and 3-(trifluoromethyl)benzonitrile (7.90 mL, 59.4 mmol, 3.0 equiv) (Note 11) are added via cannula into the flask resulting in a blue-green suspension. The flask is tightly closed with a glass stopper and the suspension is stirred in a pre-heated oil bath at 110 °C for 26 h (Note 12). The reaction mixture is allowed to cool to room temperature, EtOAc (20 mL) is added and filtered through a short pad of sand approx. 0.5 cm x 2.5 cm) and Celite (approx. 3 cm x 2.5 cm), prepacked with EtOAc (Note 13). The solid is washed thoroughly with EtOAc (4 × 30 mL) and the combined filtrates are concentrated to approx. 30 mL by rotary evaporation (40 °C water bath, 200-70 mmHg) to provide a blue-green suspension. N,N-Dimethylformamide and 3-(trifluoromethyl)benzonitrile were distilled off by Kugelrohr distillation resulting in a blue-green viscous paste (Notes 14, 15, 16 and 7). The crude product is then purified by flash column chromatography (Note 17 and 18). The combined eluent is concentrated by rotary evaporation (40 °C water bath, 700-70 mmHg) and then submitted to high vacuum (0.07 mmHg) overnight at room temperature (Note 19) to furnish 5.60–5.68 g (78–79%) of methyl 5-methyl-1-phenyl-3-(3-(trifluoromethyl)phenyl)-1H-pyrazole-4-carboxylate as a white solid (Note 20).

Notes

1. The reaction is less effective in the presence of moisture, presumably due to partial hydrolysis of enaminone, which reduces the yield of the product. With exclusion of moisture, the reaction yields the same amount of product under an atmosphere of either argon or dry air. To obtain dry air after evacuating the flask, the flask is filled with air passed through a guard tube filled with calcium chloride."

2. $Zn(ClO_4)_2 \cdot 6H_2O$ is purchased from Aldrich and used as received.

3. $MgSO_{42}$ is purchased from Fischer and stored in an oven at 120 °C.

4. Methyl acetoacetate is purchased from Alfa Aesar and used as received. Aniline is purchased from Alfa Aesar, distilled under reduced pressure, and stored in a brown Schlenk bottle (protected from direct sunlight) under an argon atmosphere.

5. The consumption of the starting material was monitored by TLC analysis on Merck silica gel 60 F_{254} plates (0.25 mm, aluminum plate, visualized with 254 nm UV lamp) using 1% Et_3N and 10% ethyl acetate in n-pentane as an eluent. The TLC is pre-treated with a solution of 1% Et_3N in n-pentane and dried (to prevent the hydrolysis of product to starting material). Aniline had an $R_f = 0.19$, methyl acetoacetate had an $R_f = 0.23$ and (Z)-methyl 3-(phenylamino)but-2-enoate had an $R_f = 0.52$ (UV active).

6. GC-MS analysis of the crude material detected only the peak corresponding to the product ($t_R = 11.2$ min (Note 7)) corresponding to the only spot ($R_f = 0.52$) on the TLC plate (n-pentane/ethyl acetate/$Et_3N = 90:10:1$, UV active).

7. GC-MS spectra were recorded on an Agilent Technologies 7890A GC-system with an Agilent 5975C VL MSD or an Agilent 5975 inert Mass Selective Detector (EI) and a HP-5MS column (0.25 mm × 30 m, Film: 0.25 µm). The major signals are quoted in m/z with the relative intensity in parentheses. The methods used start with the injection temperature T_0; after holding this temperature for 3 min, the column is heated to temperature T_1 (ramp) and this temperature is held for an additional time t: Method: $T_0 = 50$ °C, $T_1 = 290$ °C, ramp = 40 °C/min, t = 4 min.

8. The re-crystallization of the crude product was carried out by dissolving the crude mixture in minimum amount of dichloromethane and then adding n-hexane slowly into the mixture. The resulting turbid solution is kept tightly closed at lower temperature around 4 °C (in

refrigerator) for 24 h. The mixture is allowed to come to room temperature and the needle-shaped colorless crystals are filtered using a filter paper, washed with cold *n*-hexane (2 x 10 mL) and dried *in vacuo*.

9. (Z)-Methyl 3-(phenylamino)but-2-enoate has the following physicochemical and spectroscopic properties: R_f = 0.52 (*n*-pentane/ethyl acetate/Et_3N = 90:10:1); ^1H NMR (400 MHz, $CDCl_3$) δ: 2.00 (s, 3 H), 3.69 (s, 3 H), 4.70 (s, 1 H), 7.07 – 7.14 (m, 2 H), 7.16 (m, 1 H), 7.32 (m, 2 H), 10.35 (br s, 1 H,); ^{13}C NMR (100 MHz, $CDCl_3$) δ: 20.43, 50.39, 85.71, 124.62, 125.14, 129.19, 139.41, 159.24, 170.84; Exact Mass ESI-MS: calculated *m*/*z* for $[C_{11}H_{13}NO_2H]^+$: 192.1025; found: 192.1018; ATR-FTIR (cm^{-1}): 3249, 3060, 2993, 2952, 1651, 1589, 1483, 1436, 1383, 1356, 1260, 1260, 1056, 1000, 950, 912, 823, 786, 758, 729, 704, 668, 616, 551; Anal calcd for $C_{11}H_{13}NO_2$: C, 69.09; H, 6.85; N, 7.32; Found: C, 69.06; H, 6.86; N, 7.32; mp = 45.7–47.5 °C.

10. Copper(II) acetate is obtained by heating $Cu(OAc)_2(H_2O)$, purchased from Aldrich, at 100 °C under high vacuum (approx. 0.1 mmHg) for 60 h with occasional shaking. (Stirring should be avoided to limit the reduction of the particle size.)

11. *N,N*-Dimethylformamide (99.8%) was purchased from Acros and the solvent was stored over activated 4Å molecular sieves under an atmosphere of argon. 3-(Trifluoromethyl)benzonitrile was purchased from Acros and used as received.

12. The consumption of the starting material was monitored by TLC analysis on Merck silica gel 60 F_{254} plates (0.25 mm, aluminum plate, visualized with 254 nm UV lamp) using 10% ethyl acetate in *n*-pentane as an eluent. (Z)-Methyl 3-(phenylamino)but-2-enoate had an R_f = 0.44 (UV active) and methyl 5-methyl-1-phenyl-3-(3-(trifluoromethyl)phenyl)-1*H*-pyrazole-4-carboxylate had an R_f = 0.19 (UV active).

13. A large quantitiy of brown-red metallic precipitate that is typical for Cu^I/Cu^0 species is removed.

14. *N,N*-Dimethylformamide and 3-(trifluoromethyl)benzonitrile were removed from the crude reaction mixture by Kugelrohr distillation. The crude mixture was rotated in a round-bottomed cylindrical bulb and a slow gradient of reduced pressure from 5 mmHg to 0.1 mmHg is applied at 25 °C. The vapors of *N,N*-dimethylformamide and 3-(trifluoromethyl)benzonitrile condensed together in a bulb cooled using a dry ice and acetone bath.

15. The submitters reported that removal of *N,N*-dimethylformamide by an aqueous workup leads to a lower yield of the product (75 vs 81%) presumably due to the loss of the product in the aqueous phase. Thus, Kugelrohr distillation is preferred.

16. GC-MS analysis (Note 7) of the crude material detected only a single peak corresponding to the product (t_R = 10.2 min), corresponding to the only spot (R_f = 0.19) on the TLC plate (*n*-pentane/ethyl acetate = 9:1, UV active).

17. The submitters purchased silica gel (0.040-0.063) from Merck.

18. The crude mixture was dissolved in CH_2Cl_2 (approx. 10 mL), adsorbed on silica (approx. 15 g) and then charged onto a column (diameter = 5 cm, height = 15 cm) of 100 g (300 mL) of silica gel. The column was eluted with a gradient of *n*-pentane/EtOAc = 19:1 (1 L) to 3:2 (3 L) and 40-mL fractions were collected.

19. This represents a convenient way to remove trace impurities of solvents, furnishing analytically pure product as a white solid.

20. Methyl 5-methyl-1-phenyl-3-(3-(trifluoromethyl)phenyl)-1*H*- pyrazole-4-carboxylate (**3ad**) has the following physicochemical and spectroscopic properties: R_f = 0.19 (*n*-pentane/ethyl acetate = 9:1); ^1H NMR (400 MHz, CDCl$_3$) δ: 2.60 (s, 3 H), 3.77 (s, 3 H), 7.48 – 7.54 (m, 6 H), 7.63–7.65 (m, 1 H), 7.89 (m, 1 H), 7.98 (m, 1 H); ^{13}C NMR (100 MHz, CDCl$_3$) δ: 12.8, 51.2, 110.5, 124.3 (q, $^1J_{CF3}$ = 270.1 Hz), 125.0 (q, $^3J_{CF3}$ = 3.8 Hz), 125.9, 126.6 (q, $^3J_{CF3}$ = 3.9 Hz), 128.2, 129.0, 129.4, 130.2 (q, $^2J_{CF3}$ = 32.1 Hz), 132.9, 134.0, 138.6, 145.5, 152.2, 164.3; ^{19}F NMR (376 MHz, CDCl$_3$) δ: –62.53 (s, CF$_3$); Exact Mass ESI-MS: calculated *m/z* for $[C_{19}H_{15}F_3N_2O_2H]^+$: 361.1164, found: 361.1158; ATR-FTIR (cm^{-1}): 3065, 2951, 2362, 1710, 1597, 1540, 1502, 1417, 1322, 1255, 1166, 1069, 995, 808, 766, 701, 684, 662; Anal calcd for $C_{19}H_{15}F_3N_2O_2$: C, 63.33; H, 4.20; N, 7.77; Found: C, 63.31; H, 4.35; N, 7.82; mp = 103.1–104.9 °C.

Handling and Disposal of Hazardous Chemicals

The procedures in this article are intended for use only by persons with prior training in experimental organic chemistry. All hazardous materials should be handled using the standard procedures for work with chemicals described in references such as "Prudent Practices in the Laboratory" (The National Academies Press, Washington, D.C., 2011 www.nap.edu). All chemical waste should be disposed of in accordance with local regulations. For general guidelines for the management of chemical waste, see Chapter 8 of Prudent Practices.

These procedures must be conducted at one's own risk. *Organic Syntheses, Inc.*, its Editors, and its Board of Directors do not warrant or guarantee the safety of individuals using these procedures and hereby disclaim any liability for any injuries or damages claimed to have resulted from or related in any way to the procedures herein.

Discussion

Pyrazoles belong to an important class of heterocycles. Although rarely found in natural products, they are motifs of many biologically important compounds such as Celecoxib, Sildenafil and Fipronil.[3] The classical methods for the syntheses of the pyrazole moiety involve either the condensation of hydrazine derivatives with 1,3-dicarbonyl compounds, the 1,3-dipolar [3 + 2] cycloadditions or C–N and/or C–C cross coupling of the preformed pyrazoles.[4] However, these methods face synthetic limitations owing to either the use of carcinogenic hydrazines, generation of regioisomeric mixtures of products and/or the limited substrate scope.

Recently, we have developed an efficient and highly modular synthesis of tetrasubstituted pyrazoles by copper-mediated coupling of enaminones and nitriles via an oxidative C-C/N-N bond formation cascade.[5] From the optimized reaction conditions, it was found that $Cu(OAc)_2$ is required for this transformation, the reaction is sensitive to moisture, however, the reaction can be run conveniently under an atmosphere of air (or argon) in good yields.

Initially, the substrate scope of the reaction was studied using nitrile as the solvent for the synthesis of pyrazoles, limiting this method to the use of liquid nitriles only. Later, using *N,N*-dimethylformamide as the solvent, the amount of the nitriles can be reduced. Under the optimized reaction conditions, the substrate scope has been found to be impressively broad. With (Z)-methyl 3-(phenylamino)but-2-enoate (**1a**) as the coupling partner, various nitriles - aromatic, heteroaromatic as well as aliphatic nitriles - can be coupled in good yields (**3aa-3ao**). However, for aliphatic nitriles the reaction is comparatively more efficient under neat reaction conditions

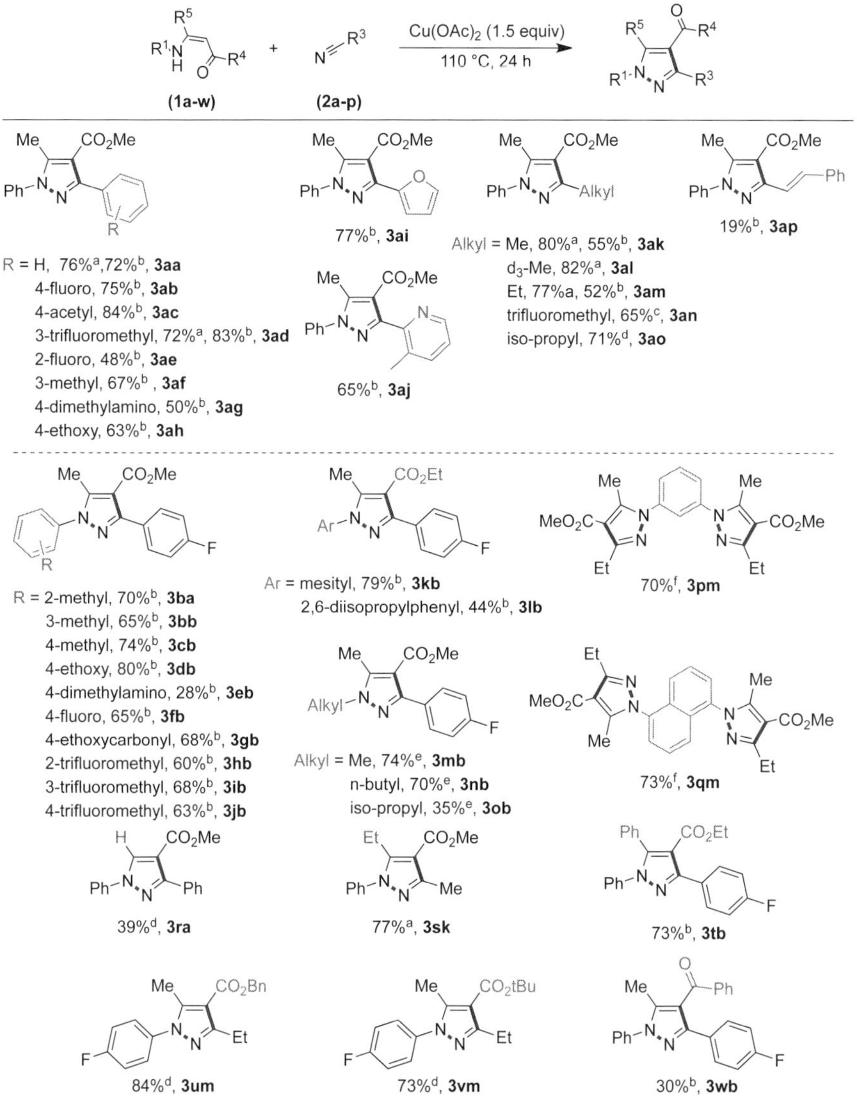

Scheme 1. Copper-mediated coupling of enaminones with nitriles. Reactions were run on a 1 mmol scale.[5] [a]**1** (1.0 equiv), Cu(OAc)$_2$ (1.5 equiv) and **2** (0.66 M), 110 °C, 24 h, air (closed tube). [b]**1** (1.0 equiv), Cu(OAc)$_2$ (1.5 equiv), **2** (3.0 equiv) and N,N-dimethylformamide (1 M), 110 °C, 24 h, air (closed tube). [c]Reaction conditions (B) after adding all the reagents and the solvent to a Schlenk tube (10 mL, 0.5 mmol scale), the tube was evacuated at 0 °C and then refilled with trifluoroacetonitrile gas. [d]Same as (A), **2** (0.33 M). [e]Same as (B), **2** (7.0 equiv) and N,N-dimethylacetamide (1 M). [f]**1** (1.0 mmol), Cu(OAc)$_2$ (6.0 equiv) and **2** (6.0 mL), 120 °C, 24 h, air (closed tube).

(**3ak-3ao**). To our delight, the gaseous nitrile trifluoroacetonitrile has also been coupled in good yield (**3an**). A styryl group can also be introduced at the 3-position of the pyrazole moiety, allowing further modifications at this position (**3ap**).

With 4-fluorobenzonitrile as the coupling partner the scope of the reaction with respect to the enaminone was studied. At 1-position of pyrazole moiety, various electron-rich as well as electron-poor groups bearing different functional groups were well tolerated (**3ba-3wb**). The sterically demanding groups such as mesityl and diisopropylphenyl could also be introduced in good yield (**3kb** and **3lb**). In addition to this, aliphatic substituents at R^1 position provided product in good yield (**3mb** and **3ob**). Bis-pyrazoles derivatives can also be obtained in good yield (**3pm** and **3qm**). The substituents at the 5-position of pyrazole products can also be varied from alkyl, aryl to hydrogen (**3pm** and **3qm**). R^4 can be varied providing pyrazole esters and ketone (**3aa-3wb**).

While studying the practicality and applicability of this efficient and modular transformation, the reaction was carried out on a 20 mmol scale. For the reaction conditions using *N,N*-dimethylformamide as the solvent, the reaction can be carried out at a 19.8 mmol scale, with **1a** and 3-(trifluoromethyl)benzonitrile (**2d**) as the coupling partner providing 79% yield of the product (as compared to 83% obtained on a 1 mmol scale), after elongating the reaction time to 26 h and distilling off excess of *N,N*-dimethylformamide and 3-(trifluoromethyl)benzonitrile by Kugelrohr distillation.

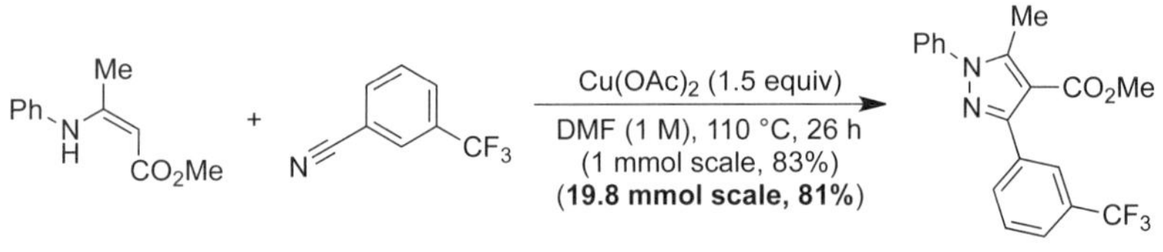

Scheme 2. Copper-mediated synthesis of tetrasubstituted pyrazole

In conclusion, we have developed an efficient and modular synthesis of pyrazoles by copper-mediated coupling of enaminones with nitriles with high regioselectivity and broad substrate scope. The convenient up-scaling of the reaction to 20 mmol scale shows the practicality of this transformation and its applicability in organic synthesis.

Org. Synth. **2014**, *91*, 211-220　　　**218**　　　DOI: 10.15227/orgsyn.091.0211

References

1. Organisch-Chemisches Institut der Westfälischen Wilhelms-Universität Münster, Corrensstrasse 40, 48149 Münster (Germany), E-mail: glorius@uni-muenster.de. The Int. NRW Graduate School of Chemistry (M.S.) and the European Research Council (ERC) under the European Community's Seventh Framework Program (FP7 2007-2013)/ERC grant agreement no. 25936 are gratefully acknowledged for financial support.
2. Bartoli, G.; Bosco, M.; Locatelli, M.; Marcantoni, E.; Melchiorre, P.; Sambri, L. *Synlett* **2004**, 239.
3. Yet, L.; *Comprehensive Heterocyclic Chemistry III, Vol. 4* (Eds.: Katritzky, A. R.; Ramsden, C. A.; Scriven, E. F. V.; Taylor R. J. K.), Elsevier, **2008**, 1.
4. Recent reviews on synthesis of pyrazoles: (a) Janin, Y. L. *Chem. Rev.*, **2012**, *112*, 3924; (b) Fustero, S.; Roselló, M. S.; Barrio, P.; Simón-Fuentes, A. *Chem. Rev.* **2011**, *111*, 6984.
5. (a) Suri, M.; Jousseaume, T.; Neumann, J. J.; Glorius, F. *Green Chem.* **2012**, *14*, 2193; (b) Neumann, J. J.; Suri, M.; Glorius, F. *Angew. Chem. Int. Ed.* **2010**, *49*, 7790; *Angew. Chem.* **2010**, *122*, 7957.

Appendix
Chemical Abstracts Nomenclature (Registry Number)

Zinc perchlorate hexahydrate; (10025-64-6)
Magnesium sulfate; (7487-88-9)
Methyl acetoacetate; (105-45-3)
Aniline; (62-53-3)
(Z)-Methyl 3-(phenylamino)but-2-enoate; (4916-22-7)
Copper(II) acetate monohydrate; (6046-93-1)
Copper(II) acetate, anhydrous; (142-71-2)
Methyl 5-methyl-1-phenyl-3-(3-(trifluoromethyl)phenyl)-1*H*-pyrazole-4-carboxylate; (1259438-02-0)
N,N-Dimethylformamide; (68-12-2)
3-(Trifluoromethyl)benzonitrile; (368-77-4)

Frank Glorius was educated in chemistry at the Universität Hannover, Stanford University (Prof. Paul A. Wender), Max-Planck-Institut für Kohlenforschung and Universität Basel (Prof. Andreas Pfaltz) and Harvard University (Prof. David A. Evans). In 2001 he began his independent research career at the Max-Planck-Institut für Kohlenforschung in Germany (Mentor: Prof. Alois Fürstner). In 2004 he became Assoc. Prof. at the Philipps-Universität Marburg and since 2007 he is a Full Prof. for Organic Chemistry at the Westfälische Wilhelms-Universität Münster, Germany. His research program focuses on the development of new concepts for catalysis and their implementation in organic synthesis.

Mamta Suri obtained her bachelor's degree in chemistry in 2007 from St. Stephen's College at University of Delhi (India). She moved to Indian Institute of Technology Guwahati (India) and completed her master's degree (2009) in the group of Prof. T. Punniyamurthy there. She then joined the group of Prof. F. Glorius at Westfälische Wilhelms-Universität Münster (Germany) in the Int. NRW Graduate School of Chemistry, where she obtained her doctoral degree in April 2013. Currently, she is working at BASF SE at Ludwigshafen, Germany.

Lan-Ting Xu received her BS degree from West China School of Pharmacy, Sichuan University in 2008, and her Ph.D. degree from Fudan University in 2013, under the supervision of Dawei Ma. She is now a MSD China R&D Postdoc Research Fellow in Shanghai Institute of Organic Chemistry. Her research interests include copper-catalyzed coupling reactions, metal-catalyzed direct C-H functionalization and heterocycle synthesis.

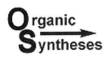

Synthesis of 1,3-Dimethyl-3-(p-tolyl)-1H-pyrrolo[3,2-c]pyridin-2(3H)-one by Cu(II)-Mediated Direct Oxidative Coupling

Chandan Dey and E. Peter Kündig[*1]

Department of Organic Chemistry, University of Geneva, 30 Quai Ernest Ansermet, 1211 Geneva 4, Switzerland

Checked by Yen-Ku Wu and Viresh H. Rawal

A.

HO₂C [structure] 1

1) SOCl₂ (2.0 equiv), reflux, 3h

2) 4-(methylamino)pyridine
Et₃N (2.0 equiv), DCM, rt, 15 h
88%

[structure] 2

B.

[structure] 2

CuCl₂ (2.2 equiv)

NaOtBu (5.0 equiv)
PhMe, reflux, 12 h
74%

[structure] 3

Procedure

Caution! Thionyl chloride reacts violently with water and releases dangerous gases.

A. *N-Methyl-N-(pyridin-4-yl)-2-(p-tolyl)propanamide (2).* A 50-mL two-necked, round-bottomed flask equipped with a 12 mm x 2.5 mm octagonal Teflon-coated magnetic stir bar, a reflux condenser fitted with a gas inlet with a positive pressure of N₂, and a glass stopper (Note 1) is charged with 2-(p-tolyl)propionic acid **1** (Note 2) (5.70 g, 34.7 mmol, 1.0 equiv) and thionyl chloride (Note 3) (5.1 mL, 69.4 mmol, 2.0 equiv). The reaction mixture is heated to reflux for 3 h (Note 4). After cooling to room temperature, excess thionyl chloride is removed under vacuum to give the crude acid chloride (Note 5). Separately, a dry, 250-mL two-necked, round-bottomed flask (Note 1) equipped with a 4.0 cm long, oval-shaped Teflon-coated magnetic stir bar, a nitrogen inlet having a positive pressure of N_2, and a rubber septum is charged with 4-(methylamino)pyridine (Note 6) (3.75 g, 34.7 mmol, 1.0 equiv), followed by dry dichloromethane (40 mL) (Note 7), which is added by syringe. The resulting solution is cooled to 0 °C (ice bath, external temperature) and treated with dry triethylamine (Note 8) (9.7 mL, 69.4 mmol, 2.0 equiv), which is added by syringe. The acid chloride formed earlier is diluted with 30 mL dichloromethane (Note 7) and the resulting solution is withdrawn by syringe and added dropwise (Note 9) to the flask containing 4-(methylamino)pyridine and triethylamine at 0 °C. The 50-mL round-bottomed flask is rinsed with 10 mL dichloromethane (Note 7), and the rinsate is added to the 250-mL flask. The ice bath is removed and the reaction mixture is stirred at 25 °C for 15 h and then transferred to a 250-mL separatory funnel. The 250-mL flask is rinsed with 20 mL dichloromethane and the solution is added to the separatory funnel. The organic layer is washed with water (50 mL) and the separated aqueous phase is back-extracted with dichloromethane (2 x 50 mL). The combined organic layers are dried over anhydrous $MgSO_4$ (10 g), filtered through a 150-mL medium porosity fritted glass funnel, and concentrated with a rotary evaporator (20 mmHg, 25 °C). The crude brown oil is dissolved in 10 mL EtOAc and is charged on a column (inner diameter = 6 cm, l = 15 cm) of 160 g of silica gel (Note 10) and subjected to flash chromatography (Note 11) using EtOAc as eluent. A first 170 mL fraction is collected in a conical flask, and subsequent fractions are collected in test tubes (Note 12). Product 2 is identified by TLC (EtOAc, $R_f = 0.24$, UV and $KMnO_4$ active) (Note 13) in test tubes 10-32. Fractions containing the product are concentrated by rotary evaporator (20 mmHg, 25 °C), and the resulting product is placed under high vacuum (1.0 mmHg, 25 °C, 2 h) to give 2 (7.77 g, 88% yield) as a white solid that is indefinitely stable at room temperature (Notes 14 and 15).

B. *1,3-Dimethyl-3-(p-tolyl)-1H-pyrrolo[3,2-c]pyridin-2(3H)-one (3)*. A 500-mL two-necked, round-bottomed flask equipped with an oval-shaped 4.0 cm Teflon-coated magnetic stir bar, a nitrogen inlet with a positive pressure of N_2, and a rubber septum (Note 1) is charged with amide 2 (6.36 g, 25.0 mmol, 1.0 equiv). The flask is placed under vacuum (1.0 mmHg) for 5 min and then back-filled with nitrogen. This sequence of vacuum and back-filling is repeated two times. Anhydrous $CuCl_2$ (Note 16)

(7.4 g, 55.0 mmol, 2.2 equiv) and NaOtBu (Note 17) (12.0 g, 125.0 mmol, 5.0 equiv) are sequentially added to the flask containing amide **2** (Note 18). Dry toluene (250 mL) (Note 19) is added by syringe under nitrogen atmosphere while stirring (Note 20), then the rubber septum is replaced with a dry reflux condenser. The nitrogen inlet in the second neck of the flask is transferred to the top of the reflux condenser and the flask opening is sealed with a glass stopper. The reaction mixture is heated to a gentle reflux for 12 h (Note 21), after which no starting material is evident (Note 22). After cooling to room temperature, the reaction mixture is diluted with 150 mL EtOAc and passed through a pad of Celite (Note 23). The flask is rinsed several times with EtOAc (6 x 150 mL) and the washings are also passed through the Celite pad (Note 24). The combined filtrate (ca. 1300 mL) is divided into two equal parts. The first half of the filtrate (ca. 650 mL) is transferred to a 1-L separatory funnel, to which brine (200 mL) and EtOAc (100 mL) are added. After extraction, the organic layer is collected and the aqueous layer is back-extracted with EtOAc (3 x 200 mL). The organic layers are combined. This sequence of extraction is repeated with the second half of the filtrate. The combined organic layers are concentrated with a rotary evaporator (20 mmHg, 25 °C) to afford a brown oil. The crude product is dissolved in 10 mL of EtOAc and charged on a column (inner diameter = 6 cm, l = 16.5 cm) of 170 g of silica gel (Note 10) and subjected to flash chromatography (Note 11) using EtOAc as the eluent. The first 180 mL of eluent is collected in a conical flask. Subsequent fractions are collected in test tubes (Note 12). Product **3** is found in test tubes 25-86 (Note 13). Fractions containing the product are concentrated by rotary evaporation (20 mmHg, 25 °C), and the resulting product placed under high vacuum (1.0 mmHg, 25 °C, 3 h) to afford **3** (4.67 g, 74% yield) as a pale yellow solid that is indefinitely stable at room temperature (Notes 25, 26 and 27).

Notes

1. All reaction vessels were dried under vacuum (1.0 mmHg) with the aid of a heat-gun for 3-4 min and maintained under nitrogen atmosphere during the course of the reaction.
2. 2-(*p*-Tolyl)propionic acid (97%) was purchased from Sigma-Aldrich and was used as received.

3. Thionyl chloride (>99%) was purchased from Sigma-Aldrich and was used as received.

4. The temperature of the oil bath was kept at 84 °C.

5. Initially, the vacuum (1.0 mmHg) was applied slowly to the stirred, crude mixture at 0 °C for 5–10 min to avoid bumping and then the flask was kept at room temperature for 1.5 h under vacuum (1.0 mmHg).

6. 4-(Methylamino)pyridine (99%) was purchased from Alfa Aesar and was used as received.

7. Methylene chloride (Optima grade) was purchased from Fisher Scientific and was dried using an alumina column-based system (Innovative Technologies PureSolv).

8. Triethylamine (Sigma-Aldrich, >99%) was distilled over KOH prior to use.

9. If the acid chloride solution was added too fast, strong fuming occurred. Normally, 0.5 mL/min was a reasonable rate of addition.

10. SiliCycle Silia*Flash*® P60 silica gel, 40-63 μm (230-400 mesh).

11. Flash chromatography was performed using 1.3-1.4 bar air pressure. Caution: a thick walled column and protective shield should be used.

12. Fractions were collected in 150 x 20 mm test tubes, allowing the collection of 28–30 mL in each test tube.

13. TLC was performed on TLC Silica gel 60 F_{254} plates.

14. Analytical data for **2**: ^{1}H NMR (500 MHz, CDCl$_3$) δ: 1.40 (d, J = 7.0 Hz, 3 H), 2.30 (s, 3 H), 3.25 (s, 3 H), 3.71 (q, J = 6.5 Hz, 1 H), 6.95 (d, J = 7.5 Hz, 2 H), 7.00 (br d, J = 5.4 Hz, 2 H), 7.05 (d, J = 8.0 Hz, 2 H), 8.57–8.58 (m, 2 H); ^{13}C NMR (125 MHz, CDCl$_3$) δ: 20.4, 20.8, 36.9, 43.1, 121.5, 126.9, 129.2, 136.3, 138.0, 150.9, 151.0, 173.3; IR (cast, cm^{-1}): 2975, 2930, 1668, 1587, 1513, 1496, 1375, 1274, 1126, 1063, 1024, 826; HRMS (ESI): calcd. for C$_{16}$H$_{19}$N$_2$O ([M+H]$^+$): 255.1492, found: 255.1488; mp 91–92 °C; Anal. calcd. for C$_{16}$H$_{18}$N$_2$O: C, 75.56; H, 7.13; N, 11.01, found: C, 75.47; H, 7.21; N, 11.03.

15. On half scale, the checkers obtained **2** in 87% yield. The submitters reported 83% yield for the full-scale reaction.

16. Anhydrous CuCl$_2$ (98%) was purchased from Strem Chemicals and used as received. It was stored in a desiccator between uses. The color of anhydrous CuCl$_2$ is dark brown. The submitters note that if the reagent becomes hydrated due to exposure to moisture and the color turns greenish, it can be dehydrated again by heating at 80 °C under vacuum (0.07 mmHg) for 12 h.

17. The checkers purchased sodium *tert*-butoxide (99.9%) from Sigma Aldrich and used as received; it was stored in a desiccator between uses. The submitter purchased sodium *tert*-butoxide (98%) from Acros Organics, sublimed it at 190 °C under vacuum (0.007 mmHg), and stored in a glove box (N$_2$ atmosphere) for use in this step.

18. The checkers carried out this step without a glove box. CuCl$_2$ and NaO*t*Bu were weighed under air atmosphere and quickly added to the round-bottomed flask under a positive flow of nitrogen. The submitters carried out this step as follows: The round-bottomed flask was transferred to a glove box (N$_2$ atmosphere). After adding CuCl$_2$ and NaO*t*Bu (sublimed), the round-bottomed flask was taken from the glove box and was connected to the Schlenk line.

19. Toluene (Optima grade) was purchased from Fisher Scientific and was dried by using an alumina column-based system (Innovative Technologies PureSolv).

20. The pale brown reaction mixture was stirred for 10 min at rt before heating was started.

21. After stirring at room temperature, the round-bottomed flask was placed in an oil bath and the temperature was set to 118 °C. It took 20–25 min for the oil to reach the preset temperature.

22. Consumption of the starting material can be checked by either TLC or ^{1}H NMR. In case of TLC, both the starting material and reaction mixture was spotted on a TLC plate and developed two times with EtOAc to get separation (TLC: EtOAc, Rf$_{SM}$ = 0.24, Rf$_{product}$ = 0.16, both UV and KMnO$_4$ active). For NMR analysis, a 0.2 mL aliquot of the reaction mixture was withdrawn under a positive flow of nitrogen and was passed through a small pad of Celite, followed by removal of solvent under vacuum (1.0 mmHg). ^{1}H NMR showed the disappearance of the resonance (quartet) at 3.71 ppm. This indicated complete consumption of the starting material. The color of the reaction mixture was deep brown. The submitters found the reaction is complete in 7 h.

23. Celite (inner diameter = 6.2 cm, l = 3 cm) was tightly packed on a glass frit (P3). The reaction mixture was added slowly to the Celite pad. Due to the presence of large amounts of inorganic salts, the upper layer of Celite needed gentle scratching for smooth filtration.

24. The Celite was washed until the filtrate became colorless and TLC showed no further product to be present.

25. Analytical data for **3**: ^{1}H NMR (500 MHz, CDCl$_3$) δ: 1.80 (s, 3 H), 2.29 (s, 3 H), 3.21 (s, 3 H), 6.88 (d, *J* = 5.5 Hz, 1 H), 7.12 (d, *J* = 8.0 Hz, 2 H),

7.21–7.19 (m, 2 H), 8.35 (s, 1 H), 8.52 (d, J = 5.0 Hz, 1 H); ^{13}C NMR (125 MHz, CDCl$_3$) δ: 20.7, 23.7, 26.2, 50.3, 103.8, 126.1, 129.2, 130.1, 136.4, 137.2, 144.3, 149.7, 150.2, 179.0; IR (cast, cm^{-1}): 3028, 2972, 2931, 1728, 1603, 1512, 1496, 1373, 1338, 1110, 1018, 818; HRMS (ESI): calcd. for C$_{16}$H$_{17}$N$_2$O ([M+H]$^+$): 253.1335, found: 253.1333; mp 103–104 °C; Anal. calcd. for C$_{16}$H$_{16}$N$_2$O: C, 76.16; H, 6.39; N, 11.10, found: C, 76.09; H, 6.39; N, 11.11.

26. On half scale, the checkers obtained **3** in 75% yield.

27. The submitters, who had stored the reagents in a glove box and set up the reaction there (Note 18), reported 87% yield for the full-scale reaction.

Handling and Disposal of Hazardous Chemicals

The procedures in this article are intended for use only by persons with prior training in experimental organic chemistry. All hazardous materials should be handled using the standard procedures for work with chemicals described in references such as "Prudent Practices in the Laboratory" (The National Academies Press, Washington, D.C., 2011 www.nap.edu). All chemical waste should be disposed of in accordance with local regulations. For general guidelines for the management of chemical waste, see Chapter 8 of Prudent Practices.

These procedures must be conducted at one's own risk. *Organic Syntheses, Inc.*, its Editors, and its Board of Directors do not warrant or guarantee the safety of individuals using these procedures and hereby disclaim any liability for any injuries or damages claimed to have resulted from or related in any way to the procedures herein.

Discussion

In 2009, this group reported a novel, efficient, and high atom economic protocol for the synthesis of 3,3-disubstituted oxindoles, which relied on the direct oxidative coupling of an aromatic Csp^2–H and a Csp^3–H centers.[2] The reactions were carried out with readily synthesized anilides in the presence of CuCl$_2$ as the oxidant and NaO*t*Bu as a base in DMF at 110 °C. Shortly following this report, Taylor and coworkers published a similar protocol leading to oxindoles with ester, nitrile and phosphonate functions in the 3-position.[3]

A likely mechanism of this reaction is shown in Scheme 1. The amidyl radical formed by one electron oxidation of the amide enolate cyclizes onto the aromatic ring. The resulting cyclohexadienyl radical is oxidized to form a σ-complex which readily aromatizes to the oxindole product.[2,3] Secondary isotope effect measurements indicates that Csp²–H bond breaking is not involved in the rate-determining step and a radical pathway is also indicated by the finding that a cyclopropyl-substituted substrate furnished a conjugated diene as the only product.[2,3a] An intramolecular radical cyclization reaction is the key step of this transformation. The base/oxidant procedure captivates by its simplicity.

Scheme 1. Proposed pathway for the direct C–H coupling reaction

The CuCl₂-mediated protocol was extended to the synthesis of 3,3-disubstituted aza-oxindoles.[4] Compounds containing the aza-oxindole structural motif exhibit biological activities such as oral anti-inflammatory activity, and potent TrkA kinase and JAK 3 kinase inhibition.[5] Unlike the synthesis of oxindole, there are fewer protocols available in the literature for this class of compounds. Examples include the oxidation of azaindoles,[6a,b] radical cyclization reactions,[6c] Pauson-Khand type [2+2+1] cycloadditions,[6d] Pd-catalyzed intramolecular α-pyridination of amides,[6e] photo cyclizations,[6f,g] and cyclization of aminopyridineacetic acids.[6h] All the aforementioned methods require a specifically functionalized precursor; for instance the presence of an *ortho*-pyridyl halogen, an α-xanthate group or an α-hydroxy group, or a preexisting bicyclic ring system. The CuCl$_2$-mediated oxidative radical coupling protocol[4] for aza-oxindole synthesis does not entail such structural specificity. The reactions with the pyridyl amides were carried out in the presence of CuCl$_2$ and NaO*t*Bu in toluene at 110 °C.[4] The key step of this transformation is a Minisci reaction, an intramolecular radical cyclization on the pyridine ring. Substrates with varying the position of nitrogen in the pyridine ring afforded aza-oxindoles in fair to excellent yields (Entry 1-11, Table 1). In general, yields with *para*-pyridyl substrates were considerably higher than with *ortho*-pyridyl substrates (Entry 1-3 *vs* 4-7, Table 1). The oxidative coupling reaction of *meta*-pyridyl amide substrate afforded two regioisomeric products (Entry 8-11, Table 1). The regioisomer resulting from the addition at *ortho* to the pyridyl N is largely favored (Entry 8, Table 1) or is the exclusive product (Entry 9-11, Table 1).[4] In the synthetic protocol for aza-oxindole detailed here, the concentration of the reaction was raised by a factor of two compared to the small scale preparation previously reported (0.1M *vs* 0.05M).[4]

In the course of these studies, a new route to aza-oxindoles, proceeding *via* a base promoted Truce-Smiles rearrangement, was found.[7]

Table 1. Selected examples of CuCl$_2$ mediated aza-oxindole synthesis[a]

Reaction scheme:

Substrate (X = N or CH, with R^4, O, N–R^2, R^3) →

2.2 equiv CuCl$_2$
5.0 equiv NaOtBu

PhMe, 110 °C
X = N or CH

→ Product

Entry	Substrate	Product	Yield(%)
1			68
2			57
3			64
4			94
5			79
6			99

Entry	Substrate	Product	Yield(%)
7			99
8		84:16	95
9			82
10			88
11			71

[a] 0.4-0.8 mmol scale in PhMe (0.05M). Yields given are those of isolated pure products.[4]

References

1. Department of Organic Chemistry, University of Geneva, 30 Quai Ernest Ansermet, 1211 Geneva 4, Switzerland. We thank the Swiss National Science Foundation and the University of Geneva for financial support.
2. Jia, Y. X.; Kündig, E. P. *Angew. Chem. Int. Ed.* **2009**, *48*, 1636.
3. (a) Perry, A.; Taylor, R. J. K. *Chem. Commun.* **2009**, 3249. (b) Pugh, D. S.; Klein, J. E. M. N.; Perry, A. Taylor, R. J. K. *Synlett* **2010**, 934. (c) Klein, J.

E. M. N.; Perry, A.; Pugh, D. S.; Taylor, R. J. K. *Org. Lett.* **2010**, *12*, 3446. (d) Pugh, D. S.; Taylor, R. J. K. *Org. Synth.* **2012**, *89*, 438.

4. Dey, C.; Kündig, E. P. *Chem. Commun.* **2012**, *48*, 3064.

5. (a) Ting, P. C.; Kaminski, J. J.; Sherlock, M. H.; Tom, W. C.; Lee, J. F.; Bryant, R. W.; Watnick, A. S.; Mcphail, A. T. *J. Med. Chem.* **1990**, *33*, 2697. (b) Kumar, V.; Dority, J. A.; Bacon, E. R.; Singh, B.; Lesher, G. Y. *J. Org. Chem.* **1992**, *57*, 6995. (c) Cheung, M.; Hunter, R. N.; Peel, M. R.; Lackey, K. E. *Heterocycles* **2001**, *55*, 1583. (d) Adams, C.; Aldous, D. J.; Amendola, S.; Bamborough, P.; Bright, C.; Crowe, S.; Eastwood, P.; Fenton, G.; Foster, M.; Harrison, T. K. P.; King, S.; Lai, J.; Lawrence, C.; Letallec, J. P.; McCarthy, C.; Moorcroft, N.; Page, K.; Rao, S.; Redford, J.; Sadiq, S.; Smith, K.; Souness, J. E.; Thurairatnam, S.; Vine, M.; Wyman, B. *Bioorg. Med. Chem. Lett.* **2003**, *13*, 3105. (e) Wood, E. R.; Kuyper, L.; Petrov, K. G.; Hunter, R. N.; Harris, P. A.; Lackey, K. *Bioorg. Med. Chem. Lett.* **2004**, *14*, 953.

6. (a) Marfat, A.; Carta, M. P. *Tetrahedron Lett.* **1987**, *28*, 4027. (b) Robinson, R. P.; Donahue, K. M. *J. Org. Chem.* **1991**, *56*, 4805. (c) Bacqué, E.; Qacemi, M. E.; Zard, S. Z. *Org. Lett.* **2004**, *6*, 3671. (d) Saito, T.; Furukawa, N.; Otani, T. *Org. Biomol. Chem.* **2010**, *8*, 1126. (e) Ackermann, L.; Vicente, R.; Hofmann, N. *Org. Lett.* **2009**, *11*, 4274. (f) Goehring, R. R.; Sachdeva, Y. P.; Pisipati, J. S.; Sleevi, M. C.; Wolfe, J. F. *J. Am. Chem. Soc.* **1985**, *107*, 435. (g) Wolfe, J. F.; Sleevi, M. C.; Goehring, R. R. *J. Am. Chem. Soc.* **1980**, *102*, 3646. (h) Okuda, S.; Robison, M. M. *J. Am. Chem. Soc.* **1959**, *81*, 740.

7. Dey, C.; Katayev, D.; Ylijoki, K. E. O.; Kündig, E. P. *Chem. Commun.* **2012**, *48*, 10957.

Appendix
Chemical Abstracts Nomenclature (Registry Number)

2-(*p*-Tolyl)propionic acid; (938-94-3)
Thionyl chloride, (7719-09-7)
4-(Methylamino)pyridine; (1121-58-0)
N-Methyl-*N*-(pyridin-4-yl)-2-(*p*-tolyl)propanamide, (1364651-81-9)
Copper (II) chloride; (7447-39-4)
Sodium *tert*-butoxide; (865-48-5)
1,3-Dimethyl-3-(*p*-tolyl)-1*H*-pyrrolo[3,2-*c*]pyridin-2(3*H*)-one; (1364652-24-3)

Peter Kündig graduated from the Federal Institute of Technology (ETH) in Zurich and then moved to the University of Toronto where he obtained his Ph.D. (G.A.S. Ozin). Following a postdoctoral stay at the University of Bristol (P. Timms), he started his own research at the University of Geneva, focusing on synthetic and mechanistic organometallic chemistry and on metal mediated and catalyzed reactions in organic synthesis. He held from 1990-2012 a Chair in Organic Chemistry. His awards include the Werner Prize of the Swiss Chemical Society (1986) and the EUCHEMS award (2007). Kündig was chair of the ChemComm Editorial Board (2007-11) and presently is President of the Swiss Chemical Society.

Chandan Dey was born in Midnapore, West Bengal, India in 1984. He obtained a B.Sc. (Hons.) degree in Chemistry from Midnapore college, India in 2006. In 2008, he received a M.Sc. degree from Indian Institute of Technology-Bombay, India. He joined the group of Prof. E. Peter Kündig at University of Geneva, Switzerland where he completed his doctoral study in 2012.

Yen-Ku Wu was born in Taipei, Taiwan. He received his B.Sc. in chemistry in 2004 and M.Sc. with Professor Hsing-Jang Liu in 2006 from National Tsing Hua University. In 2008, he started his Ph.D. studies working on non-conventional Nazarov chemistry under the supervision of Professor F. G. West at the University of Alberta. After obtaining his Ph.D. in 2013, he moved to the University of Chicago, where he is currently a postdoctoral scholar in the laboratories of Professor Viresh H. Rawal.

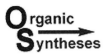

Preparation of (R)-4-Cyclohexyl-2,3-butadien-1-ol

Juntao Ye and Shengming Ma[*1]

State Key Laboratory of Organometallic Chemistry, Shanghai Institute of Organic Chemistry, Chinese Academy of Sciences, 345 Lingling Lu, Shanghai, China

Checked by Mingyao Wu and Dawei Ma

A.

$$\text{1 (OTBS)} + \text{cyclohexyl-CHO} \xrightarrow[\text{toluene, 130 °C}]{\substack{\text{pyrrolidine-CPh}_2\text{OH} \\ \text{ZnBr}_2 \text{ (0.75 equiv)}}} \text{(R)-2}$$

B.

$$\text{(R)-2 (CH}_2\text{OTBS)} \xrightarrow[\text{THF, 0 °C, rt}]{\text{TBAF·3H}_2\text{O}} \text{(R)-3 (CH}_2\text{OH)}$$

Procedure

A. *(R)-tert-Butyldimethyl(4-cyclohexyl-2,3-butadien-1-yloxy)silane* ((R)-2). An oven-dried, three-necked 500-mL round-bottomed flask containing a magnetic stir bar (Notes 1 and 2) is connected to a vacuum line via a one-stopcock adapter in the left neck. A solid addition funnel is placed in the middle neck and a glass stopper is placed in the right neck. Zinc bromide (ZnBr₂) powder (17.3 g, 76.6 mmol) (Notes 3 and 4) is added to the flask via the solid addition funnel under argon atmosphere. After addition, the solid addition funnel is removed and replaced with a glass stopper. The flask is dried under vacuum (1 mmHg) with a heat gun for approximately 1 min. After cooling to room temperature, the middle neck is equipped with a Dean-Stark trap (Note 5). The glass stopper is removed and (S)-α,α-diphenylprolinol (25.9 g, 100 mmol, 1.00 equiv) (Notes 6 and 7) and *tert*-butyldimethyl(2-propynyloxy)silane (1)[2] (18.9 g, 110 mmol, 1.1 equiv) (Note 8) are added sequentially with stirring under argon atmosphere at room temperature. The suspension is allowed to stir for 5 min at room temperature (Note 9) before 150 mL of freshly distilled toluene (Note 7) is added via a 50-mL syringe under an argon atmosphere. Freshly distilled cyclohexanecarboxaldehyde (16.8 g, 150 mmol, 1.5 equiv) (Note 7) is added over the course of one min via a syringe. The flask containing the aldehyde is rinsed with 50 mL of toluene, which in turn is added to the reaction flask via syringe under the argon atmosphere. The resulting mixture is stirred for 3 min at room temperature, during which time 8 mL of freshly distilled toluene is added to fill the Dean-Stark trap. The trap is connected to an argon source under slight positive pressure, which is released via a bubbler to the atmosphere. The flask is then placed in a pre-heated oil bath of 130 °C for 11 h with stirring (Notes 10 and 11). After cooling to room temperature, the supernatant is transferred into a 500-mL flask and the precipitate is discarded after rinsing with 30 mL of ethyl ether. After concentration by rotary evaporation (15 mmHg with a water bath of 30 °C) (Note 12), flash chromatography on silica gel (eluent: petroleum ether/ethyl ether = 40:1) afforded 20.8 g of the crude product (R)-2 as an orange-red liquid (Notes 13, 14, and 15).

B. *(R)-4-Cyclohexyl-2,3-butadien-1-ol ((R)-3)*. The crude product (R)-2 prepared above (20.80 g) is dissolved in 200 mL of THF (Note 7) in a 500-mL round-bottomed flask that contains a magnetic stir bar and is equipped with a solid addition funnel. The flask is placed in an ice-water bath and stirred while open to the atmosphere. To this solution is added TBAF·3H$_2$O (32.0 g, 101 mmol) (Note 7) in one portion via a solid addition funnel. After the addition is complete, the ice-water bath is removed and the resulting mixture is allowed to stir at room temperature. After 2 h, the reaction is complete as monitored by TLC (Note 10). The reaction mixture is poured into ice water (150 mL) in a 1-L beaker followed by addition of diethyl ether (300 mL). The mixture is transferred to a separatory funnel, the organic layer is separated, and the aqueous layer is extracted with ethyl ether (3 × 50 mL). The combined organic layer is washed with brine (100 mL) and dried over anhydrous Na$_2$SO$_4$ (15 g). After evaporation (15 mmHg with a water bath of 30 °C), the residue is purified by chromatography on silica gel (eluent: petroleum ether/ethyl acetate = 20:1 → 15:1) (Note 16) to afford 9.42 g of (R)-3 (62% combined yield over steps A and B) (Notes 17 and 18) as a light yellow liquid with 99% ee (Notes 19, 20, and 21).

Notes

1. All glassware was thoroughly washed and dried in an oven at 120 °C. Teflon-coated magnetic stirring bars were washed with acetone and dried.

2. Efficient stirring is crucial for the reaction; the submitters used an oval-shaped (40 × 15 mm) magnetic stirring bar.

3. Anhydrous $ZnBr_2$ was stored under Ar or N_2 in a desiccator. While the material was weighed out open to the atmosphere, it should be transferred quickly into the flask using a solid addition funnel under argon atmosphere.

4. The submitters found that powder form of $ZnBr_2$ (>98%) purchased from TCI facilitated more efficient stirring. Granular crystal form of $ZnBr_2$ (99.9%), purchased from Alfa Aesar, was also used after being ground into powder with a mortar and pestle avoiding exposure to moisture by applying Ar or N_2 atmosphere.

5. Dean-Stark trap (205 × 130 mm, capacity: 5-mL) was purchased from Beijing Synthware Glass Co., Ltd, China. A larger size of Dean-Stark trap was found to be less efficient for water-removal. The outer surface of the Dean-Stark trap should be fully wrapped with cotton to improve the water-removing performance.

6. Since the starting materials are solid, the stirring apparatus should be turned on from the very beginning to ensure efficient stirring while the flask is in the oil bath.

7. (S)- α,α-Diphenylprolinol (98%) was purchased from Shanghai Darui Fine Chemical Co., Ltd., China. Toluene (≥99.5%) was purchased from Shanghai Experimental Reagent Co., Ltd., China and distilled over sodium wire with benzophenone under an atmosphere of argon before use. Tetrahydrofuran (≥99%) was purchased from Shanghai Experimental Reagent Co., Ltd, China; Ethyl acetate (≥99.5%), petroleum ether (60-90 °C, ≥99%), were purchased from Suzhou JIMCEL H&N Electronic Material Co., Ltd, China. $ZnBr_2$ (>98%) was purchased from TCI, Ltd. or Alfa Aesar. Cyclohexanecarboxaldehyde (≥98%) and tetrabutylammonium fluoride trihydrate (≥98%) were purchased from Shanghai Darui Fine Chemical Co., Ltd., China. Cyclohexanecarboxaldehyde was distilled before use while all other chemicals were used as received.

8. Butyldimethyl(2-propynyloxy)silane **1** (97%) is available from Sigma-Aldrich or can be prepared from the reaction of propargyl alcohol and TBSCl according to Radha's procedure.[2] When prepared, the submitters purified **1** by distillation under reduced pressure (68–74 °C/20 mmHg). Anhydrous Na_2SO_4 (1 g of Na_2SO_4/ 10 g of **1**) was added to the distillate to remove trace amounts of water, which is critical for the success of the reaction. The compound could be stored at 5-8 °C for >12 h before use.

9. The submitters found that stirring the suspension of $ZnBr_2$, (*S*)- α,α-diphenylprolinol, and *tert*-butyldimethyl(2-propynyloxy)silane **1** for 5 min at room temp is crucial for success of a larger scale reaction.

10. TLC analysis during the reaction: Step A: R_f of compound **1** = 0.61 (eluent : petroleum ether); R_f of compound **2** = 0.72 (eluent : petroleum ether); Step B: R_f of compound **3** = 0.41 (eluent : petroleum ether/EtOAc (10/1)). All samples were visualized using an aqueous solution of $KMnO_4$.

11. The solution in the flask boils vigorously within the first 1-2 h and refluxes gently thereafter. Approximately 2 mL of water were collected in the Dean-Stark trap when the reaction was complete. The color of the solution changed from light yellow to orange yellow and finally to dark brown over the course of the reaction.

12. After removing most of the toluene under vacuum (15 mmHg with a water bath of 30 °C) and cooling to room temperature, $CHCl_3$ (20 mL) was added to the mixture to dissolve some insoluble substances, Methylene bromide (CH_2Br_2) (1.74 g, 10 mmol) was then added via a syringe and mixed thoroughly to serve as the internal standard. Analysis of the crude reaction mixture by ^{1}H NMR showed that the NMR yields of the reaction ranged from 64-69% in different runs.

13. The purpose of this manipulation is to remove the polar imine by-product,[3] which may inhibit the efficient deprotection of the TBS group in the subsequent step. Therefore, after this operation, all the fractions containing the products can be collected together; minor impurities in these fractions will not affect its use in the next step.

14. After evaporation with a rotary evaporator (15 mmHg with a water bath of 30 °C), the remaining toluene can be removed under vacuum (1 mmHg) at room temperature before chromatography purification.

15. The column (diameter = 80 mm) was packed with 300 g of silica gel (10-40 μm): a mixture of 2.4 L of petroleum ether (60–90 °C) and 60 mL of ethyl ether was used as the eluent (~100 mL for each). The obtained

crude product was used directly in the next step without further treatment. Chromatographic separation on silica gel with an eluent of petroleum ether was used to obtain the relatively pure product (R)-**2**, which exhibits the following characteristics: ^{1}H NMR (400 MHz, CDCl$_3$) δ: 0.08 (s, 6 H), 0.90 (s, 9 H), 1.05–1.29 (m, 5 H), 1.58–1.76 (m, 5 H), 1.97–2.00 (m, 1 H), 4.15–4.18 (m, 2 H), 5.15–5.25 (m, 2 H); ^{13}C NMR (100 MHz, CDCl$_3$) δ: –5.11, –5.09, 18.3, 26.0, 26.1, 33.1, 37.1, 62.2, 92.5, 98.3, 202.3; MS (EI) m/z (%): 209 (M$^+$-^{t}Bu, 80.74), 127 (59.05), 75 (100); IR (neat): v = 2927, 2854, 1962, 1448, 1362, 1255, 1092, 1005 cm^{-1}; HRMS calcd for C$_{10}$H$_{30}$OSi [M$^+$]: 266.2066, found: 266.2064.

16. The column (diameter = 80 mm) was packed with 390 g of silica gel (10–40 μm): a mixture of 4.0 L of petroleum ether (60–90 °C) and 0.2 L of ethyl acetate was first used as the eluent (~400 mL for each fraction); then a mixture of 3.8 L of petroleum ether (60–90 °C) and 0.25 L of ethyl acetate was used (~100 mL for each fraction). After evaporation of the solvents, 9.42 g of the product (R)-**3** was obtained as a light yellow liquid.

17. The purity of the product (R)-4-cyclohexyl-2,3-butadien-1-ol ((R)-**3**) is 99% as determined by GC; GC conditions: hp-5 (30 m × 0.32 mm × 0.25 μm); oven: 50 °C for 2 min, then 20 °C/min, 250 °C for 10 min; inject: 320 °C; FID: 320 °C; split: 80:1; N$_2$: 20 mL/min.

18. The yield was calculated based on (S)-α,α-diphenylprolinol since it is the limiting reagent (1 equiv).

19. A second full-scale run provided 9.82 g (65%) for the two steps.

20. Determination of the enantiomeric excess of the product (R)-4-cyclohexyl-2,3-butadien-1-ol ((R)-**3**) required the synthesis of the corresponding racemic 4-cyclohexyl-2,3-butadien-1-ol (±)-**3** (Scheme 1).[4a]

Scheme 1. Preparation of racemic 4-cyclohexyl-2,3-butadien-1-ol (±)-3

To a flame-dried Schlenk tube that contained a stir bar were added CuBr (7.4 mg, 0.05 mmol) and activated 4 Å molecular sieves

(301.7 mg). Toluene (2 mL), *tert*-butyldimethyl(2-propynyloxy)silane **1** (170.5 mg, 1.0 mmol), cyclohexanecarboxaldehyde (124.0 mg, 1.1 mmol) and pyrrolidine (78.3 mg, 1.1 mmol) were then added sequentially under an argon atmosphere. The solution was then stirred at 25 °C until completion of the reaction as monitored by TLC (12 h). The crude reaction mixture was filtrated through a short pad of silica gel eluted with ether (20 mL). After evaporation, the crude product was used in the next step without further treatment. To another Schlenk tube that contained a stir bar were added anhydrous ground ZnI$_2$ (147.0 mg, 0.45 mmol) and NaI (75.2 mg, 0.5 mmol). The Schlenk tube was heated with a heating gun under vacuum (1 mmHg) for about 1 min, and the flask was refilled with argon. The above crude product was then dissolved in toluene (5 mL) and transferred to the Schlenk tube via a syringe under argon atmosphere. The Schlenk tube was then equipped with a condenser and placed in a pre-heated oil bath at 110 °C with stirring. After 6 h, the reaction was judged as complete by TLC, after which the crude reaction mixture was filtrated through a short pad of silica gel (diameter: 30 mm, height: 20 mm) with ether (20 mL) as the eluent. After evaporation, the crude product was then dissolved in THF (3 mL) and treated at 0 °C with TBAF·3H$_2$O (316 mg, 1.0 mmol). The resulting mixture was allowed to warm to room temperature naturally with stirring. After 2 h, the reaction was complete, as determined by TLC, and H$_2$O (10 mL) and ether (10 mL) were then added. The organic layer was separated and the aqueous layer was extracted with ether (3 × 10 mL). The combined organic layer was dried over anhydrous Na$_2$SO$_4$. After filtration and evaporation, the residue was purified by flash chromatography (eluent: petroleum ether/ethyl acetate = 10/1) to afford 4-cyclohexyl-2,3-butadien-1-ol (±)-**3** (113.0 mg, 74%) as a liquid.

21. The enantiomeric excess was determined by chiral HPLC (conditions: Chiralcel AS-H column, eluent: hexane/*i*-PrOH = 98/2, flow rate: 0.7 mL/min). The peaks were visualized at 214 nm with retention times of 17.2 (major isomer) and 20.3 min (minor isomer). Enantioenriched (R)-4-cyclohexyl-2,3-butadien-1-ol ((R)-**3**) exhibits the following characteristics: $[\alpha]_D^{28}$ = − 97.1 (*c* = 1.01, CHCl$_3$); ^{1}H NMR (500 MHz, CDCl$_3$) δ: 1.00–1.32 (m, 5 H), 1.62–1.77 (m, 6 H), 1.98–2.05 (m, 1 H), 4.10 (s, 2 H), 5.27–5.38 (m, 2 H); ^{13}C NMR (125 MHz, CDCl$_3$) δ: 25.8, 25.9, 32.8, 32.9, 36.9, 60.7, 92.4, 99.4, 202.0; MS (EI) *m/z* (%): 152 (M$^+$, 0.70), 55 (100); IR (neat): *v* = 3331, 2924, 2851, 1961, 1448, 1302, 1258, 1214, 1062, 1012 cm^{-1}; HRMS calcd for C$_{10}$H$_{16}$O [M$^+$]: 152.1201, found: 152.1198.

Handling and Disposal of Hazardous Chemicals

The procedures in this article are intended for use only by persons with prior training in experimental organic chemistry. All hazardous materials should be handled using the standard procedures for work with chemicals described in references such as "Prudent Practices in the Laboratory" (The National Academies Press, Washington, D.C., 2011 www.nap.edu).   All chemical waste should be disposed of in accordance with local regulations. For general guidelines for the management of chemical waste, see Chapter 8 of Prudent Practices.

These procedures must be conducted at one's own risk.   *Organic Syntheses, Inc.*, its Editors, and its Board of Directors do not warrant or guarantee the safety of individuals using these procedures and hereby disclaim any liability for any injuries or damages claimed to have resulted from or related in any way to the procedures herein.

Discussion

Allenes have become an important class of compounds for the purposes of organic synthesis, medicinal chemistry, and materials science,[5] therefore, methods for the efficient synthesis of allenes from easily available starting materials are highly desirable.[6] In 1979, Crabbé *et al.* reported the first CuBr-mediated one-pot protocol for the synthesis of terminal allenes from terminal alkynes and formaldehyde in the presence of diisopropylamine.[3a] Based on this method, we have developed a modified procedure that proceeds in higher yields by utilizing CuI and dicyclohexylamine.[3b] However, it should be noted that the reaction is limited to paraformaldehyde: no allene is formed when other aldehydes are used. To address such a limitation, we have recently developed an efficient ZnI_2-mediated[3c] or CuI-catalyzed[3d] one-pot protocol for the synthesis of 1,3-disubstituted allenes from 1-alkynes, aldehydes, and morpholine (Scheme 2) or di(*n*-alkyl)amine (Scheme 3). Notably, functionalities such as halide, hydroxyl, and amine groups can all be tolerated in these transformations.

Org. Synth. **2014**, *91*, 233-247   **239**   DOI: 10.15227/orgsyn.091.0233

$R^1\!\!-\!\!\equiv$ + $R^2\!-\!CHO$ + [morpholine] $\xrightarrow[\text{toluene, 130 or 150 °C}]{ZnI_2 \text{ (0.8 equiv)}}$ allene product with R^2 and R^1

(1.0 mmol)　(1.8 mmol)

(1.4 mmol)

R^1 = n-C_8H_{17}, R^2 = Ph, 7.2 h, 65%
R^1 = n-C_8H_{17}, R^2 = p-ClC_6H_4, 6.7 h, 52%
R^1 = p-$O_2NC_6H_4CH_2OCH_2$, R^2 = i-Pr, 22 h, 56%
R^1 = p-$O_2NC_6H_4CH_2OCH_2$, R^2 = n-Bu, 6 h, 42%

R^1 = p-$O_2NC_6H_4CH_2O(CH_2)_2$, R^2 = Ph, 9.2 h, 51%
R^1 = n-$C_7H_{15}CH(OH)$, R^2 = Ph, 7.8 h, 62%
R^1 = n-$C_7H_{15}CH(NHTs)$, R^2 = Ph, 6.4 h, 59%
R^1 = n-$C_7H_{15}CH(NHTs)$, R^2 = i-Pr, 9.6 h, 49%

Scheme 2. ZnI$_2$-promoted one-pot synthesis of 1,3-disubstituted allenes

$R^1\!\!-\!\!\equiv$ + $R^2\!-\!CHO$ + $R^3\!\!\diagdown_{\!\!N}\!\!\diagup^{R^3}_{H}$ $\xrightarrow[\text{dioxane, 130 - 150 °C}]{CuI \text{ (10 or 20 mol%)}}$ allene product with R^2 and R^1

(1.0 mmol)　(1.6 mmol)

(1.4 mmol)

R^1 = PhCH(OH), R^2 = n-Pr, R^3 = n-Bu, 12 h, 73%
R^1 = n-$C_7H_{15}CH(OH)$, R^2 = n-Pr, R^3 = n-Bu, 36 h, 60%
R^1 = n-$C_7H_{15}CH(NHTs)$, R^2 = n-Pr, R^3 = n-Bu, 48 h, 54%
R^1 = BnOCH$_2$, R^2 = n-Pr, R^3 = n-Bu, 28 h, 55%
R^1 = PhCH(OH), R^2 = Et, R^3 = i-Bu, 18 h, 49%
R^1 = PhCH(OH), R^2 = Cy, R^3 = i-Bu, 18 h, 39%
R^1 = PhCH(OH), R^2 = n-C_5H_{11}, R^3 = i-Bu, 12 h, 55%
R^1 = HOCH$_2$, R^2 = n-C_7H_{15}, R^3 = i-Bu, 8 h, 41%

Scheme 3. CuI-catalyzed one-pot synthesis of 1,3-disubstituted allenes

After numerous unsuccessful attempts, we have recently developed the synthesis of optically active allenes via two different strategies involving a) the utilization of a chiral amines, and b) the utilization of chiral ligands (Scheme 4).[4]

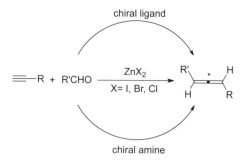

Scheme 4. Two approaches for the synthesis of optically active allenes

Among the allene family, axially chiral α-allenols are of particular interest as they are not only versatile building blocks for the synthesis of different types of heterocycles with a central chirality, but they are also the most basic starting materials for the synthesis of allenyl amines, malonates, thiols, aldehydes or ketones, and carboxylic acids with an axial chirality.[7] However, enantioselective synthesis of α-allenols is still challenging due to the fact that such an axial chirality spreads over three carbon atoms and there is a free hydroxyl group. Traditionally, a tedious, low-yielding route to axially chiral primary α-allenols from optically active propargylic alcohols using inconvenient and potentially hazardous chemicals such as *n*-BuLi and LiAlH₄ has been used.[8] Thus, a highly efficient approach to such primary α-allenols is in great demand.

Tertiary and secondary α-allenols may be prepared via the catalytic formation of optically active propargylic amines (Scheme 5 and 6).[4a] However, this approach is not suitable for the synthesis of such primary alcohols in high ees and yields.

R¹, R² = -(CH₂)₅-, R³ = Cy, 68%, 96% ee
R¹, R² = -(CH₂)₄-, R³ = Cy, 69%, 95% ee
R¹, R² = Me, Me, R³ = Cy, 71%, 95% ee
R¹, R² = Et, Et, R³ = Cy, 77%, 96% ee
R¹, R² = -(CH₂)₅-, R³ = *i*-Pr, 50%, 95% ee
R¹, R² = -(CH₂)₅-, R³ = *i*-Bu, 75%, 90% ee
R¹, R² = -(CH₂)₅-, R³ = *n*-C₇H₁₅, 54%, 92% ee
R¹, R² = -(CH₂)₅-, R³ = Ph, 78%, 93% ee

Scheme 5. Chiral ligand approach for the synthesis of tertiary α-allenols

Scheme 6. Chiral ligand approach for the synthesis of four diastereoisomers of axially chiral secondary α-allenols

This submission details the procedure based on the second approach shown in Scheme 4: a highly efficient $ZnBr_2$-mediated[4,9] multi-gram synthesis of highly enantioenriched axially chiral α-allenols from TBS-protected propargylic alcohols, aldehydes, and the inexpensive commercially available secondary amine (S)-α,α-diphenylprolinol followed by desilylation. By utilizing this strategy, axially chiral primary and secondary α-allenols may all be prepared efficiently in moderate to good yields with 96-99% ee or de (Scheme 7).

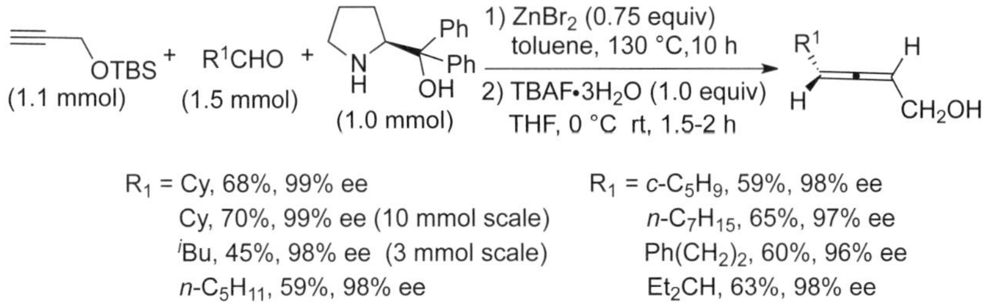
(1.1 mmol)
(1.5 mmol)
(1.0 mmol)
1) ZnBr2 (0.75 equiv) toluene, 130 °C,10 h
2) TBAF•3H2O (1.0 equiv) THF, 0 °C rt, 1.5-2 h
R1 = Cy, 68%, 99% ee
Cy, 70%, 99% ee (10 mmol scale)
iBu, 45%, 98% ee (3 mmol scale)
n-C5H11, 59%, 98% ee
R1 = c-C5H9, 59%, 98% ee
n-C7H15, 65%, 97% ee
Ph(CH2)2, 60%, 96% ee
Et2CH, 63%, 98% ee

Scheme 7. Chiral amine approach for the synthesis of axially chiral primary α-allenols

When (*R*)-α,α-diphenylprolinol is used as the chiral amine, the corresponding (*S*)-α-allenols are conveniently prepared with the same level of enantioselectivity. Therefore, all four diastereoisomers of axially chiral secondary α-allenols can be prepared with high stereoselectively by simply adjusting the absolute configurations of the central chiralities in the TBS-protected propargylic alcohols and α,α-diphenylprolinol (eq. 1-4, Scheme 8).

The primary enantioenriched α-allenols obtained through this method are very useful and may be transformed into a wide variety of not-readily-available, yet synthetically useful, optically active heterocyclic compounds with central chirality or into functionalized allenes with axial chirality without loss of enantiomeric purity (Scheme 9).

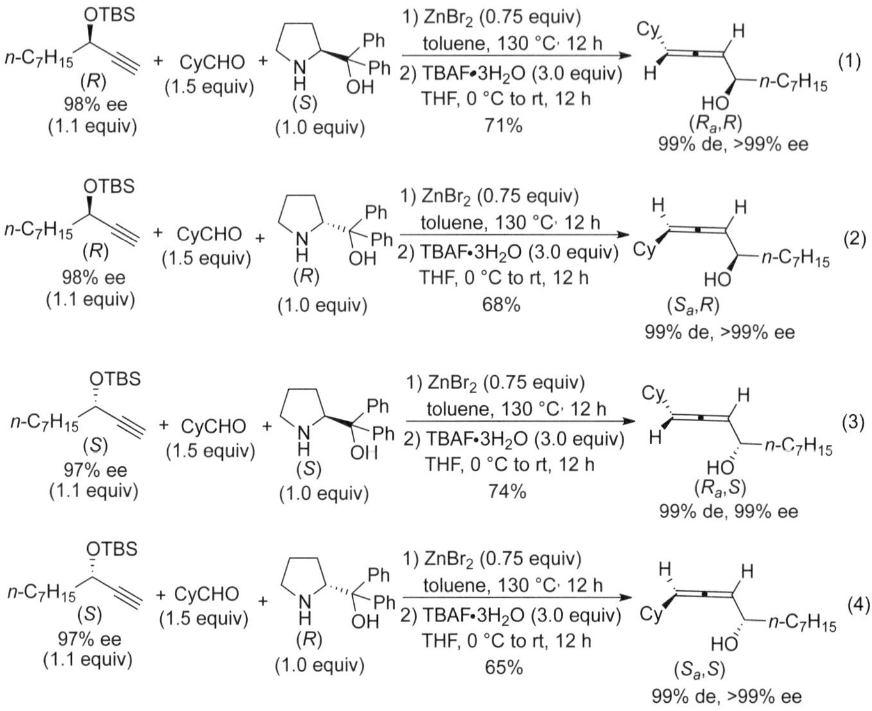

Scheme 8. Chiral amine approach for the synthesis of four diastereoisomers of axially chiral secondary α-allenols

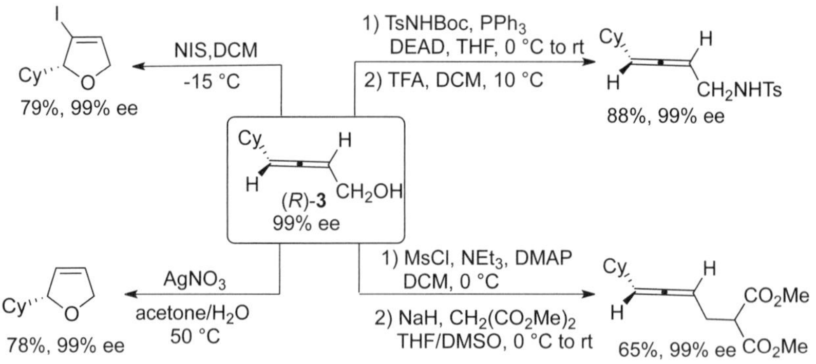

Scheme 9. Transformations of axially chiral α–allenol **(R)-3**

References

1. State Key Laboratory of Organometallic Chemistry, Shanghai Institute of Organic Chemistry, Chinese Academy of Sciences, 345 Lingling Lu, Shanghai, China. E-mail: masm@sioc.ac.cn.
2. Falck, J. R.; He, A.; Fukui, H.; Tsutsui, H.; Radha, A. *Angew. Chem. Int. Ed.* **2007**, *46*, 4527.
3. (a) Crabbé, P.; Fillion, H.; André, D.; Luche, J.-L. *J. Chem. Soc., Chem. Commun.* **1979**, 859. (b) Kuang, J.; Ma, S. *J. Org. Chem.* **2009**, *74*, 1763. (c) Kuang, J.; Ma, S. *J. Am. Chem. Soc.* **2010**, *132*, 1786. (d) Kuang, J.; Luo, H.; Ma, S. *Adv. Synth. Catal.* **2012**, *354*, 933.
4. (a) Ye, J.; Li, S.; Chen, B.; Fan, W.; Kuang, J.; Liu, J.; Liu, Y.; Miao, B.; Wan, B.; Wang, Y.; Xie, X.; Yu, Q.; Yuan, W.; Ma, S. *Org. Lett.* **2012**, *14*, 1346. Recently, Periasamy, *et al.* reported a "two-step" procedure of ZnBr$_2$-promoted synthesis of axially chiral 1,3-disubstituted allenes: (b) Periasamy, M.; Sanjeevakumar, N.; Dalai, M.; Gurubrahamam, R.; Reddy, P. O. *Org. Lett.* **2012**, *14*, 2932. In their paper, they reported that stirring the solution of ZnBr$_2$, (S)-α,α-diphenylprolinol, and terminal alkynes in toluene for 10 min at 120 °C followed by the addition of aldehydes at room temperature and then gradually heating the solution to 120 °C are very important for the high ees and yields for 1-aryl substituted allenes. However, we have not been able to reproduce their results in terms of ee and yield: the calibrated specific optical rotations of the same allenes with similar ee values from our study and the data in their report are different.
5. For most recent reviews on the chemistry of allenes, see: (a) Ma, S. *Chem. Rev.* **2005**, *105*, 2829. (b) Ma, S. *Aldrichimica Acta* **2007**, *40*, 91. (c) Brasholz, M.; Reissig, H.-U.; Zimmer, R. *Acc. Chem. Res.* **2009**, *42*, 45. (d) Ma, S. *Acc. Chem. Res.* **2009**, *42*, 1679. (e) Alcaide, B.; Almendros, P.; Campo, T. M. d. *Chem. Eur. J.* **2010**, *16*, 5836. (f)Aubert, C.; Fensterbank, L.; Garcia, P.; Malacria, M.; Simonneau, A. *Chem. Rev.* **2011**, *111*, 1954; (g) Inagaki, F.; Kitagaki, S.; Mukai, C. *Synlett* **2011**, 594. (h) López, F.; Mascareñas, J. L. *Chem. Eur. J.* **2011**, *17*, 418. (i) Yu, S.; Ma, S. *Angew. Chem., Int. Ed.* **2012**, *51*, 3074.
6. For reviews on the synthesis of allenes, see: (a) Sydnes, L. K. *Chem. Rev.* **2003**, *103*, 1133. (b) Krause, N. Hashmi, A. S. K., Eds.; *Modern Allene Chemistry;* Wiley-VCH: Weinheim, Germany, 2004; Vol. 1 and 2. (c) Krause, N.; Hoffmann-Röder, A. *Tetrahedron* **2004**, *60*, 11671. (d)

Brummond, K. M.; Deforrest, J. E. *Synthesis* **2007**, 795. (e) Ogasawara, M. *Tetrahedron: Asymmetry* **2009**, *20*, 259. (f) Yu, S.; Ma, S. *Chem. Commun.* **2011**, *47*, 5384.

7. For selected reports on the reactions of α-allenols, see: (a) Olsson, L.-I.; Claesson, A. *Synthesis* **1979**, 743. (b) Nikam, S. S.; Chu, K. H.; Wang, K. K. *J. Org. Chem.* **1986**, *51*, 745. (c) Marshall, J. A.; Wang, X. *J. Org. Chem.* **1990**, *55*, 2995. (d) Marshall, J. A.; Pinney, K. G. *J. Org. Chem.* **1993**, *58*, 7180. (e) Marshall, J. A.; Sehon, C. A. *J. Org. Chem.* **1995**, *60*, 5966. (f) Ma, S.; Gao, W. *Tetrahedron Lett.* **2000**, *41*, 8933. (g) Hoffmann-Röder, A.; Krause, N. *Org. Lett.* **2001**, *3*, 2537. (h) Friesen, R. W.; Blouin, M. *J. Org. Chem.* **1993**, *58*, 1653. (i) Ma, S.; Zhao, S. *J. Am. Chem. Soc.* **1999**, *121*, 7943. (j) Fu, C.; Li, J.; Ma, S. *Chem. Commun.* **2005**, 4119. (k) Li, J.; Fu, C.; Chen, G.; Chai, G.; Ma, S. *Adv. Synth. Catal.* **2008**, *350*, 1376. (l) Deng, Y.; Jin, X.; Ma, S. *J. Org. Chem.* **2007**, *72*, 5901. (m) Deng, Y.; Yu, Y.; Ma, S. *J. Org. Chem.* **2008**, *73*, 585.

8. For early leading references, see: (a) Olsson, L.-I.; Claesson, A. *Acta Chem. Scand.* **1977**, *B31*, 614. (b) Claesson, A.; Olsson, L.-I. *J. Am. Chem.Soc.* **1979**, *101*, 7302. (c) Smith, R. A.; White, R. L.; Krantz, A. *J. Med. Chem.* **1988**, *31*, 1558. For an alternative eight-step approach, see: (d) Stichler-Bonaparte, J.; Kruth, H.; Lunkwitz, R.; Tschierske, C. *Liebigs Ann.* **1996**, 1375.

9. Ye, J.; Fan, W.; Ma, S. *Chem. Eur. J.* **2013**, *19*, 716.

Appendix
Chemical Abstracts Nomenclature (Registry Number)

2-Propyn-1-ol; (107-19-7)
Silane, chloro(1,1-dimethylethyl)dimethyl-; (18162-48-6)
1*H*-Imidazole; (288-32-4)
Silane, (1,1-dimethylethyl)dimethyl(2-propyn-1-yloxy)-; (76782-82-6)
Zinc bromide; (7699-45-8)
Cyclohexanecarboxaldehyde; (2043-61-0)
2-Pyrrolidinemethanol, α,α-diphenyl-, (2S)-; (112068-01-6)
Butanaminium, *N,N,N*-tributyl-, fluoride, hydrate (1:1:3) (87749-50-6)
Pyrrolidine; (123-75-1)
Copper(I) bromide; (7787-70-4)
Zinc iodide; (10139-47-6)
Sodium iodide; (7681-82-5)
(±)-2,3-Butadien-1-ol, 4-cyclohexyl-; (153489-62-4)

Prof. Shengming Ma was born in 1965 in Zhejiang, China. He graduated from Hangzhou University (1986) and received his Ph.D. degree from Shanghai Institute of Organic Chemistry (1990). He became an assistant professor of SIOC in 1991. After his postdoctoral appointments at ETH with Prof. Venanzi and Purdue University with Prof. Negishi from 1992–1997, he joined the faculty of SIOC in 1997. From February 2003 to September 2007, he was jointly appointed by SIOC and Zhejiang University. He works for East China Normal University and SIOC since October 2007.

Juntao Ye was born in Hubei, China. He received his B.S. degree in chemistry from Huazhong University of Science and Technology (HUST) in 2008. He obtained his Ph.D. degree in 2013 under the supervision of Prof. Shengming Ma. His doctoral research was focused on the synthesis and cyclization reactions of allenes. Currently, he is a postdoctoral fellow in Prof. Mark Lautens' group at the University of Toronto.

Mingyao Wu was born in Hubei, China. He received his B.S. degree in chemistry from Wuhan University in 2008. He earned his Ph.D. in Shanghai Institute of Organic Chemistry (SIOC) in 2013 under the supervision of Prof. Dawei Ma, working on total synthesis of complex natural products.

Discussion Addendum for:

Formation of γ-Keto Esters from β-Keto Esters: Methyl 5,5-dimethyl-4-oxohexanoate

Yashoda N. D. Bhogadhi and Charles K. Zercher[1*]

Department of Chemistry, University of New Hampshire, Durham, NH 03824

Original article: Ronsheim, M. D.; Hilgenkamp, R. K.; Zercher, C. K. Org. Synth. ***2002***, *79, 146–153.*

![Reaction scheme: A β-keto ester (tert-butyl group with two carbonyls and OMe) reacts with Et₂Zn (3 equiv), CH₂I₂ (3 equiv) in CH₂Cl₂, 0 °C, 45 min, 89–94% to give a γ-keto ester product.]

Et₂Zn (3 equiv), CH₂I₂ (3 equiv)

CH₂Cl₂, 0 °C, 45 min

89–94%

Methods for the preparation of donor-acceptor (push-pull) cyclopropanes for the purpose of incorporating a single carbon between two carbonyl groups have been developed by a number of research groups (**Scheme 1**). Bieräugel described the cyclopropanation of a β-keto ester-derived enamine, which upon hydrolysis provided homologated material.[2] Attempts by Saigo to mimic these results with silyl enol ethers were inefficient and provided mixtures of products when using zinc carbenoids, although reactions based on copper-carbenoids derived from diazoesters were more efficient.[3] A radical-based method for homologation of β-dicarbonyls was reported by Dowd, although the one-carbon insertion was limited to α-substituted β-dicarbonyl starting materials.[4] A complementary strategy for the formation of donor-acceptor cyclopropanes was reported by Reissig, in which methyl ketone-derived enol ethers were reacted with stabilized rhodium carbenoids.[5] The resulting cyclopropanes could be converted cleanly to γ-keto esters through hydrolysis and used as nucleophilic species in tandem reaction processes.

Org. Synth. **2014**, *91*, 248-259
DOI: 10.15227/orgsyn.091.0248

Published on the Web 8/6/2014
© 2014 Organic Syntheses, Inc.

Bieraugel[2]

Saigo[3]

Dowd[4]

Reissig[5]

Scheme 1. Chain extension reactions through donor-acceptor cyclopropanes

A one-pot zinc carbenoid-mediated homologation reaction was reported by Zercher and co-workers.[6] Treatment of unfunctionalized β-keto esters with the Furukawa-modified Simmons-Smith reagent,[7] generated from an equimolar ratio of diethylzinc and diiodomethane, provided rapid and efficient access to γ-keto esters **(Scheme 2)**. Labeling studies revealed that the carbenoid carbon was inserted regioselectively adjacent to the ketone functionality, an observation that suggested the intermediacy of

Scheme 2. Zinc-mediated chain extension of β-keto esters

a donor-accepter cyclopropane. Mechanistic understanding of the zinc carbenoid-mediated homologation reaction was also informed by computational investigations, NMR analyses of reactive intermediates, and studies involving reaction stoichiometry.[8] The proposed reaction mechanism is summarized in **Scheme 3**. After initial deprotonation by zinc carbenoid (or diethylzinc), the resulting enolate reacts with carbenoid to provide homoenolate **3**. Intramolecular cyclization into the more electrophilic carbonyl provides the donor-acceptor cyclopropane (**4**), which fragments with a low energy barrier to provide an organometallic intermediate. This species is structurally reminiscent of the traditional zinc-Reformatsky intermediate,[9] and the strong covalent character of the carbon zinc bond is likely responsible for the absence of its reactivity with the various alkylating electrophiles (carbenoid and ethyl iodine) present in solution.

Scheme 3. Proposed reaction mechanism for zinc-mediated chain extension

The proposed mechanism illustrates that two equivalents of diethylzinc are necessary to effect the transformation, although in practice three equivalents of diethylzinc are used to ensure that the reactions proceed to completion. If acidic protons are present in the reaction substrate, the use of additional diethylzinc might be necessary and does not hinder the homologation reaction. Neat diethylzinc usually offers superior results, although commercial solutions of diethylzinc can be used to avoid the handling of the pyrophoric neat reagent. The replacement of diethylzinc with an alternate base for the purpose of enolate formation, thereby reducing the amount of pyrophoric reagent required in the reaction, should only be undertaken with the understanding that the stability of the zinc-organometallic intermediate **5** may be compromised in the presence of an alternate counterion. Reactions performed in the presence of non-zinc counterions result in greater product diversity and lower isolated yields of the simple homologated product.[10]

The zinc carbenoid–mediated transformation described herein offers a number of advantages to many of the other donor-acceptor cyclopropane methods. For instance, easily accessible and often commercially available β-keto carboxylic acid derivatives serve as starting materials with no need for derivatization as enol ethers or enamines. Furthermore, no hydrolysis step is necessary to fragment the cyclopropane, which means that protic quenching of the reactive intermediate can be delayed until after homologation is complete and tandem reactions are performed. The zinc carbenoid-mediated homologation reaction operates on a wide variety of β-keto carboxylic acid derivatives,[11,12] and β-keto phosphonates[13] can also be transformed into the γ-keto homologues **(Scheme 4)**.

Scheme 4. Zinc carbenoid-mediated homologation of β-keto amides and β-keto phosphonates

α-Substitution present on β-keto carboxylate starting materials often results in poor yields of chain extended or ring expanded products due to further reaction of the intermediate enolate; in contrast, α-substitution on β-keto phosphonate starting materials is well tolerated. Treatment of α-carboxyester cycloketones with carbenoid provides modest yields of the ring-expanded products, with use of more electrophilic carbenoids and control of stoichiometry being key to maximizing the efficiency of the transformation **(Scheme 5).**[14]

Scheme 5. Zinc carbenoid-mediated reactions of α-substituted starting materials

Substituted carbenoids provides the means to incorporate functionality at the β-position of the γ-keto ester products. For example, the carbenoid prepared from 1,1-diiodoethane leads to regioselective incorporation of a methyl group adjacent to the ketone functionality **(Scheme 6)** while preserving enolate character developed adjacent to the ester.[15]

Scheme 6. Formation of β-substituted γ-keto ester using a substituted carbenoid

The utility of the zinc-carbenoid-mediated homologation reactions has been enhanced through its application to tandem reaction protocols. The zinc enolate, which is regioselectively incorporated adjacent to the ester functionality, can react with electrophiles to effect the tandem reactions. While most of the tandem sequences described below are applicable to the array of β-keto carboxylic acid derivatives, β-keto phosphonate substrates are often poor partners in the tandem reaction processes. While a variety of electrophiles can be used to effect tandem reactions, not all electrophiles react efficiently with the intermediate. For example, alkylation of the

Org. Synth. **2014**, *91*, 248-259

organometallic intermediate (enolate) is slow, which is consistent with its similarity to the traditional zinc-Reformatsky intermediate.[8] Therefore, incorporation of alkyl substituents at the α-carbon of the homologated material must rely upon indirect methods, some of which are highlighted below. Fortunately, many electrophiles react efficiently with the organometallic intermediate to provide access to a diverse array of α-substituted γ-keto esters.

Exposure of the organometallic intermediate **5** to aldehydes or ketones provide tandem aldol products.[16] Use of ester or amide starting materials provides selective access to *syn*-aldol products with diastereocontrol as high as 20:1 **(Scheme 7a)**. The syn selectivity is attributed to the influence of the γ-keto functionality, which is believed to favor formation of an intermediate Z-enolate.[17] Reaction through a closed transition state would then account for the *syn*-selectivity. Tandem homologation–aldol reactions on substrates that possess an Evans auxiliary result in formation of *anti*-aldol products with excellent enolate facial selectivity **(Scheme 7b)**. The anti-selectivity is consistent with the stereochemistry observed when excess Lewis acid is used in the reaction of oxazolidinone-derived Z-enolates.[18] Due to the use of three or more equivalents of diethylzinc in the homologation portion of the reaction, excess zinc(II) is likely responsible for the aldol reaction proceeding through an open transition state.

a)

a) Et$_2$Zn,CH$_2$I$_2$
b) PhCHO
97%

dr = 12:1

b)

a) Et$_2$Zn,CH$_2$I$_2$
b) PhCHO
68%

R = *i*-Pr

R = *i*-Pr

Scheme 7. Tandem chain extension-aldol reactions

The products of tandem homologation aldol reactions exist in hemiacetal forms, which are attractive precursors to substituted tetrahydrofuranyl systems. A tandem homologation-aldol reaction, which utilized a carbenoid-derived from 1,1-diiodoethane, provided rapid access to the trisubstituted ring system, which was eventually converted to a *cis, cis*-phaseolinic acid derivative via oxidative cleavage **(Scheme 8)**. The presence of the methyl group adjacent to the enolate was responsible for controlling both enolate facial and the *anti*-aldol selectivities.[19] The aldol products can also be oxidized to provide 1,4-diketones, which have been used as precursors for heterocycle formation.[20]

a) Et₂Zn (5 equiv)
b) MeCHI₂ (5 equiv)

c) *n*-C₅H₁₁CHO (2 equiv)

51%

CAN
H₂O, MeCN

77%

Scheme 8. Formation of 3,4,5-Trisubstituted-γ-lactone

The organometallic intermediate can also be trapped with imine functionality, which provides a rapid route to substituted β-proline derivatives.[21] While the yields are modest and the chiral sulfonamide-derived imines cannot be used due to carbenoid reactivity with the sulfur, diastereocontrol can be effected through selection of the nitrogen activating group. Deprotection and reduction of the cyclic imine completes the preparation of β-proline derivatives **(Scheme 9)**. Enantiomerically-enriched products can be accessed through the use of oxazolidinone auxiliaries.

a) Et₂Zn, CH₂I₂

b)

1. TFA, CH₂Cl₂

2. NaBH₃CN
HCl, MeOH

52%
dr = 4:1

65%

Scheme 9. Formation of β-proline derivatives

The stability of the intermediate organometallic species (**5**) generated from a β-keto ester is demonstrated through the absence of its reaction with excess carbenoid, even over an extended period of time; however, the addition of catalytic trimethylsilyl chloride to the organometallic intermediate induces reaction with carbenoid to effect formation of an intermediate ester homoenolate.[22] Protonation completes the installation of a methyl group, while formation of an iodomethyl side chain through treatment with iodine[23] provides access to a versatile functional group for further manipulation (**Scheme 10**). The use of the electrophilic carbenoid for carbon-carbon bond formation, therefore, provides a route to products unavailable from direct alkylation with alkyl halides.

Scheme 10. Tandem Chain Extension-iodomethylation reaction

Direct treatment of the organometallic intermediate with iodine provides access to the isolable α-iodo-species,[17] which can be efficiently transformed to an α,β-unsaturated-γ-keto ester functionality by treatment with base.[24] This transformation has been a key step in the synthesis of numerous natural products, including (+)-brefeldin A[25] (**Scheme 11**) and patulolides A and B.[26] In addition, carbon nucleophiles can be incorporated

(+)-Brefeldin A

Scheme 11. Chain Extension-Iodination-Elimination in an Approach to (+)-Brefeldin A

into these products via regioselective conjugate addition, which provides another indirect route to the incorporation of an alkyl group at the α-carbon.[27]

Exposure of the organometallic intermediate to anhydrides is effective for the production of α-acylated products. The products produced from tandem homologation-acylation reactions are β-keto esters, which are appropriate substrates for heterocycle formation through Paul-Knorr reactions.[20] In addition, the acylated products are appropriate β-keto ester substrates for which a second homologation reaction would effect formation of double chain extended products. One example of a tandem chain extension-acylation reaction involves treatment of methyl acetoacetate with carbenoid, followed by addition of methoxymaleic anhydride, which generates an acylated product that spontaneously cyclizes to generate a spiroketal. Members of the papyracillic acid family of natural products were successfully prepared through the use of this strategy (Scheme 12).[28]

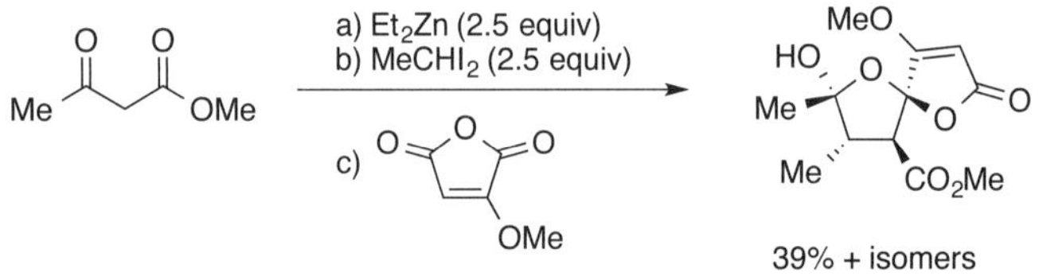

Scheme 12. One-step synthesis of spiroketal core of papyracillic acid

The reactions of β-diketones differ significantly from the reactions of β-keto esters, amides, or phosphonates. While the installation of the carbenoid carbon between the two ketone groups is believed to proceed through fragmentation of a donor-acceptor cyclopropane, the resulting intermediate is a reactive enolate as opposed to the stable organometallic formed from β-keto esters. The enolate reacts with additional equivalents of the carbenoid to form chain-extended cyclopropanol products (Scheme 13).[6, 29] The absence of similar

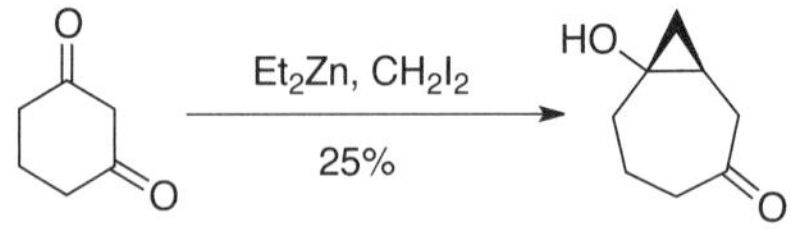

Scheme 13. Chain extension-cyclopropanation of β-diketone

Org. Synth. **2014**, *91*, 248-259

cyclopropanol products when starting with β-keto esters reflects the differences between zinc enolates of ketones and zinc enolates of carboxylic acid derivatives. Strongly electrophilic carboxylic acid derivatives, such as acylated oxazolidinones, also provide access to cyclopropanated products when exposed to carbenoid for extended reaction times.[12, 23]

References

1. Department of Chemistry, University of New Hampshire, Durham, NH 03824; chuck.zercher@unh.edu; The support of the National Institutes of Health (R15 GM060967-02) is acknowledged.
2. Bieräugel, H.; Akkerman, J. M.; Lapierre Armond, J. C.; Padit, U. K. *Tetrahedron Lett.* **1974**, 2817–2820.
3. (a) Saigo, K.; Kurihara, H.; Miura, H.; Hongu, A.; Kubota, N.; Nohira, H.; Hasegawa, M. *Synth. Commun.* **1984**, *14*, 787–796. (b) Saigo, K.; Yamashita, T.; Hongu, A.; Hasegawa, M. *Synth. Commun.* **1985**, *15*, 715–721.
4. (a) Dowd, P.; Choi, S. C. *J. Am. Chem. Soc.* **1987**, *109*, 3493–3494. (b) Dowd, P.; Choi, S. C. *J. Am. Chem. Soc.* **1987**, *109*, 6548–6549. (c) Dowd, P.; Choi, S. C. *Tetrahedron Lett.* **1989**, *30*, 6129–6132. (d) Dowd, P.; Choi, S. C. *Tetrahedron* **1989**, *45*, 77–90.
5. (a) Reichelt, I.; Reissig, H. U. *Chem. Ber.l* **1983**, *116*, 3895–3914. (b) Kunkel, E.; Reichelt, I.; Reissig, H. U. *Liebigs Ann. Chem.* **1984**, 512–530. (c) Reichelt, I.; Reissig, H. U. *Liebigs Ann. Chem.* **1984**, 531–551. (d) Grimm, E. L.; Reissig, H. U. *J. Org. Chem.* **1985**, *50*, 242–244.
6. Brogan, J. B.; Zercher, C. K. *J. Org. Chem.* **1997**, *62*, 6444–6446.
7. (a) Furukawa, J.; Kawabata, N.; Nishimur, J. *Tetrahedron Lett.* **1966**, 3353. (b) Simmons, H. E.; Smith, R. D. *J. Am. Chem. Soc.* **1959**, *81*, 4256–4264.
8. Eger, W. A.; Zercher, C. K.; Williams, C. M. *J. Org. Chem.* **2010**, *75*, 7322–7331.
9. (a) Reformatsky, S. *Chem. Ber.* **1887**, *20*, 1210–1211. (b) Orsini, F.; Pelizzoni, F.; Ricca, G. *Tetrahedron Lett.* **1982**, *23*, 3945–3948.
10. Hilgenkamp, R., M. S. Thesis, University of New Hampshire, **2000**.
11. Hilgenkamp, R.; Zercher, C. K. *Tetrahedron* **2001**, *57*, 8793–8800.
12. Lin, W.; Theberge, C. R.; Henderson, T. J.; Zercher, C. K.; Jasinski, J. P.; Butcher, R. J. *J. Org. Chem.* **2009**, *74*, 645–651.

13. Verbicky, C. A.; Zercher, C. K. *J. Org. Chem.* **2000,** *65,* 5615–5622.

14. Schwartz, B. D.; Tilly, D. P.; Heim, R.; Wiedemann, S.; Williams, C. M.; Bernhardt, P. V. *Eur. J. Org. Chem.* **2006,** 3181–3192.

15. Lin, W. M.; McGinness, R. J.; Wilson, E. C.; Zercher, C. K. *Synthesis* **2007,** 2404–2408.

16. Lai, S.; Zercher, C. K.; Jasinski, J. P.; Reid, S. N.; Staples, R. J. *Org. Lett.* **2001,** *3,* 4169–4171.

17. Aiken, K. S.; Eger, W. A.; Williams, C. M.; Spencer, C. M.; Zercher, C. K. *J. Org. Chem.* **2012,** *77,* 5942–5955.

18. Heathcock, C. H., *Asymmetric Synthesis*, Morrison, J. D., Ed. Academic Press Inc.: London, 1983; Vol. 3, pp 111–212.

19. Jacobine, A. M.; Lin, W.; Walls, B.; Zercher, C. K. *J. Org. Chem.* **2008,** *73,* 7409–7412.

20. Minetto, G.; Raveglia, L. F.; Sega, A.; Taddei, M. *Eur. J. Org. Chem.* **2005,** 5277–5288.

21. Jacobine, A. M.; Puchlopek, A.; Zercher, C. K.; Briggs, J. B.; Jasinski, J. P.; Butcher, R. J. *Tetrahedron* **2012,** *68,* 7799–7805.

22. Hilgenkamp, R.; Zercher, C. K. *Org. Lett.* **2001,** *3,* 3037–3040.

23. Pu, Q.; Wilson, E.; Zercher, C. K. *Tetrahedron* **2008,** *64,* 8045–8051.

24. Ronsheim, M. D.; Zercher, C. K. *J. Org. Chem.* **2003,** *68,* 4535–4538.

25. Lin, W.; Zercher, C. K. *J. Org. Chem.* **2007,** *72,* 4390–4395.

26. Ronsheim, M. D.; Zercher, C. K. *J. Org. Chem.* **2003,** *68,* 1878–1885.

27. (a) Deziel, R.; Plante, R.; Caron, V.; Grenier, L.; Llinas-Brunet, M. *J. Org. Chem.* **1996,** *61,* 2901–2903. (b) Captain, L. F.; Xia, X.; Liotta, D. C. *Tetrahedron Lett.* **1996,** *37,* 4293–4296.

28. Mazzone, J. R.; Zercher, C. K. *J. Org. Chem.* **2012,** *77,* 9171–9178.

29. Xue, S.; Li, L.-Z.; Liu, Y.-K.; Guo, Q.-X. *J. Org. Chem.* **2006,** *71,* 215–218.

Yashoda Bhogadhi was born in Avanigadda, India. She received her Bachelor's degree in Pharmacy from Andhra University, India. She then moved to United States where she obtained her Master's degree in Chemistry in 2009 from University of South Dakota. She is currently pursuing her Ph.D. in Organic Chemistry at University of New Hampshire under the supervision of Prof. Charles K. Zercher. Her current graduate research focuses on the synthesis of natural products CJ-12,954 and CJ-13,014 and investigation of diastereoselectivity in the formation of novel hydroxycyclopropyl peptide isosteres.

Chuck Zercher received his B.A. in Chemistry in 1981 from Messiah College. He earned a M.S. in Medicinal Chemistry from SUNY Buffalo under the direction of Leo Fedor and a Ph.D. in Chemistry from Notre Dame under the direction of Prof. Marvin Miller in 1989. Post-doctoral research under the direction with Prof. Paul Wender at Stanford University preceded his move to the University of New Hampshire where he is Robert E. and Gloria G. Lyle Professor of Chemistry. His research interests are in the areas of methods development and total synthesis.

Ni-catalyzed Reductive Cleavage of Methyl 3-Methoxy-2-Naphthoate

Josep Cornella,[†] Cayetana Zarate,[†] and Ruben Martin*[†§]

[†] Institute of Chemical Research of Catalonia (ICIQ), Av Països Catalans 16, 43007, Tarragona, Spain

[§] Catalan Institution for Research and Advanced Studies (ICREA), Passeig Lluïs Companys, 23, 08010, Barcelona, Spain

Checked by Hande Gunduz, Richard P. Loach, and Mohammad Movassaghi

A. [naphthalene-CO₂H / OH] → [naphthalene-CO₂Me / OMe] **1**
  4.0 equiv K₂CO₃
  5.2 equiv MeI
  DMF, 40 °C

B. **1** [naphthalene-CO₂Me / OMe] → [naphthalene-CO₂Me / H] **2**
  5.0% Ni(COD)₂
  10.0% PCy₃
  1.0 equiv TMDSO
  PhCH₃, 110 °C

Procedure

A. *Methyl 3-methoxy-2-naphthoate (1)*. An oven-dried 500 mL round-bottomed flask containing 3-hydroxy-2-naphthoic acid (10.1 g, 53.4 mmol, 1 equiv) (Note 1) and K_2CO_3 (29.5 g, 213.6 mmol, 4.0 equiv) (Note 2) is equipped with an oval magnetic stir bar (34 mm x 12 mm), argon inlet and a rubber septum. The flask is evacuated and back-filled with argon (this sequence is repeated three times over a period of 5 min) and the argon atmosphere is maintained throughout the reaction using an argon manifold system. The flask is charged through the septum via syringe with anhydrous DMF (120 mL) (Note 3), and iodomethane (17.3 mL, 277.7 mmol, 5.2 equiv) (Note 4) is added dropwise over a period of 2 min. The flask is then immersed in a pre-heated oil bath (Note 5) at 40 °C and the mixture is stirred for 14 h. After this time, the flask is allowed to reach room temperature. The septum is removed and saturated NH$_4$Cl solution (250 mL) is added (Note 6). The mixture is transferred to a 500 mL separatory funnel and extracted with EtOAc (5 x 50 mL) (Note 7). The combined organic layers are dried over anhydrous MgSO$_4$ (30 g) (Note 8), filtered, rinsed with additional 30 mL of EtOAc and concentrated by rotary evaporation (from 760 mmHg to 36 mmHg, 42 °C). The residue is purified by column chromatography on silica gel (Note 9), obtaining the title compound (**1**) as a colorless oil (11.1 g, 96%) (Notes 10 and 11).

B. *Methyl 2-naphthoate (2)*. An oven-dried 100 mL pressure tube equipped with a Teflon screw-cap and a magnetic stir bar (13 mm x 5 mm) is brought into a nitrogen-filled glovebox, charged with methyl 3-methoxy-2-naphthoate (**1**) (5.13 g, 23.7 mmol, 1 equiv), bis(cyclooctadiene)nickel(0) (Ni(COD)$_2$) (0.33 g, 1.18 mmol, 0.05 equiv) (Note 12), tricyclohexylphosphine (PCy$_3$) (0.66 g, 2.37 mmol, 0.10 equiv) (Note 13) and toluene (47.5 mL) (Note 14). After addition of 1,1,3,3-tetramethyldisiloxane (TMDSO) (4.20 mL, 23.8 mmol, 1.0 equiv) (Note 15), the pressure tube is closed with the Teflon screw-cap, brought back to the fume hood and immersed in a pre-heated oil bath at 110 °C. After 17 h, the resulting black solution is allowed to cool to room temperature, is filtered through a plug of Celite® (36 g) (Note 16) in a filter funnel (60 mm diameter) and is eluted with 130 mL of EtOAc (Note 17). The filtrate is concentrated by rotary evaporation (from 760 mmHg to 45 mmHg) and purified by column chromatography on silica gel (Note 18), obtaining the title compound as a white solid (3.9 g, 89% yield) (Notes 19 and 20).

Notes

1. 3-Hydroxy-2-naphthoic acid (98%) was purchased form Aldrich and used as received.
2. Potassium carbonate (>99%) was purchased from Sigma-Aldrich and used as received without any further purification.
3. DMF anhydrous (content in H$_2$O < 10 ppm) was dried from an Instrument Solvent Purification System (J. C. Meyer Solvent Systems).
4. Iodomethane (MeI, >99%) was purchased from Sigma-Aldrich and kept in the fridge at 4 °C. It was used as received.

5. Oil Bath: silicone oil ò=0.97, was purchased from VWR and used as received (working temperature from –40 °C to +200 °C).

6. NH₄Cl was purchased from VWR. The solution was prepared using 30 g of NH₄Cl in 900 mL of distilled water.

7. Ethyl acetate was purchased from Sigma-Aldrich and used as received.

8. Anhydrous magnesium sulfate was purchased from VWR and used as received.

9. Column chromatography was performed on 170 g of silica gel (230-400 mesh SiliaFlash®P60), purchased from Silicycle. It was wet packed in a 5 cm diameter column using hexanes/ethyl acetate (95/5) and the crude material was loaded into the column by means of a Pasteur pipette (the remaining residue was loaded in the minimal amount of hexanes/ethyl acetate (95/5)). Fractions of 40 mL were collected at 0.5 mL/s rate, eluting with hexanes/ethyl acetate (95/5). All fractions containing the desired product were combined, concentrated by rotary evaporation (from 760 mmHg to 38 mmHg) and dried overnight at 1.1 mmHg.

10. A second reaction on identical scale also provided 11.1 g (96%) of pure product.

11. Compound **1** has the following physical properties: R_f 0.27 (hexanes/ethyl acetate (90:10)); ^{1}H NMR (400 MHz, CDCl₃) δ: 3.97 (s, 3 H), 4.02 (s, 3 H), 7.22 (s, 1 H), 7.39 (ddd, J = 8.1, 6.9, 1.2 Hz, 1 H), 7.53 (ddd, J = 8.2, 6.9, 1.2 Hz, 1 H), 7.75 (dd, J = 8.3, 0.7 Hz, 1 H), 7.83 (dd, J = 8.3, 0.7 Hz, 1 H), 8.32 (s, 1 H); ^{13}C NMR (100 MHz, CDCl₃) δ: 51.87, 55.56, 106.46, 121.45, 124.07, 126.15, 127.19, 128.07, 128.33, 132.40, 135.78, 155.37, 166.34. IR (neat, cm⁻¹): 1732, 1632, 1281, 1210, 1074. HRMS Calcd for $C_{13}H_{12}O_3$: 217.0859, Found 217.0844. Anal Calcd for $C_{13}H_{12}O_3$: C, 72.21; H, 5.59; found: C, 71.95; H, 5.55. The submitters report that the product was isolated as a white solid with mp = 48–50 °C. Checkers determined the purity of the isolated compound **1** by GC-MS analysis (Column: Agilent Technologies, HP-5ms, 30 m x 0.250 mm, 0.25 μm. Max temp 350 °C; GC equipment: Agilent Technologies, 7890 A - Model number: G1033A; Detector: Agilent Technologies 5975C inert MSD with triple axis detector - Model number: G3171A). GC Temperature Program: Initial 50 °C 1 min, Ramp 40 °C/min up to 250 °C. Hold for 3 min. Total run time: 9 min. Front Inlet (SS Inlet) flow: 19.00 mL/min Column flow: 1.000 mL/min. Retention time: 6.834 min. Submitters also determined the purity of the isolated compound **1** by GC analysis (Column: Agilent Technologies,

HP-5ms, 30 m x 0.250 mm, 0.25 μm, 19091S-433. Max temp 325 °C; GC equipment: Agilent Technologies, 7890 A - Model number: G3440A; Detector: Agilent Technologies 5975C inert MSD with triple axis detector - Model number: G3171A). GC Temperature Program: Initial 50 °C for 1 min, Ramp 40 °C/min up to 250 °C. Hold for 3 min. Total run time: 9 min. Column flow: 1.00 mL/min. Retention time: 7.099 min.

12. Bis(cyclooctadiene)nickel(0) (Ni(COD)$_2$, >98%) was purchased from Strem Chemicals and stored at –35 °C in the glove-box. It was used without further purification.

13. Tricyclohexylphosphine (PCy$_3$, 97%) was purchased from Strem Chemicals and kept in the glovebox. It was used as received.

14. Toluene anhydrous (content in H$_2$O < 10 ppm) was dried from an Instrument Solvent Purification System (J. C. Meyer Solvent Systems).

15. 1,1,3,3-Tetramethyldisiloxane (97%) was purchased from Aldrich and used as received.

16. Celite®545 coarse was purchased from EMD Chemicals and used as received.

17. A filter plate (90 mm x 65 mm) was loaded with 36 g of Celite® and packed with 50 mL EtOAc. Extra 130 mL of EtOAc were used to elute the reaction mixture. The filtrate was collected in a 1 L round-bottomed flask.

18. Column chromatography was performed on 80 g of silica gel (230-400 mesh SiliaFlash®P60), purchased from Silicycle. It was wet packed in a 5 cm diameter column using hexanes/ethyl acetate (99/1) and the crude material was loaded into the column by means of a Pasteur pipette. (The remaining residue was loaded in the minimal amount of hexanes/ethyl acetate (99/1)). Fractions of 12 mL were collected eluting with 1 L hexanes/ethyl acetate (99/1), 0.5 L of hexanes/ethyl acetate (97/3) and finally 0.5 L hexanes/ethyl acetate (95/5). All fractions containing the desired product were combined, concentrated by rotary evaporation (from 760 mmHg to 45 mmHg) and dried for 30 min at 0.03 mmHg.

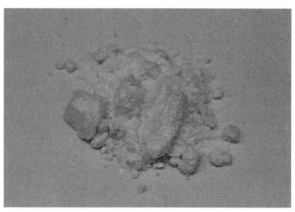

19. A second run on similar scale provided 4.5 g (82%) of pure product.

20. Compound **2** has the following physical properties: R_f 0.27 (hexanes/ethyl acetate (99:1)) mp = 73.7–74.5 °C; ^{1}H NMR (400 MHz, CDCl$_3$) δ: 3.99 (s, 3 H), 7.52–7.63 (m, 2 H), 7.87 (dt, J = 8.1, 0.7 Hz, 2 H), 7.95 (dt, J = 8.6, 0.7 Hz, 1 H), 8.08 (dd, J = 8.6, 1.7 Hz, 1 H), 8.63 (s, 1 H); ^{13}C NMR (100 MHz, CDCl$_3$) δ: 52.12, 125.15, 126.55, 127.33, 127.68, 128.07, 128.14, 129.27, 130.99, 132.42, 135.44, 167.15; IR (neat, cm^{-1}): 1711, 1233, 1201, 1154, 1130, 1100; HRMS [M + H] Calcd for C$_{12}$H$_{10}$O$_2$: 187.0759, Found 187.0750; Anal. calcd for C$_{12}$H$_{10}$O$_2$: C, 77.40; H, 5.41; found: C, 76.83; H, 5.36. Checkers determined the purity of the isolated compound **2** by GC-MS analysis (Column: Agilent Technologies, HP-5ms, 30 m x 0.250 mm, 0.25 μm. Max temp 350 °C; GC equipment: Agilent Technologies, 7890 A - Model number: G1033A; Detector: Agilent Technologies 5975C inert MSD with triple axis detector - Model number: G3171A). GC Temperature Program: Initial 50 °C 1 min, Ramp 40 °C/min up to 250 °C. Hold for 3 min. Total run time: 9 min. Front Inlet (SS Inlet) flow: 19.00 mL/min Column flow: 1.00 mL/min. Retention time: 6.073 min. Submitters also determined the purity of the isolated compound **2** by GC-MS analysis (Column: Agilent Technologies, HP-5ms, 30 m x 0.250 mm, 0.25 μm, 19091S-433. Max temp 325 °C; GC equipment: Agilent Technologies, 7890 A - Model number: G3440A; Detector: Agilent Technologies 5975C inert MSD with triple axis detector - Model number: G3171A). GC Temperature Program: Initial 50 °C for 1 min, Ramp 40 °C/min up to 250 °C. Hold for 3 min. Total run time: 9 min. Column flow: 1.000 mL/min. Retention time: 6.316 min.

Handling and Disposal of Hazardous Chemicals

The procedures in this article are intended for use only by persons with prior training in experimental organic chemistry. All hazardous materials should be handled using the standard procedures for work with chemicals described in references such as "Prudent Practices in the Laboratory" (The National Academies Press, Washington, D.C., 2011 www.nap.edu). All chemical waste should be disposed of in accordance with local regulations. For general guidelines for the management of chemical waste, see Chapter 8 of Prudent Practices.

These procedures must be conducted at one's own risk. *Organic Syntheses, Inc.*, its Editors, and its Board of Directors do not warrant or guarantee the safety of individuals using these procedures and hereby disclaim any liability for any injuries or damages claimed to have resulted from or related in any way to the procedures herein.

Discussion

Aryl methyl ethers are highly ubiquitous in nature. The importance of such motifs in organic synthesis is primarily associated to their unique role as directing groups, thus allowing a wide number of powerful synthetic transformations (Scheme 1, path a). Among these, particularly interesting are the *ortho*-metalation;[2] electrophilic aromatic substitution;[3] Friedel-Crafts-type reactions;[4] *ortho*[5], *meta*[6] or *para*[7] C–H bond-functionalization reactions; etc. Importantly, all these transformations occur with an exquisite chemo- and regioselectivity control. In sharp contrast, unbiased and unactivated arenes are known to promote similar types of reactions, but in most instances harsh reaction conditions are required, thus making these transformations not particularly attractive from a synthetic standpoint due to regio- and chemoselectivity issues (Scheme 1, path b).[8] In recent years, elegant approaches based upon C–H bond-functionalization reactions have overcome most of these drawbacks. Despite the advances realized, an *ortho*-directing group is typically required, and these procedures are not yet attractive enough as the cleavage of such groups still constitutes a tremendous challenge.[9]

Scheme 1. Aryl ethers as temporary protecting groups

Despite formidable advances in the cross-coupling arena, this field of expertise is largely dominated by the use of organic halides as coupling counterparts. To such end, chemists have been challenged to come up with more flexible strategies from easily available precursors that do not require the use of halide waste. Recently, metal-catalyzed C–O bond-cleavage has become a powerful alternative to commonly employed organic halides.[10] Among their advantages is that such C–O electrophiles can be derived from simple and cheap phenols, thus enhancing the flexibility in catalytic design. Among the different C–O electrophiles, the utilization of simple aryl methyl ethers constitutes the best alternative because aryl methyl ethers are the simplest derivatives from all phenol series. Unfortunately, however, the high activation energy of the C–OMe bond constitutes a serious drawback to be overcome. Recently, we developed a new strategy based upon a metal-catalyzed reductive cleavage of C–OMe bonds as a means to use aryl methyl ethers as temporary removable directing groups (Scheme 1, path c).[11,12] In this manner, such a technique represents a powerful alternative for arene functionalization that is difficult to accomplish using common synthetic organic techniques.

Scheme showing reaction conditions: Ni(COD)₂ (5 mol%), PCy₃ (10 mol%), TMDSO (1 equiv.), Toluene, 110 °C, converting aryl-OMe to aryl-H.

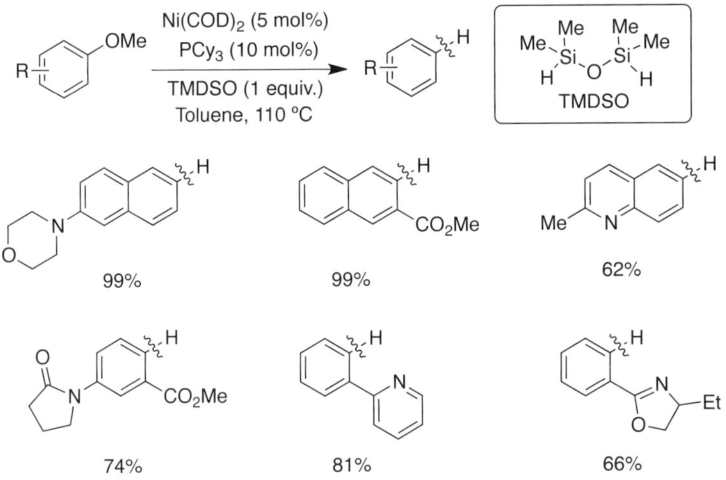

TMDSO structure shown.

Products:

99% 99% 62%

74% 81% 66%

Scheme 2. Scope for the Ni-catalyzed reductive cleavage of C-OMe bonds

The Ni-catalyzed reductive cleavage was distinguished by its wide scope, including challenging substrate combinations. The protocol turned out to be widely tolerant of functional groups as silyl groups, esters, amides, acetals, amines and heterocycles remained intact (Scheme 2). Not surprisingly, extended π-systems were much more reactive than anisole derivatives, a common requisite in many C–O bond cleavage reactions. Although tentative, we postulate that such observation is in line with the known ability of extended π-systems to bind the Ni center in a η²-fashion via the Dewar-Chatt-Duncanson model, thus retaining, unlike a regular arene, certain aromaticity that provides an extra stabilization. Interestingly, we found that the presence of an electron-withdrawing directing group in an *ortho*-position allowed for the cleavage of C–OMe bonds in anisole derivatives.

Scheme 3. Site-selectivity in Ni-catalyzed reductive cleavage of C–OMe bonds

Our protocol could also be amenable for site-selectivity, as C–OMe bonds could be reductively cleaved in the presence of other C–OMe bonds, likely due to subtle steric as well as electronic differences (Scheme 3). Importantly, selectivity was also observed in highly complex molecules such as estradiol or quinine-type derivatives, thus reinforcing the notion that these transformations can potentially be used as a manifold for natural product diversity (Scheme 3). While one might question the synthetic applicability of a reductive cleavage event, we demonstrated that our technique might be useful for building up molecular complexity using the aryl methyl ether as a temporary protecting group. Indeed, we found that 2-naphthol could be easily transformed into three different regioisomeric naphthalenes by using the known ability of C–O bonds to direct functionalization at different positions on the arene followed by our Ni-catalyzed reductive cleavage event. These results can hardly be underestimated; indeed, it would be rather difficult to imagine a synthetic technique capable of rapidly accessing different regioisomers from a common precursor.

Scheme 4. Synthetic applicability

Most recently, we have shed light on the mechanism of this rather appealing transformation, by concluding that the commonly accepted Ni(0)/Ni(II) mechanism via oxidative addition into the C-OMe is not operating.[13] Indeed, we gathered evidence, both experimentally and computationally, that putative catalytically active Ni(I) intermediates come into play. Such species are likely generated via the comproportionation event of an initially formed Ni(II) species.

In conclusion, we have found a functional group tolerant Ni-catalyzed reductive cleavage of aryl methyl ethers that occurs with a wide substrate scope. Practicality aside, we believe such a reaction might open up new perspectives in the field of C–O bond-cleavage, a less-explored but vibrant area of expertise, and we expect that this method will find immediate application in organic synthesis.

References

1. Institute of Chemical Research of Catalonia (ICIQ); Av. Països Catalans, 16, 43007, Tarragona, Spain. Catalan Institution for Research and Advanced Studies (ICREA), Passeig Lluís Companys, 23, 08010, Spain. E-mail: rmartinromo@iciq.es; We thank ICIQ Foundation, MICINN (CTQ2012-34054) and the European Research Council (ERC-277883) for financial support. J. Cornella thank European Union for a Marie Curie fellowship (FP7-PEOPLE-2012-IEF-328381).

2. (a) Clayden, J. *Organolithiums: Selectivity for Synthesis*; VCH: Weinheim, 2002; (b) Snieckus, V. *Chem. Rev.* **1990**, *90*, 879.
3. Taylor, R. *Electrophilic Aromatic Substitution*; Wiley: New York, 1995.
4. Klumpp, G. W. *Reactivity in Organic Chemistry*; Wiley: New York, 1982, pp 227–378.
5. For selected examples, see: (a) Hennings, D. D.; Iwasa, S.; Rawal, V. H. *J. Org. Chem.* **1997**, *62*, 2; (b) Satoh, T.; Kawamura, Y.; Miura, M.; Nomura, M. *Angew. Chem. Int. Ed.* **1997**, *36*, 1740; (c) Huang, C.; Gevorgyan, V. *J. Am. Chem. Soc.* **2009**, *131*, 10844; (d) Zhao, X.; Yeung, C. S.; Dong, V. M. *J. Am. Chem. Soc.* **2010**, *132*, 5837; (e) Ackermann, L.; Diers, E.; Manvar, A. *Org. Lett.* **2012**, *14*, 1154.
6. Dai, H.-X.; Li, G.; Zhang, X.-G.; Stepan, A. F.; Yu, J.-Q. *J. Am. Chem. Soc.* **2013**, *135*, 7567.
7. Cian, C.-L.; Phipps, R. J.; Brandt, J. R.; Meyer, F.-M.; Gaunt, M. J. *Angew. Chem. Int. Ed.* **2011**, *50*, 458.
8. Astruc, D. *Modern Arene Chemistry*; Wiley-VCH: Weinheim, 2002.
9. Hoveyda, A. H.; Evans, D. E.; Fu, G. C. *Chem. Rev.* **1993**, *93*, 1307.
10. For reviews, see: Tehetena, M.; Garg, N. K. *Org. Process Res. Dev.* **2013**, *17*, 29; (b) Yamaguchi, J.; Muto, K.; Itami, K. *Eur. J. Org. Chem.* **2013**, 19; (c) Correa, A.; Cornella, J.; Martin, R. *Angew. Chem. Int. Ed.* **2013**, *52*, 1878; (d) Rosen, B. M.; Quasdorf, K. W.; Wilkson, D. A.; Zhang, N.; Resmerita, A.-M.; Garg, N. K.; Percec, B. *Chem. Rev.* **2011**, *111*, 1346; (e) Yu, D.-G.; Li, B.-J.; Shi, Z.-J. *Acc. Chem. Res.* **2010**, *43*, 1486.
11. Alvarez-Bercedo, P.; Martin, R. *J. Am. Chem. Soc.* **2010**, *132*, 17353.
12. For other related approaches, see: (a) Sergeev, A. G.; Hartwig, J. F. *Science* **2011**, *332*, 439; (b) Tobisu, M.; Yamakawa, K.; Shimasaki, T.; Chatani, N. *Chem. Commun.* **2011**, 2946.
13. Cornella, J.; Gomez-Bengoa, E.; Martin, R. *J. Am. Chem. Soc.* **2013**, *135*, 1997.

Appendix
Chemical Abstracts Nomenclature (Registry Number)

Methyl 3-methoxy-2-naphthoate (13041-60-6)
3-Hydroxy-2-naphthoic acid (92-70-6)
K_2CO_3: Potassium carbonate (584-08-7)
MeI: Iodomethane; methyl iodide (74-88-4)

Methyl 2-naphthoate (2459-25-8)
Ni(COD)₂: Bis(cyclooctadiene)nickel(0) (1295-35-8)
PCy₃: tricyclohexylphosphine (2622-14-2)
TMDSO: 1,1,3,3-Tetramethyldisiloxane (3277-26-7)

Ruben Martin received his PhD in 2003 from the University of Barcelona with Prof. Antoni Riera. In 2004, he moved to the Max-Planck Institut für Kohlenforschung as a Humboldt postdoctoral fellow with Prof. Alois Fürstner. In 2005 he then undertook further postdoctoral studies at MIT with Prof. Stephen L. Buchwald as a MEC-Fulbright fellow. In September 2008 he initiated his independent career as an assistant professor at the Institute of Chemical Research of Catalonia (ICIQ) and in July 2013 he was promoted to associate professor. His research is focused on the development of new metal-catalyzed activation of inert bonds.

Dr. Josep Cornellà was born in Barcelona and graduated with distinction from the University of Barcelona (Spain) with a M.Sc in Organic Chemistry in 2008. Subsequently, he joined the group of Prof. Igor Larrosa at Queen Mary University of London (United Kingdom). In early 2012, he earned his PhD working on the development of metal-catalyzed decarboxylative transformations. In 2012, he was granted a Marie Curie Fellowship at Prof. Ruben Martin's group at ICIQ and he is currently studying the discovery of novel transformations involving metal-catalyzed C–O bond activation and CO₂ fixation.

Cayetana Zárate received her B.Sc. from the University of Valladolid in 2012 with Extraordinary Award. During her B.Sc. she was an undergraduate fellow for two years carrying out research in gold-catalyzed reactions at Prof. Espinet´s group. In October 2012 she began graduate studies under the supervision of Prof. Ruben Martin at ICIQ, where she earned her M.Sc from the University Rovira i Virgili. She is currently pursuing her PhD studies at Prof. Ruben Martin´s group in the area of metal-catalyzed C-O bond-activation.

Richard Loach was born in Birmingham (U.K.) and graduated from Imperial College, London in 2003 with a B.Sc in Chemistry. In 2007 he joined the research group of Professor John Boukouvalas at Laval University in Québec (Canada). In 2013, he earned his PhD for his studies on the total syntheses of novel γ-hydroxybutenolide natural products. In 2014, he was granted a FRQNT fellowship to pursue his postdoctoral research in Professor Mohammad Movassaghi's group at MIT. He is currently working on the total synthesis of alkaloid natural products.

Hande Gunduz was born in Istanbul (Turkey) and received her B.Sc. from Istanbul Technical University in 2009. In the same year she began her Ph.D. studies in the research group of Professor Naciye Talinli, focusing on the use of dioxinone chemistry for the synthesis of anticancer compounds. In 2013, she was granted a Fullbright Fellowship and joined Professor Mohammad Movassaghi's research group at MIT as a visiting graduate student. She worked on the total synthesis of alkaloid natural products.

 **O**rganic **S**yntheses

Indium-Catalyzed Heteroaryl–Heteroaryl Bond Formation through Nucleophilic Aromatic Substitution: Preparation of 2-Methyl-3-(thien-2-yl)-1H-indole

Yuta Nagase and Teruhisa Tsuchimoto[*1]

Department of Applied Chemistry, School of Science and Technology, Meiji University, Higashimita, Tama-ku, Kawasaki 214-8571, Japan

Checked by Joyce Leung and John Wood

Procedure

2-Methyl-3-(thien-2-yl)-1H-indole. A 100-mL Schlenk tube equipped with a Teflon-coated, egg-shaped magnetic stir bar (9.5 x 19 mm) (Note 1) is charged with indium(III) trifluoromethanesulfonate (360 mg, 0.64 mmol, 2 mol %) (Note 2). The tube fitted with a glass stopper is slowly heated to 150 °C over a period of 1 h by an oil bath with stirring under vacuum (0.50 mmHg) (Note 3). After heating at 150 °C for an additional 1 h, the tube is cooled down to room temperature and filled with nitrogen (Note 4). The glass stopper is replaced with a rubber septum under a flow of nitrogen, and the tube is evacuated and refilled with nitrogen three times. Dry 1,4-dioxane (16 mL) (Note 5) and dry toluene (1.3 mL) (Note 6) are added to the tube by syringes, and the resulting mixture is stirred at room temperature for approximately 10 min until the indium salt is dissolved. 2-Methyl-1H-indole (4.20 g, 32.0 mmol, 1.0 equiv) (Note 7) is added under a flow of nitrogen by temporarily removing the septum. 2-Methoxythiophene (4.02 g, 35.2 mmol, 1.1 equiv) (Note 7) is added by syringe, and the septum is replaced with the glass stopper. The reaction mixture is stirred at an oil bath temperature of 85 °C for 5 h (Note 8). After cooling to room temperature, a saturated aqueous solution of NaHCO$_3$ (20 mL) is added to the mixture, which is then stirred for 5 min to form a white solid. The resulting mixture is filtered through a glass funnel with a cotton plug to remove the solid, which is then washed with EtOAc (10 mL). The filtrate is transferred to a 250-mL separatory funnel. The organic layer is retained, and the separated aqueous layer is extracted with EtOAc (3 x 30 mL). The combined organic layers are washed with brine (20 mL) and dried over anhydrous sodium sulfate. After filtration through a glass funnel with a cotton plug to remove the drying agent that is washed with EtOAc (10 mL), the solvent is removed by rotary evaporation (50 mmHg, 40 °C water bath) and then under oil pump vacuum (0.50 mmHg, room temperature) for 20 min (Note 9). The crude product is re-dissolved in chloroform (30 mL) (Note 10), and silica gel (16 g) (Note 11) is added to the solution. Chloroform is removed by rotary evaporation (50 mmHg, 40 °C water bath), and the mixture is further dried under oil pump vacuum (0.50 mmHg, room temperature) for 15 min (Note 12) to obtain a free-flowing powder, which is placed at the top of silica gel (310 g) pre-eluted with 600 mL of 25% Et$_2$O in hexanes in a column (7.5 cm diameter). After covering the surface of the silica gel with sand (Note 13), elution with 2.4 L of 25% Et$_2$O in hexanes gave fractions containing the desired product, which can be visualized on thin-layer chromatography (TLC) (Note 14). The fractions are combined, concentrated by rotary evaporation (50 mmHg, 40 °C water bath), and then dried under oil pump vacuum (0.50 mmHg, room temperature) (Note 15) for 2 h to afford 2-methyl-3-(thien-2-yl)-1H-indole (5.66–5.80 g, 83–85% yield) as a white crystalline solid (Note 16).

Notes

1. The submitters used octagon-shaped magnetic stir bar (1.5 cm x 6.5 mm).
2. Indium(III) trifluoromethanesulfonate [In(OTf)$_3$, Tf = SO$_2$CF$_3$] was purchased from Strem Chemical, Inc., and should be stored in a

desiccator due to its highly hygroscopic nature. The submitters purchased indium (III) trifluoromethanesulfonate from Sigma Aldrich.

3. The submitters heated indium (III) trifluoromethanesulfonate at 150 °C under vacuum (0.083 mmHg).

4. The submitters conducted the reaction under the atmosphere of argon.

5. 1,4-Dioxane (CHROMASOLV® Plus, for HPLC, ≥99.5%) was purchased from Sigma Aldrich and was used as received without further purification.

6. Toluene was dried using a solvent purification system manufactured by SG Water U.S.A., LLC. The submitters purchased toluene (99.5+%) from Wako Pure Chemical Industries, Ltd. and distilled under argon from calcium chloride just prior to use.

7. 2-Methyl-1H-indole (>99.0%) and 2-methoxythiophene (>98.0%) were purchased from Tokyo Chemical Industry Co., Ltd. and used as received.

8. The progress of the reaction can be monitored by thin-layer chromatography (TLC) using Silicycle glass-backed extra hard layer, 60 Å plates (indicator F-254, 250 μm). A solution of 25% Et$_2$O in hexanes was used as the development solvent and p-anisaldehyde stain was used to visualized the spots of both reactants (2-methyl-1H-indole: R$_f$ = 0.35; 2-methoxythiophene: R$_f$ = 0.70) and products (R$_f$ = 0.30). The submitters monitored the reaction progress by gas chromatography (GC). Complete conversion of 2-methoxythiophene was observed at the reaction time of 5 h. GC analyses were performed on a Shimadzu model GC-2014 instrument equipped with a capillary column of InertCap 5 (5% phenyl polysilphenylene-siloxane, 30 m x 0.25 mm x 0.25 μm) using nitrogen as carrier gas (1.78 mL/min) and measured under the following conditions: injector temperature: 280 °C; detector temperature: 300 °C; initial temperature: 50 °C for 2 min; rate of temperature rise: 30 °C/min; final temperature: 280 °C for 21 min. The retention times of the reagents and the product were as follows: 2.06 min for 2-methoxythiophene; 7.66 min for 2-methyl-1H-indole; 11.17 min for 2-methyl-3-(thien-2-yl)-1H-indole.

9. According to submitter's procedure, solvent was removed by rotary evaporator (1.5 mmHg, 30 °C) and then under oil pump (0.38 mmHg, room temperature) for 2 h.

10. Chloroform (Approx. 0.75% EtOH, HPLC) was purchased from Fisher and was used as received. The submitters purchased chloroform (>99.0%) from Junsei Chemical Co., Ltd. which was used as received.

11. Column Chromatography was performed with the indicated solvents using Silicycle SiliaFlash® P60 (230–400 mesh) silica gel. The submitters purchased Silica gel (Silica gel 60N, spherical neutral, particle size 40–50 μm) from Kanto Chemical Co., Inc. and used as received.

12. According to the submitter's procedure, chloroform was removed by rotary evaporator (1.5 mmHg, 30 °C) and then under oil pump (0.38 mmHg, room temperature) for 1 h.

13. The submitters covered the silica gel with anhydrous sodium sulfate.

14. TLC plates were eluted with 25% Et$_2$O in hexanes, and a spot of the desired product (R$_f$ = 0.30) was visualized with 254 nm UV light and by staining with p-anisaldehyde stain. Fractions were collected in 25 x 200 mm Pyrex glass test tubes. According to submitter's procedure, TLC (TLC silica gel 60G F254) was purchased from Merck KGaA. A phosphomolybdic acid/ethanol solution was used as the TLC stain and desired product spot has an R$_f$ value of 0.32. The desired product was obtained in fractions 20–39 (1.3 L).

15. According to submitter's procedure, combined fractions were concentrated by rotary evaporator (1.5 mmHg, 30 °C) and then dried under oil pump (0.38 mmHg, room temperature) for 2 h.

16. 2-Methyl-3-(thien-2-yl)-1H-indole has the following physical and spectroscopic properties: mp 80–81.0 °C; ^{1}H NMR (500 MHz, CDCl$_3$) δ: 2.59 (s, 3 H), 7.15–7.21 (m, 4 H), 7.31–7.33 (m, 2 H), 7.84 (d, J = 10.0 Hz, 1 H), 7.96 (bs, 1 H); ^{13}C NMR (125 MHz, CDCl$_3$) δ: 13.2, 108.0, 110.4, 119.2, 120.4, 122.0, 123.5, 124.6, 127.4, 127.9, 132.5, 135.1, 137.4; IR (neat): 3395, 3051, 2923, 2853, 1725, 1561, 1455, 1446, 743 cm^{-1}; HRMS (ESI) Calcd for C$_{13}$H$_{10}$NS: [M-H], 212.0534. Found: m/z 212.0557; Anal. Calcd. for C$_{13}$H$_{11}$NS: C, 73.20; H, 5.20; N, 6.57; S, 15.03. Found: C, 73.10; H, 5.25; N, 6.52; S, 15.09.

Handling and Disposal of Hazardous Chemicals

The procedures in this article are intended for use only by persons with prior training in experimental organic chemistry. All hazardous materials should be handled using the standard procedures for work with chemicals described in references such as "Prudent Practices in the Laboratory" (The National Academies Press, Washington, D.C., 2011 www.nap.edu). All chemical waste should be disposed of in accordance with local regulations.

For general guidelines for the management of chemical waste, see Chapter 8 of Prudent Practices.

These procedures must be conducted at one's own risk. *Organic Syntheses, Inc.*, its Editors, and its Board of Directors do not warrant or guarantee the safety of individuals using these procedures and hereby disclaim any liability for any injuries or damages claimed to have resulted from or related in any way to the procedures herein.

Discussion

A heteroaryl–heteroaryl unit constitutes a core structure in many important organic molecules such as optoelectronic materials,[2] liquid crystals,[3] biologically active compounds,[4] and ligands for asymmetric catalysis.[5] Over the past 35 years, transition metal-catalyzed cross-coupling reactions have contributed mainly to constructing (hetero)aryl–(hetero)aryl bonds.[6] On the other hand, nucleophilic aromatic substitution (S_NAr) has actually been studied to make such biaryl units since the 1940s.[7] However, in order to synthesize biaryls by the S_NAr reaction, two aryl substrates with entirely opposite electronic demands must be arranged. Thus, electron-rich aryl nucleophiles with highly electropositive metals (e.g. Li^+, Mg^{2+}, Zn^{2+}) and/or electron-poor aryl electrophiles with one or more strong electron-withdrawing groups (EWGs; e.g. CF_3, NO_2, CN, CO_2R) are necessary for each substrate.[8] In addition, more than a stoichiometric amount of promoter is often required.[8c,p,q,r] In sharp contrast to such conventional approaches, we have achieved, for the first time, S_NAr-based catalytic heteroaryl–heteroaryl bond formation without using both the heteroarylmetal nucleophile and EWGs-substituted heteroaryl electrophile.[9]

Synthesis of 2-methyl-3-(thien-2-yl)-1*H*-indole on a 0.25 mmol scale[9] was carried out with 2-methyl-1*H*-indole and 2-methoxythiophene in a 1.3:1 molar ratio. The desired product could be isolated as a pure form by column chromatography on silica gel, but not from a reaction mixture performed on a large scale over 30 mmol, where the desired product contaminated with 2-methyl-1*H*-indole was obtained. In order to obtain the pure product in such a large-scale synthesis, the use of a slight excess molar amount of not 2-methyl-1*H*-indole but 2-methoxythiophene is important.

The indium-catalyzed S_NAr reaction is applicable to a range of indoles with OMe, alkyl, Br, Ph, and/or p-MeOC$_6$H$_4$ groups as shown in Table 1,

Table 1. Indium-catalyzed S_NAr reaction of indoles with 2- or 3-methoxy-thiophene[a]

82% yield

84% yield

78% yield

66% yield

42% yield

47% yield

80% yield

77% yield

46% yield[b]

42% yield

81% yield

70% yield

[a] Yields of isolated products are shown here. Futher details on reaction conditions for each reaction are provided in Supporting Information of reference 9. [b] In(ONf)₃ (Nf = SO₂C₄F₉; 10 mol%) instead of In(OTf)₃ was used.

which shows results of reactions performed on a 0.25 or 0.40 mmol scale. It should be noted that the OMe groups on the indole substrates does not participate in the substitution reaction, thus indicating remarkable chemoselectivities. 3-Methoxythiophene also reacted with indoles to give the corresponding thienylindoles. Results of reactions using other substrates,[9] for instance, pyrroles as nucleophiles and 2,5-dimethoxythiophene, 5,5'-dimethoxy-2,2'-bithiophene, 2-pyridyl triflate and 3-acetoxyindole as electrophiles are presented. Possible reaction mechanisms are also discussed.[9]

As disclosed above, transition-metal-catalyzed cross-coupling reaction is, in general, an effective tool to connect two heteroaryl molecules. However, the cross-coupling by using indolylmetals seems to be impractical because pre-synthesis of indolylmetals requires multi-steps.[6a,10] Accordingly, our strategy will be helpful to prepare such type of heteroaryl compounds.

References

1. Department of Applied Chemistry, School of Science and Technology, Meiji University, Higashimita, Tama-ku, Kawasaki 214-8571, Japan. E-mail: tsuchimo@isc.meiji.ac.jp. Financial support by a Grant-in-Aid for Scientific Research (No. 19750083) from the Ministry of Education, Culture, Sports, Science and Technology is highly acknowledged. Y.N. thanks the JSPS Research Fellowship for Young Scientist.

2. For selected recent examples, see: (a) Fourati, M. A.; Maris, T.; Skene, W. G.; Bazuin, C. G.; Prud'homme, R. E. *J. Phys. Chem. B* **2011**, *115*, 12362–12369. (b) Kim, J.; Kwon, Y. S.; Shin, W. S.; Moon, S.-J.; Park, T. *Macromolecules* **2011**, *44*, 1909–1919. (c) Vijayakumar, C.; Saeki, A.; Seki, S. *Chem. Asian J.* **2012**, *7*, 1845–1852.

3. For selected recent studies, see: (a) Miyajima, D.; Araoka, F.; Takezoe, H.; Kim, J.; Kato, K.; Takata, M.; Aida, T. *Angew. Chem., Int. Ed.* **2011**, *50*, 7865–7869. (b) Yasuda, T.; Shimizu, T.; Liu, F.; Ungar, G.; Kato, T. *J. Am. Chem. Soc.* **2011**, *133*, 13437–13444.

4. For selected recent reports, see: (a) Nehrbass-Stuedli, A.; Boykin, D.; Tidwell, R. R.; Brun, R. *Antimicrob. Agents Chemother.* **2011**, *55*, 3439–3445. (b) Beaulieu, P. L.; Gillard, J.; Jolicoeur, E.; Duan, J.; Garneau, M.; Kukolj, G.; Poupart, M.-A. *Bioorg. Med. Chem. Lett.* **2011**, *21*, 3658–3663. (c) La Regina, G.; Bai, R.; Rensen, W. M.; Di Cesare, E.; Coluccia, A.; Piscitelli, F.; Famiglini, V.; Reggio, A.; Nalli, M.; Pelliccia, S.; Da Pozzo, E.; Costa, B.; Granata, I.; Porta, A.; Maresca, B.; Soriani, A.; Iannitto, M. L.; Santoni, A.; Li, J.; Cona, M. M.; Chen, F.; Ni, Y.; Brancale, A.; Dondio, G.; Vultaggio, S.; Varasi, M.; Mercurio, C.; Martini, C.; Hamel, E.; Lavia, P.; Novellino, E.; Silvestri, R. *J. Med. Chem.* **2013**, *56*, 123–149. (d) Perspicace, E.; Jouan-Hureaux, V.; Ragno, R.; Ballante, F.; Sartini, S.; La Motta, C.; Da Settimo, F.; Chen, B.; Kirsch, G.; Schneider, S.; Faivre, B.; Hesse, S. *Eur. J. Med. Chem.* **2013**, *63*, 765–781.

5. For selected reviews, see: (a) Au-Yeung, T. T.-L.; Chan. A. S. C. *Coord. Chem. Rev.* **2004**, *248*, 2151–2164. (b) Shimizu, H.; Nagasaki, I.; Saito, T. *Tetrahedron* **2005**, *61*, 5405–5432.

6. For representative reviews, see: (a) Hassan, J.; Sévignon, M.; Gozzi, C.; Schulz, E.; Lemaire, M. *Chem. Rev.* **2002**, *102*, 1359–1469. (b) de Meijere, A., Diederich, F. *Metal-Catalyzed Cross-Coupling Reactions*, 2nd ed.; Wiley-VCH: Weinheim, 2004. (c) Satoh, T.; Miura, M. *Chem. Lett.* **2007**, *36*, 200–205. (d) Han, W.; Ofial, A. R. *Synlett* **2011**, 1951–1955. (e) Zhao, D.; You, J.; Hu, C. *Chem. Eur. J.* **2011**, *17*, 5466–5492. (f) Bugaut, X.; Glorius, F. *Angew. Chem., Int. Ed.* **2011**, *50*, 7479–7481. For a brief historical note on the cross-coupling reaction, see: (g) Tamao, K.; Hiyama, T.; Negishi, E.-i. *J. Organomet. Chem.* **2002**, *653*, 1–4.

7. Smith, M. B.; March, J. *March's Advanced Organic Chemistry: Reactions, Mechanisms, and Structure*, 6th ed.; Wiley-Interscience: Hoboken, 2007, pp. 481–482 and 853–857.

8. (a) Fuson, R. C.; Wassmundt, F. W. *J. Am. Chem. Soc.* **1956**, *78*, 5409–5413. (b) Wilson, J. M.; Cram, D. J. *J. Org. Chem.* **1984**, *49*, 4930–4943. (c) Stahly, G. P. *J. Org. Chem.* **1985**, *50*, 3091–3094. (d) Cram, D. J.; Bryant, J. A.; Doxsee, K. M. *Chem. Lett.* **1987**, *16*, 19–22. (e) Shindo, M.; Koga, K.; Tomioka, K. *J. Am. Chem. Soc.* **1992**, *114*, 8732–8733. (f) Reuter, D. C.; Flippin, L. A.; McIntosh, J.; Caroon, J. M. Hammaker, J. *Tetrahedron Lett.* **1994**, *35*, 4899–4902. (g) Kamikawa, K.; Uemura, M. *Tetrahedron Lett.* **1996**, *37*, 6359–6362. (h) Hattori, T.; Suzuki, M.; Tomita, N.; Takeda, A.; Miyano, S. *J. Chem. Soc., Perkin Trans. 1* **1997**, 1117–1123. (i) Norman, D. P. G.; Bunnell, A. E.; Stabler, S. R.; Flippin, L. A. *J. Org. Chem.* **1999**, *64*, 9301–9306. (j) Boisnard, S.; Neuville, L.; Bois-Choussy, M.; Zhu, J. *Org. Lett.* **2000**, *2*, 2459–2462. (k) Moosa, B. A.; Abu Safieh, K. A.; El-Abadelah, M. M. *Heterocycles* **2002**, *57*, 1831–1840. (l) Hattori, T.; Takeda, A.; Yamabe, O.; Miyano, S. *Tetrahedron* **2002**, *58*, 233–238. (m) Hattori, T.; Shimazumi, Y.; Goto, H.; Yamabe, O.; Morohashi, N.; Kawai, W.; Miyano, S. *J. Org. Chem.* **2003**, *68*, 2099–2108. (n) Hattori, T.; Iwato, H.; Natori, K.; Miyano, S. *Tetrahedron: Asymmetry* **2004**, *15*, 881–887. (o) Al-Hiari, Y. M.; Qaisi, A. M.; El-Abadelah, M. M.; Voelter, W. *Monatsh. Chem.* **2006**, *137*, 243–248. (p) Cecchi, M.; Micoli, A.; Giomi, D. *Tetrahedron* **2006**, *62*, 12281–12287. (q) De Rosa, M.; Arnold, D.; Medved', M. *Tetrahedron Lett.* **2007**, *48*, 3991–3994. (r) Yu, C.; Jiao, L.; Tan, X.; Wang, J.; Xu, Y.; Wu, Y.; Yang, G.; Wang, Z.; Hao, E. *Angew. Chem., Int. Ed.* **2012**, *51*, 7688–7691. (s) Aissaoui, R.; Nourry, A.; Coquel, A.; Dao, T. T. H.; Derdour, A.; Helesbeux, J.-J.; Duval, O.; Castanet, A.-S.; Mortier, J.

J. Org. Chem. **2012**, *77*, 718–724. (t) Xiong, Y.; Wu, J.; Xiao, S.; Xiao, J.; Cao, S. *J. Org. Chem.* **2013**, *78*, 4599–4603. See also the following reviews: (u) Bringmann, G.; Walter, R.; Weirich, R. *Angew. Chem., Int. Ed. Engl.* **1990**, *29*, 977–991. (v) Gant, T. G.; Meyers, A. I. *Tetrahedron* **1994**, *50*, 2297–2360. (w) Hattori, T.; Miyano, S. *J. Synth. Org. Chem., Jpn.* **1997**, *55*, 121–131.

9. Tsuchimoto, T.; Iwabuchi, M.; Nagase, Y.; Oki, K.; Takahashi, H. *Angew. Chem., Int. Ed.* **2011**, *50*, 1375–1379.

10. For examples, see: (a) Kawasaki, I; Yamashita, M.; Ohta, S. *Chem. Pharm. Bull.* **1996**, *44*, 1831–1839. (b) Denhart, D. J.; Deskus, J. A.; Ditta, J. L.; Gao, Q.; King, H. D.; Kozlowski, E. S.; Meng, Z.; LaPaglia, M. A.; Mattson, G. K.; Molski, T. F.; Taber, M. T.; Lodge, N. J.; Mattson, R. J.; Macor, J. E. *Bioorg. Med. Chem. Lett.* **2009**, *19*, 4031–4033. (c) Bourderioux, A.; Ouach, A.; Bénéteau, V.; Mérour, J.-Y.; Routier, S. *Synthesis* **2010**, 783–790.

Appendix
Chemical Abstracts Nomenclature (Registry Number)

2-Methyl-1*H*-indole; (95-20-5)
2-Methoxythiophene; (16839-97-7)
Indium(III) trifluoromethanesulfonate; (128008-30-0)
1,4-Dioxane; (123-91-1)
2-Methyl-3-(thien-2-yl)-1*H*-indole; (1159415-23-2)

Yuta Nagase was born in 1985 in Ishikawa, Japan. He received his B.Eng. degree in 2008 and his M.Eng. degree in 2010 from Meiji University. Since 2010, he has started his Ph.D. study under supervision of Associate Professor Teruhisa Tsuchimoto. He has been a JSPS research fellow since 2012. His research interests focus on development of new Lewis acid-catalyzed reaction with indoles as nucleophiles.

Teruhisa Tsuchimoto was born in 1970 in Gifu, Japan. He received his Ph.D. degree in 1997 from Tokyo Institute of Technology under the supervision of Professor Tamejiro Hiyama. After working with Professor Peter Wipf at University of Pittsburgh as a postdoctoral research fellow, he returned to Japan in 1998 to join Professor Shirakawa's group at JAIST as an Assistant Professor. In 2006, he moved to Meiji University as an Associate Professor. He is a recipient of the Incentive Award in Synthetic Organic Chemistry, Japan in 2009. His research interests cover development of new synthetic methodologies based on Lewis acid-catalyzed activation of hydrocarbon functional groups, and their application to synthesis of novel optoelectronic materials.

Joyce Leung was born and raised in Hong Kong. She received her B.S. in Chemical Biology from University of California, Berkeley in 2007. Then she worked as a Research Associate at Nanosyn, Inc. for a year, and began her doctoral studies at University of Texas at Austin in 2008. She obtained her Ph.D. in 2013 under the supervision of Professor Michael J. Krische. Her doctoral research focused on transition metal catalyzed carbon-carbon bond formation method development. She is currently conducting postdoctoral research on natural product synthesis under the supervision of Professor John L. Wood at Baylor University.

Synthesis of 2,3-Disubstituted Benzofurans by the Palladium-Catalyzed Coupling of 2-Iodoanisoles and Terminal Alkynes, Followed by Electrophilic Cyclization: 3-Iodo-2-phenylbenzofuran

Tuanli Yao,[§1]* Dawei Yue,[¶2] and Richard C. Larock[¶2]*

[§]University of Kansas Specialized Chemistry Center, University of Kansas, Lawrence, KS 66047; [¶]Department of Chemistry, Iowa State University, Ames, IA 50011

Checked by Michael Rombola and Viresh H. Rawal

A.

B.

Procedure

A. *2-(Phenylethynyl)anisole (3)*. A 250-mL, one-necked round-bottomed flask, equipped with a 15 mm × 32 mm ellipsoid-shaped magnetic stirring bar, is charged with 2-iodoanisole (**1**, 7.02 g, 30.0 mmol, 1.0 equiv) (Note 1), phenylacetylene (**2**, 36.0 mmol, 3.68 g, 1.2 equiv) (Note 2), triethylamine (60 mL) (Note 3) and bis(triphenylphosphine)palladium(II) dichloride (0.6 mmol, 0.421 g, 0.02 equiv) (Note 4). After stirring for 5 min, copper(I) iodide (0.3 mmol, 0.057 g, 0.01 equiv) (Note 5) is added and stirring is continued for another 2 min (Note 6). The flask is then capped with a rubber septum, into which is inserted a nitrogen inlet, and subjected to

Org. Synth. **2014**, *91*, 283-292

283

Published on the Web 9/12/2014

DOI: 10.15227/orgsyn.091.0283

© 2014 Organic Syntheses, Inc.

three cycles of evacuation and refilling with nitrogen. While being maintained under a slight positive pressure of nitrogen, the mixture is allowed to stir at room temperature for 3 h (Notes 7 and 8). The resulting dark gray solution is filtered through a medium porosity fritted glass funnel and EtOAc (4 x 25 mL) is used to rinse the flask and wash the filter cake. The combined organic phases are concentrated by rotary evaporation (28 °C, 150 to 8 mmHg) to give a dark brown oil (Note 9). The crude product is purified by flash column chromatography on silica gel (Note 10) to afford alkyne **3** as an amber oil (Note 11) (6.19–6.24 g, 99%).

B. *3-Iodo-2-phenylbenzofuran* **(4)**. A 500-mL, three-necked round-bottomed flask, equipped with a 15 mm × 32 mm ellipsoid-shaped magnetic stirring bar, is charged with 2-(phenylethynyl)anisole (**3**, 5.15 g, 24.7 mmol, 1.0 equiv) and dichloromethane (250 mL) (Note 12). Two necks are capped with rubber septa, and to the third neck is attached a 50 mL screw feed solid addition funnel. The side port of the addition funnel is capped with a rubber septum. A nitrogen inlet is inserted into the middle septum and the flask is subjected to three cycles of evacuation and refilling with nitrogen. Solid iodine (I_2) (12.6 g, 49.6 mmol, 2.0 equiv) (Note 13) is added *via* a solid addition funnel over 20 min. The addition funnel is replaced with a rubber septum and the reaction is allowed to stir at room temperature under nitrogen for 5 h (Notes 14 and 15). The dark purple solution is transferred to a 500-mL separatory funnel. The original round-bottomed flask is rinsed with dichloromethane (2 x 25 mL) and the rinse is also transferred to the 500-mL separatory funnel. A saturated aqueous $Na_2S_2O_3$ solution (150 mL) (Note 16) is added. After shaking for 1 min, the layers are separated. The organic phase is dried over anhydrous sodium sulfate (Na_2SO_4) (Note 17), filtered through a medium porosity fritted glass funnel, which was rinsed with dichloromethane (2 x 25 mL), and concentrated by rotary evaporation (28 °C, 150 to 8 mmHg) to give a brown oil. The residue is purified by flash column chromatography on silica gel (Note 18) to afford benzofuran **4** as a beige solid (Note 19) (6.85–6.88 g, 87% yield).

Notes

1. 2-Iodoanisole (**1**, 98%) was purchased from Sigma-Aldrich and used as received.

DOI: 10.15227/orgsyn.091.0283

2. Phenylacetylene (**2**, 98%) was purchased from Sigma-Aldrich and used as received.

3. Triethylamine (Et₃N, ≥ 99%) was purchased from Sigma-Aldrich and used as received.

4. Bis(triphenylphosphine)palladium(II) dichloride (PdCl₂(PPh₃)₂, 98%) was purchased from Sigma-Aldrich and used as received.

5. Copper(I) iodide (CuI, 98%) was purchased from Sigma-Aldrich and used as received.

6. The color of the reaction changes to dark green and a considerable amount of triethylamine hydroiodide salt precipitates.

7. The speed of stirring might need to be adjusted, since precipitation of the triethylamine hydroiodide salt could impede efficient stirring of the solution.

8. Completeness of the reaction is judged by the disappearance of 2-iodoanisole by thin-layer chromatography (TLC), performed on glass-backed pre-coated silica gel plates (250 μm, Merck Millipore) with a UV254 indicator, using 33:1 hexane/ethyl acetate as the eluent (R$_f$ of 2-iodoanisole = 0.54; Rf of the product **3** = 0.23). The product is visualized with a 254 nm UV lamp.

9. The crude product is dried on a rotary evaporator (28 °C, 7.5 mmHg) for 45 min.

10. Column chromatography was performed on a 4 cm diameter column, dry-packed with 110 g of silica gel (SiliaFlash® P60 230 × 400 mesh, 60Å), and eluted with 33:1 hexanes/ethyl acetate (1 L). Fifty 20 mL fractions were collected. Fractions 19-42 contained the desired product and were concentrated by rotary evaporation (28 °C, 7.5 mmHg) and dried under vacuum (3 mmHg) for 24 h while occasionally rotating the flask.

11. The physical properties of **3** are: R$_f$ = 0.23 (TLC, Note 8); ¹H NMR (500 MHz, CDCl₃) δ: 3.92 (s, 3 H), 6.90 (d, J = 8.5, 1 H), 6.95 (t, J = 7.5, 1 H), 7.30–7.37 (m, 4 H), 7.51 (d, J = 7.5 Hz, 1 H), 7.56–7.58 (m, 2 H); ¹³C NMR (126 MHz, CDCl₃) δ: 56.0, 86.0, 93.6, 111.0, 112.7, 120.7, 123.8, 128.3, 128.4, 129.9, 131.9, 133.8, 160.2; IR (neat) 3059, 2835, 2216, 1593, 1574 cm⁻¹; Anal. Calcd. for C₁₅H₁₂O: C, 86.51; H, 5.81. Found: C, 86.54; H, 5.76.

12. Dichloromethane (DCM, 99.9%) was purchased from Fisher Scientific and used as received.

13. The checkers purchased iodine (I₂, 1 – 3 mm beads, 99.7%) from Sigma-Aldrich and used it as received.

---

14. The completeness of the reaction is judged by the disappearance of 2-(phenylethynyl)anisole by thin-layer chromatography performed on glass-backed pre-coated silica gel plates (250 μm, Merck Millipore) with a UV254 indicator, using 33:1 hexane/ethyl acetate as the eluent (R_f of 2-(phenylethynyl)anisole = 0.23; R_f of the product **4** = 0.53). The product is visualized with a 254 nm UV lamp.

15. A minor side product is present and possesses a very similar R_f to 2-(phenylethynyl)anisole upon TLC analysis.

16. The checkers purchased sodium thiosulfate pentahydrate ($Na_2S_2O_3 \bullet 5H_2O$, Certified ACS) from Fisher Scientific and used it as received.

17. Anhydrous sodium sulfate (Na_2SO_4, $\geq$ 99%) was purchased from Fisher Scientific and used as received. To ensure proper dryness, 35 g of Na_2SO_4 was added to the organic phase and the resulting mixture was kept at room temperature for 10 min with occasional swirling.

18. Column chromatography is performed on a 4 cm diameter column, dry-packed with 95 g of silica gel (SiliaFlash® P60230 × 400 mesh, 60Å) and eluted with 33:1 hexanes/ethyl acetate. Fifteen 20 mL fractions are collected. Fractions 8–14 contained the desired product. They were combined, concentrated by rotary evaporation (28 °C, 150 to 8 mmHg), dried under vacuum (3 mmHg) at 23 °C for 6 h and kept in the refrigerator (1 °C) for 3 days. The solid obtained was ground up and dried under vacuum (4 mbar) at 23 °C for 2 h until a constant weight (6.85–6.88 g) was obtained.

19. The physical properties of **4** are: mp 40–42 °C; R_f = 0.53 (TLC, Note 14); ^{1}H NMR (500 MHz, CDCl$_3$) δ: 7.31–7.38 (m, 2 H), 7.42–7.52 (m, 5 H), 8.18–8.19 (m, 2 H); ^{13}C NMR (126 MHz, CDCl$_3$) δ: 61.4, 111.4, 122.1, 123.7, 125.9, 127.7, 128.7, 129.4, 130.2, 132.7, 153.3, 154.1; IR (CHCl$_3$) 3062, 1590, 1453 cm^{-1}; HRMS m/z calcd. for $C_{14}H_9IO$ [M]$^+$ 319.9698, found 319.9688. Anal. Calcd. for $C_{14}H_9OI$: C, 52.53; H, 2.83. Found: C, 52.60; H, 2.88.

Working with Hazardous Chemicals

The procedures in *Organic Syntheses* are intended for use only by persons with proper training in experimental organic chemistry. All hazardous materials should be handled using the standard procedures for work with chemicals described in references such as "Prudent Practices in the Laboratory" (The National Academies Press, Washington, D.C., 2011; the full text can be accessed free of charge at http://www.nap.edu/catalog.php?record_id=12654). All chemical waste should be disposed of in accordance with local regulations. For general guidelines for the management of chemical waste, see Chapter 8 of Prudent Practices.

In some articles in *Organic Syntheses*, chemical-specific hazards are highlighted in red "Caution Notes" within a procedure. It is important to recognize that the absence of a caution note does not imply that no significant hazards are associated with the chemicals involved in that procedure. Prior to performing a reaction, a thorough risk assessment should be carried out that includes a review of the potential hazards associated with each chemical and experimental operation on the scale that is planned for the procedure. Guidelines for carrying out a risk assessment and for analyzing the hazards associated with chemicals can be found in Chapter 4 of Prudent Practices.

The procedures described in *Organic Syntheses* are provided as published and are conducted at one's own risk. *Organic Syntheses, Inc.*, its Editors, and its Board of Directors do not warrant or guarantee the safety of individuals using these procedures and hereby disclaim any liability for any injuries or damages claimed to have resulted from or related in any way to the procedures herein.

Discussion

The benzo[*b*]furan nucleus is prevalent in a wide variety of biologically active natural and unnatural compounds.[3] There has been growing interest in developing a general and versatile synthesis of benzo[*b*]furans. A number of synthetic approaches to this class of compounds have been introduced in recent years.[4] One common approach to heterocycles has been the electrophilic cyclization of alkynes using ICl and I_2. Cacchi and co-workers have previously reported the synthesis of 3-iodobenzo[*b*]furans by the iodocyclization of alkynylphenols.[5] Unfortunately, the protecting and deprotecting steps required to synthesize the alkynylphenol are not particularly attractive synthetically. We have successfully made this overall approach more synthetically attractive by employing the corresponding methyl ethers and using I_2 and ICl as electrophiles.[6] The preparation of 3-iodo-2-phenylbenzofuran described here illustrates a general protocol for the palladium/copper-catalyzed cross-coupling of various *o*-iodoanisoles and terminal alkynes, followed by electrophilic cyclization with I_2. Since this process was first communicated in 2005, it has been subsequently employed in the synthesis of coumestans,[7] permethylated anigopreissin A[8] and XH-14[9] derivatives, inhibitors of mycobacterium protein tyrosine phosphatase B,[10] and retinoic acid receptor agonists.[11]

This approach to 2,3-disubstituted benzofurans is very versatile. The substituents attached to the triple bond can be vinylic groups or aromatic rings bearing certain types of functionality. Unfortunately, alkynes bearing an alkyl group fail to undergo electrophilic cyclization. This approach has also been successfully applied to the synthesis of furopyridines (Table 1).

Table 1. Synthesis of Benzo[b]furans by Electrophilic Cyclization[a]

Entry	Alkyne	Product	Time (h)	Yield (%)
1			3	87
2			3	80
3			3	100
4			3	95
5			12	67
6			12	80
7			12	67

[a] All reactions were run with 0.25 mmol alkyne and 2 equiv of I$_2$ in 5 mL of CH$_2$Cl$_2$ at 25 °C.

References

1. University of Kansas Specialized Chemistry Center, University of Kansas, Lawrence, KS 66047; t394y119@ku.edu. We thank the National Institute of General Medical Sciences (GM070620) for support of this research and Johnson Matthey, Inc., and Kawaken Fine Chemicals Co., Ltd., for donations of palladium salts.

2. Department of Chemistry, Iowa State University, Ames, IA 50011; larock@iastate.edu.

3. (a) Dean, F. M. *The Total Synthesis of Natural Products*; ApSimon, J., Ed.; Wiley: New York, 1973; Vol. 1, pp 467–562. (b) Cagniant, P.; Cagniant, D. In *Advances in Heterocyclic Chemistry*; Katritzky, A. R., Boulton, A. J., Eds.; Academic Press: New York, 1975; Vol. 18, pp 337–482. (c) Dean, F. M.; Sargent, M. V. In *Comprehensive Heterocyclic Chemistry*; Katritzky, A. R., Rees, C. W., Eds.; Pergamon Press: Oxford, UK, 1984; Vol. 4, pp 531–597. (d) *Comprehensive Heterocyclic Chemistry II*; Katritzky, A. R., Rees, C. W., Scriven, E. F. V., Eds.; Pergamon Press: Oxford, UK, 1996; Vol. 2, pp 259–321.

4. (a) Horaguchi, T.; Iwanami, H.; Tanaka, T.; Hasegawa, E.; Shimizu, T.; Suzuki, T.; Tanemura, K. *J. Chem. Soc., Chem. Commun.* **1991**, 44–46. (b) Horaguchi, T.; Kobayashi, H.; Miyazawa, K.; Hasegawa, E.; Shimizu, T. *J. Heterocycl. Chem.* **1990**, *27*, 935–940. (c) Boehm, T. L.; Showalter, H. D. H. *J. Org. Chem.* **1996**, *61*, 6498–6499. (d) Nicolaou, K. C.; Snyder, S. A.; Bigot, A.; Pfefferkorn, J. A. *Angew. Chem., Int. Ed.* **2000**, *39*, 1093–1096. (e) Kondo, Y.; Shiga, F.; Murata, N.; Sakamoto, T.; Yamanaka, H. *Tetrahedron* **1994**, *50*, 11803–11812. (f) Arcadi, A.; Cacchi, S.; Rosario, M. D.; Fabrizi, G.; Marinelli, F. *J. Org. Chem.* **1996**, *61*, 9280–9288. (g) Cacchi, S.; Fabrizi, G.; Moro, L. *Tetrahedron Lett.* **1998**, *39*, 5101-5104. (h) Monteiro, N.; Arnold, A.; Balme, G. *Synlett* **1998**, 1111–1113. (i) Kraus, G. A.; Zhang, N.; Verkade, J. G.; Nagarajan, M.; Kisanga, P. B. *Org. Lett.* **2000**, *2*, 2409–2410. (j) Katritzky, A. R.; Ji, Y.; Fang, Y.; Prakash, I. *J. Org. Chem.* **2001**, *66*, 5613–5615.

5. Arcadi, A.; Cacchi, S.; Fabrizi, G.; Marinelli, F.; Moro, L. *Synlett* **1999**, 1432–1434.

6. Yue, D.; Yao, T.; Larock, R. C. *J. Org. Chem.* **2005**, *70*, 10292–10296.

7. Yao, T.; Yue, D.; Larock, R. C. *J. Org. Chem.* **2005**, *70*, 9985–9989.

8. Chiummiento, L.; Funicello, M.; Lopardo, M. T.; Lupattelli, P.; Choppin, S.; Colobert, F. *Eur. J. Org. Chem.* **2012**, 188–192.

9. Bang, H. B.; Han, S. Y.; Choi, D. H.; Hwang, J. W.; Jun, J.-G. *ARKIVOC* **2009**, 112–125.

10. He, Y.; Xu, J.; Yu, Z.-H.; Gunawan, A. M.; Wu, L.; Wang, L.; Zhang, Z.-Y. *J. Med. Chem.* **2013**, *56*, 832–842.

11. Santin, E. P.; Khanwalkar, H.; Voegel, J.; Collette, P.; Mauvais, P.; Gronemeyer, H.; de Lera, A. R. *ChemMedChem* **2009**, *4*, 780–791.

Appendix
Chemical Abstracts Nomenclature (Registry Number)

2-Iodoanisole; (529-28-2)
Phenylacetylene; (536-74-3)
Triethylamine; (121-44-8)
Bis(triphenylphosphine)palladium(II) dichloride; (13965-03-2)
Copper(I) iodide; (7681-65-4)
Iodine; (7553-56-2)

Tuanli Yao earned his B.S. and M.S. degrees in chemistry from Peking University in China. He obtained his Ph.D. in 2005 from Iowa State University working with Professor Richard C. Larock. His graduate research at Iowa State focused on new approaches to heterocycles and carbocycles. After postdoctoral research with Professor Richmond Sarpong at U.C. Berkeley, he joined Deciphera Pharmaceuticals in Lawrence, Kansas. Currently, he is a Research Associate at University of Kansas Specialized Chemistry Center. His research interests include palladium catalysis and medicinal chemistry.

Richard C. Larock, University Professor and Distinguished Professor Emeritus at Iowa State University, received his B.S. at the University of California, Davis in 1967. He then joined the group of Prof. Herbert C. Brown at Purdue University, where he received his Ph.D. in 1972. He worked as an NSF Postdoctoral Fellow at Harvard University in Prof. E. J. Corey's group and joined the Iowa State University faculty in 1972. His research interests have included aryne chemistry, electrophilic cyclization, palladium catalysis, and polymer chemistry based on biorenewable resources.

Dawei Yue received his B.S. degree in Chemistry from Xiamen University in 1997, where he conducted research with Professor Huilin Wan. He then moved to the United States, where he received his Ph.D. degree from Iowa State University in 2004 under the mentorship of Professor Richard C. Larock. He was a postdoctoral fellow with Professor Michael E. Jung at the University of California, Los Angeles in 2004 and with Professor Sheng Ding at the Scripps Research Institute (2004-2006) before beginning his career in pharmaceutical industry. He is currently director of chemistry at BroadPharm, a customer-focused research and development company based in San Diego, CA.

Michael Rombola was born in 1989 in Rochester, NY. He studied as an undergraduate at Cornell University, where he completed a B.S. degree in biology in 2011. He is now pursuing his doctoral degree at the University of Chicago, working under the guidance of Professor Viresh H. Rawal. He is currently developing novel chiral diene ligands for use in asymmetric catalysis.

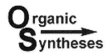

Enantioselective Preparation of (S)-5-Oxo-5,6-dihydro-2H-pyran-2-yl Benzoate

Tamas Benkovics,[1]* Adrian Ortiz, Zhiwei Guo, Animesh Goswami and Prashant Deshpande

Chemical Development, Bristol-Myers Squibb Co., 1 Squibb Drive, New Brunswick, NJ 08903

Checked by Magnus C. Eriksson, Suresh R. Kapadia and Chris H. Senanayake

A.

1 $\xrightarrow[\text{CH}_2\text{Cl}_2 \\ 0\ °\text{C to rt, 1 h}]{m\text{-CPBA}}$ 2

B.

2 $\xrightarrow[\text{t-amyl-alcohol} \\ \text{rt, 1 h}]{\text{Bz}_2\text{O, 4-DMAP}}$ 3

C.

3 $\xrightarrow[\text{n-BuOH, toluene} \\ 40\ °\text{C, 3 h}]{\text{Lipase from} \\ \textit{Candida rugosa}}$ 4

Procedure

Caution! Reactions and subsequent operations involving peracids and peroxy compounds should be run behind a safety shield. For relatively fast reactions, the rate of addition of the peroxy compound should be slow enough so that it reacts rapidly and no significant unreacted excess is allowed to build up. The reaction

> *mixture should be stirred efficiently while the peroxy compound is being added, and cooling should generally be provided and maintained since many reactions of peroxy compounds are exothermic. New or unfamiliar reactions, particularly those run at elevated temperatures, should be run first on a small scale. Reaction products should never be recovered from the final reaction mixture by distillation until all residual active oxygen compounds (including unreacted peroxy compounds) have been destroyed. Decomposition of active oxygen compounds may be accomplished by the procedure described in Korach, M.; Nielsen, D. R.; Rideout, W. H. Org. Synth. 1962, 42, 50 (Org. Synth. 1973, Coll. Vol. 5, 414).*

A. *5-Oxo-5,6-dihydro-2H-pyran* (**2**). To a clean and dry 3-L four-necked flask equipped with an overhead mechanical stirrer and a PTFE-coated temperature probe, furfuryl alcohol (**1**) (140 g, 1.40 moles, 1.05 equiv) and 1.2 L of dichloromethane are added using medium stirring rate under ambient atmosphere (Note 1). The third neck of the flask is fitted with an inlet adapter with a piece of tubing attached to a bubbler. The fourth neck is sealed with a glass stopper. The flask is cooled to 0–5 °C, then charged with *meta*-chloroperoxybenzoic acid (315 g, 1.36 moles, actual potency 74.7 w%) in five equal portions through the fourth neck, maintaining the internal temperature between 0 and 10 °C (Note 2). Once the addition of oxidant is complete, the heterogeneous mixture is warmed to room temperature by removing the ice bath and stirred at ambient temperature for 1 h (Notes 3 and 4). The reaction is subsequently cooled to 0–5 °C, and the reaction by-product, solid *meta*-chlorobenzoic acid is removed by filtration using a 150 mm Büchner funnel equipped with a Whatman grade 1 filter paper. The solids are washed with pre-cooled (0 °C) dichloromethane (250 mL) (Note 5). The combined filtrate is transferred to a 2 L flask, and one-half of the dichloromethane is removed under reduced pressure (250–350 mmHg) on a rotary evaporator at 30 °C bath temperature. Isopropyl alcohol (450 mL) is added to this solution and the distillation is continued, using vacuum as low as 30 mmHg, until the volume of the solution reached 300 mL (Note 6). After verifying the final volume with a graduated cylinder, the solution is transferred back to the clean 3-L round-bottomed flask. Using an addition funnel heptane (300 mL) is added over 15 min under moderate agitation (Note 7). Once the product crystals appear, the solution is stirred for 30 min, followed by addition of heptane (600 mL) over 15 min. The

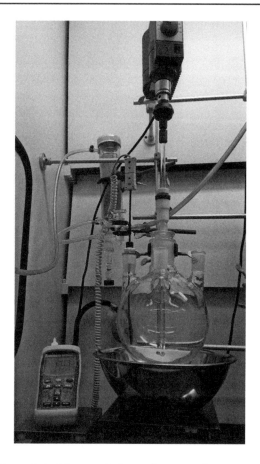

slurry is cooled to 0–5 °C, aged for 1 h, and then placed in a 0–5 °C refrigerator overnight (Note 8). The cold slurry is filtered using a 125 mm Büchner funnel equipped with a Whatman grade 1 filter paper, and the solids are subsequently washed with pre-cooled (0 °C) 3:1 heptane:isopropyl alcohol mixture (150 mL), followed by heptane (150 mL). The solid is dried on the filter for 1 h using vacuum, after which 5-oxo-5,6-dihydro-2H-pyran (**2**) 96-98 g, 62-63%) is isolated as a light yellow solid (Note 9).

B. *5-Oxo-5,6-dihydro-2H-pyran-2-yl benzoate (3)*. To a clean and dry 3-L four-necked round-bottomed flask equipped with an overhead mechanical stirrer and a PTFE-coated temperature probe, lactol **2** (100 g, 0.876 moles) and *tert*-amyl-alcohol (700 mL) is added using rapid stirring under ambient atmosphere (Note 10). The third neck of the flask is fitted with an inlet adapter with a piece of tubing attached to a bubbler. The fourth neck is sealed with a glass stopper. The light yellow slurry is heated gently to 20 °C until all lactol dissolves. To this solution is added powdered benzoic anhydride (243 g, 1.05 moles, 1.2 equiv) in one portion, followed by the addition of powdered 4-dimethylaminopyridine (3.3 g, 26.3 mmoles, 0.03 equiv) in one portion (Note 11), through the fourth neck. After addition of the solid reagents, an orange homogeneous solution appears within 15 min, and agitation is continued for 1 h (Note 12). The temperature is maintained between 20–23 °C through the use of a water bath over that time period. Solids begin to precipitate after about 45 min of stirring to form a slurry within 2–3 min. The flask is subsequently cooled to 0–5 °C, which results in a thick white slurry (Note 13). To this slurry is added water (100 mL) in one portion and the suspension is held at 0–5 °C for 1 h (Note 14). The cold slurry is transferred to a 150 mm Büchner funnel equipped with a Whatman grade 1 filter paper, and filtered by applying house vacuum (~100 mmHg). The solids are subsequently washed with pre-cooled (0 °C) 9:1 *tert*-amyl alcohol: water mixture (200 mL), followed in sequence by water (200 mL) and heptane (400 mL). The overall filtration sequence is completed in about 45 min. After drying on the filter for 12 h using house vacuum (~100 mmHg), 5-oxo-5,6-dihydro-2H-pyran-2-yl benzoate (**3**) 142–144 g, 74–75%) is isolated as a light tan solid (Note 15).

C. *(S)-5-Oxo-5,6-dihydro-2H-pyran-2-yl benzoate* (**4**). To a clean and dry 2-L Erlenmeyer equipped with a PTFE-coated 15 x 80 mm stir bar, benzoate **3** (125 g, 0.573 moles) and toluene (1 L) is charged, and the mixture is heated at 30 °C for 10 min to afford a slightly turbid orange solution. At this point, any remaining red insolubles not dissolved at this temperature are removed by filtration using a 125 mm Büchner funnel equipped with a Whatman grade 1 filter paper, which provides a clear orange solution (Note 16). To a 3-L four-necked, round-bottomed flask equipped with an overhead mechanical stirrer, a 250 mL addition funnel and a PTFE-coated temperature probe, is added lipase from *Candida rugosa* (Lipase MY) (12.5 g, 10 weight%, actual activity 30,000 u/g). The fourth neck is sealed with a glass stopper. The filtered organic layer that contains **3** is added in one portion using rapid stirring under ambient atmosphere (Note 17). The heterogeneous thin suspension is heated carefully to an internal temperature of 40 °C using a heating mantle. The heating is controlled such that the internal temperature of 40 °C is not exceeded. In a separate flask, water (6.25 g) in *n*-butanol (125 mL) is prepared, and this solution is added dropwise to the reaction flask via the 250 mL addition funnel over 30 min

(Note 18). The suspension is agitated for 6.5 h at 40 °C, and then is immediately filtered using a Büchner funnel equipped with a Whatman grade 1 filter paper (Notes 19 and 20). The clear orange filtrate is added to a separatory funnel and then washed with saturated sodium bicarbonate solution (600 mL), followed by water (600 mL). The organic layer is stirred for 20 min with anhydrous sodium sulfate. The suspension is filtered to remove the sodium sulfate and the solution stored in a flask in the refrigerator (4 °C) overnight. The organic layer is concentrated to 250 mL under reduced pressure (<100 mmHg) on a rotary evaporator at 50 °C bath temperature. Upon reaching the desired volume, *tert*-amyl alcohol (500 mL) is added to the flask, and the total volume is reduced under vacuum to 250 mL. The resulting warm solution is transferred to a 500 mL two-necked round-bottomed flask containing a magnetic stirbar and fitted with a thermocouple in one neck and a 250 mL addition funnel in the other. The solution is allowed to cool by stirring at ambient temperature, and crystals begins to precipitate at about 40 °C internal temperature. Once the internal temperature reaches about 30 °C, heptane (125 mL) is added. The suspension is then stirred at ambient temperature for 1 h, cooled to 0–5 °C in an ice/water bath and stirred for 1 h. The cold slurry is filtered on a 125 mm Büchner funnel equipped with a Whatman grade 1 filter paper using house vacuum, and the cake is subsequently washed with 1:1 *tert*-amyl alcohol/n-heptane (125 mL) followed by heptane (125 mL) (Note 21). After drying on the filter for 1 h using house-vacuum, (S)-5-oxo-5,6-dihydro-2H-pyran-2-yl benzoate (4) (44–46 g, 35–37%, >99%ee) is isolated as a light orange to light brown solid. Purity is determined by normal and chiral phase HPLC as well as NMR (Note 22).

Notes

1. 3-Chloroperoxybenzoic acid, 77%; dichloromethane, Chromasolv for HPLC; 2-propanol, Chromasolv plus for HPLC; *t*-amyl alcohol, ReagentPlus 99%; 4-dimethylamino pyridine, ReagentPlus 99% were purchased from Sigma-Aldrich. Furfuryl alcohol 98%, was purchased from Alfa Aesar; *n*-heptane, Ultra-resi analyzed, and toluene, Baker analyzed for HPLC were purchased from JT Baker; benzoic anhydride, 98% was purchased from Acros. Lipase MY was purchased from Meito–Sangyo Co, Japan. All chemicals were used as received.

2. Detailed safety studies of this reaction have been conducted to show the Achmatowitz rearrangement with *m*CPBA to be a dose-controlled exotherm. To avoid unreacted peroxides in the subsequent distillation, furfuryl alcohol should be used in slight excess. The checkers found that the addition of *m*CPBA takes about 3 h on this scale.

3. The reaction can be monitored by thin layer chromatography (TLC) with EMD 60 F_{254} pre-coated silica gel plates. The plates were eluted with a 1:1 mixture of hexanes and ethyl acetate; furfuryl alcohol (**1**) ($R_f =$ 0.6), lactol **2** ($R_f = 0.3$) and benzoate **3** ($R_f = 0.9$) can all be visualized using potassium permanganate stain prepared using 1.5 g of potassium permanganate and 10 g potassium carbonate dissolved in 200 mL water and 1.25 mL of 10 weight% of NaOH solution.

4. Conversion for all three steps can be determined by HPLC on a Imtakt Cadenza CD-C18 3 μM, 4.6 x 75 mm column using 0.05% TFA in CH_3CN:water (5:95) as solvent A and 0.05% TFA in CH_3CN:water (95:5) as solvent B. The 10 min gradient started with B = 0% and became 100% at 8 min. The wavelength of detection was 220 nm and the flow rate was 1 mL/min. Using this method, furfuryl alcohol (**1**) elutes at 3.23 min, lactol **2** elutes at 1.69 min, and benzoate **3** elutes at 6.72 min. Using this HPLC analysis, the oxidation reaction typically achieves 90–93% conversion per furfuryl alcohol.

5. The cake has a tendency to compress and form cracks during the filtration. For an effective cake wash, the solids need to be leveled with a spatula to avoid channeling. The filtration removes about 80% of the 3-chlorobenzoic acid.

6. Low dichloromethane and water contents are critical for good recovery of **2**. The main source of water contamination for this reaction is the wet *m*CPBA that can contain up to 20% water. Both dichloromethane and isopropyl alcohol can azeotrope water. In our experience, adding 450 mL isopropanol per 140 g of furfuryl alcohol has consistently provided a sufficiently dry solution, measured to be less than 1 weight % water based on Karl-Fisher titration.

7. After heptane addition, the product can oil out, but continued stirring leads to crystallization. The addition of seed crystals ensures crystallization. Seed crystals can be generated by transferring a small portion of the reaction solution to a disposable vial, scratching the glass with a metal spatula, then placing the contents under high flow of nitrogen.

8. In general, lactol **2** is the least stable of all compounds produced in this sequence. To ensure good quality **2**, the oxidation and the crystallization should be performed on the same day and the internal temperature should not be allowed to go above 30 °C during evaporation. The checkers found that the isolated yield was only slightly lower (59%) if the filtered dichloromethane solution was stored in the refrigerator overnight before concentration, crystallization and isolation of **2**.

9. Lactol **2** exhibits the following properties: ^{1}H NMR (500 MHz, CDCl$_3$) δ: 4.14 (d, J = 17.0 Hz, 1 H), 4.36 (d, J = 5.7 Hz, 1 H), 4.57 (d, J = 17.0 Hz, 1 H), 5.63 (dd, J = 5.2, 3.0 Hz, 1 H), 6.17 (d, J = 10.4 Hz, 1 H), 6.98 (dd, J = 10.4, 3.0 Hz, 1 H); ^{13}C NMR (125 MHz, CDCl$_3$) δ: 66.7, 88.3, 127.9, 146.6, 195.3. IR (film): 3308, 1663, 1624, 1278, 1090, 1008, 980, 847, 686 cm^{-1}. Anal. calcd. for C$_5$H$_6$O$_3$: C, 52.63; H, 5.30; found C, 52.66; H, 5.31. HRMS (ESI+): [M+H] calculated: 115.0390, measured: 115.0380. mp 58-60 °C. This product should be stored at or below 5 °C.

10. *tert*-Butanol can also be used as the solvent to give a similar (67–70%) yield of **3**. Once the reagents were added, no freezing of the *tert*-butanol has been observed even after holding the temperature of the reaction mixture at 0 °C.

11. Safety studies have been conducted on this transformation and have revealed that a mild exotherm occurs at the onset of the reaction. This exotherm, however, is offset by the mild endotherm associated with the dissolution of the benzoic anhydride. The internal temperature decreased to 15–16 °C following the benzoic anhydride addition and then rose to 25–26 °C within 10-15 min after the DMAP addition.

12. Using the HPLC method outlined in Note 4 for analysis, >95% conversion of **2** to **3** based on area percent has consistently been observed after 1 h reaction time.

13. Over the course of the reaction, the DMAP also decomposes the final product, turning the solution or slurry to a darker red color. If product crystals do not appear spontaneously within 30 min of the 1 h hold, seeding should be attempted to ensure high product quality. Seed crystals can be generated by the same procedure outlined in Note 7.

14. Water should only be added after the slurry was formed. Even though the slurry is more stable with water, it should still be filtered the same day.

15. Benzoate **3** exhibits the following properties: ^{1}H NMR (500 MHz, CDCl$_3$) δ: 4.29 (d, J = 17.0 Hz, 1 H), 4.61 (d, J = 17.0 Hz, 1 H), 6.34 (d, J = 10.4 Hz, 1 H), 6.75 (d, J = 3.6 Hz, 1 H), 7.06 (dd, J = 10.4, 3.6 Hz, 1 H), 7.47 (t, J = 7.8, 2 H), 7.61 (t, J = 7.5 Hz, 1 H), 8.06 (dd, J = 8.0, 1.0 Hz, 2 H); ^{13}C NMR (126 MHz, CDCl$_3$) δ: 67.70, 87.38, 128.80, 129.16, 129.21, 130.13, 134.00, 142.50, 165.22, 193.59. IR (film): 2930, 2856, 1683, 1261, 1089, 1067, 914, 704 cm^{-1}. Anal. calcd. For C$_{12}$H$_{10}$O$_4$: C, 66.05; H, 4.61; C, 66.25; H, 4.53; mp 77–79 °C; HRMS (ESI+): [M+H] calculated: 219.065, measured: 219.063. Please store this chemical at or below 5 °C.

16. Polish filtration of the reddish insolubles is critical to avoid premature degradation of the enzyme. If insolubles are still present after the first filtration, the filtration should be repeated until a transparent solution is achieved.

17. Lipases (CAS# 9001-62-1) belong to the hydrolase class of enzymes, acting on ester bonds to catalyze the hydrolysis of fats to fatty acids and glycerol. Lipase MY is a commercial lipase enzyme product obtained from *Candida rugosa* and supplied by Meito-Sangyo Company, Japan. In addition to Lipase MY, we found the reaction to be catalyzed similarly by various *Candida rugosa* lipase products, e.g. *Candida rugosa* lipases L1754 and 62316 from Sigma-Aldrich, Lipase OF from Meito-Sangyo, Lipase AY from Amano. The conditions described here are optimum for Lipase MY with an activity of 30,000 u/g. The optimum conditions are different for other *Candida rugosa* lipase products. The checkers used Lipase MY from Meito-Sangyo Company, Japan with an activity of 30,000 u/g, about 45% of the strength of the submitters batch (66,700 u/g) from the same company. With the same weight amount of lipase, the reaction time had to be approximately doubled to achieve the same outcome as the submitted protocol.

18. The optimal amount of water for this reaction was found to be between 50-100 weight% compared to the enzyme. The amount of water added to the reaction was optimized; less water led to a lower rate and higher water amounts agglomerated the enzyme in toluene, shutting down the reactivity.

19. Enantiomeric excess (ee) was determined on Chiralpak AD-3R 3 μM 4.6 x 150 mm column using 0.05% TFA in CH$_3$CN:water (5:95) as solvent A and 0.05% TFA in CH$_3$CN:water (95:5) as solvent B. The 30 min gradient started with B = 10%, became 10% at 5 min, 25% at 10 min, 50% at 15 min and 100% at 20 min. The wavelength of detection was 220 nm and

the flow rate was 1 mL/min. The desired S-enantiomer elutes at 18.3 min and the R-enantiomer elutes at 20.2 min. At 3 h, chiral HPLC typically indicates >95%ee in the crude solution when Lipase MY with 66,700 u/g is used. By the checkers, the same outcome, >95%ee, was achieved after 6 h with Lipase MY with 30,000 u/g .

20. It is critical to stop the enzyme-catalyzed transesterification reaction immediately after the ee of remaining S-enantiomer (**4**) reached the desired level (95%ee) to obtain the highest yield. Lipase MY is very selective, but not absolutely specific and transesterifies the S-enantiomer at a much lower rate than R-enantiomer. Removal of the enzyme by filtration stops this undesired reaction.

21. In all crystallization procedures examined, the enantioselectivity of product **4** is increased during crystallization. The color of the final crystals can be correlated to step 2 benzoylation; if proper seeding is performed, which results in a lighter color of **2**, the color carried through to step 3 is also less dark.

22. Benzoate **4** exhibits the same properties as benzoate **3** (Note 15). The optical rotation of the final product was determined to be $[\alpha]_D = +252$ (C = 2.0, CHCl$_3$). The product should be stored at or below 5 °C.

Working with Hazardous Chemicals

The procedures in *Organic Syntheses* are intended for use only by persons with proper training in experimental organic chemistry. All hazardous materials should be handled using the standard procedures for work with chemicals described in references such as "Prudent Practices in the Laboratory" (The National Academies Press, Washington, D.C., 2011; the full text can be accessed free of charge at http://www.nap.edu/catalog.php?record_id=12654). All chemical waste should be disposed of in accordance with local regulations. For general guidelines for the management of chemical waste, see Chapter 8 of Prudent Practices.

In some articles in *Organic Syntheses*, chemical-specific hazards are highlighted in red "Caution Notes" within a procedure. It is important to recognize that the absence of a caution note does not imply that no significant hazards are associated with the chemicals involved in that procedure. Prior to performing a reaction, a thorough risk assessment should be carried out that includes a review of the potential hazards associated with each chemical and experimental operation on the scale that is planned for the procedure. Guidelines for carrying out a risk assessment and for analyzing the hazards associated with chemicals can be found in Chapter 4 of Prudent Practices.

The procedures described in *Organic Syntheses* are provided as published and are conducted at one's own risk. *Organic Syntheses, Inc.*, its Editors, and its Board of Directors do not warrant or guarantee the safety of individuals using these procedures and hereby disclaim any liability for any injuries or damages claimed to have resulted from or related in any way to the procedures herein.

Discussion

Derivatives of lactol **2** contain a rich array of functionalities for elaboration in a diastereoselective fashion,[2,3] and have been shown to be convenient starting materials for natural product synthesis.[4,5] The methods described herein illustrate a chromatography-free synthesis of crystalline benzoate **4** in high enantioselectivity using commercially available chemicals.

Despite the rapid accessibility of **2** from furfurol via the Achmatowitz rearrangement, chiral derivatives such as **4** have been difficult to prepare, especially the *S*-enantiomeric derivatives of **2**.[6] Sugawara and co-workers reported a preparation of non-crystalline TBS ether **5**, but it required over five synthetic operations (Scheme 1).[4]

The Feringa group prepared the analogous acetate of **2**;[2,7] unfortunately, the acetate product required chromatography to isolate, and only in a highly pure form does it appear to be a low-melting solid.[8] The more crystalline benzoate **4** was also prepared by Feringa and co-workers, but racemic **3** had to be separated via chiral chromatography.[2]

Scheme 1. Sugawara synthesis of (S)-5

Highlights of our three-step synthetic sequence include isolation of intermediate **2** via crystallization as a white solid, enabling the storage of this otherwise unstable material at 0 °C for extended periods of time. Benzoylation of lactol **2** in a tertiary alcohol solvent[9] allows not only the desired reaction to out-compete the background decomposition, but product **3** also precipitates out of the reaction mixture. Finally, enzymatic transesterification of **3** using a low loading of a commercially available enzyme allows the desired S-enantiomer to be accessed without the need for chiral separation. In addition, the enantioselectivity of **4** is further upgraded during the crystallization, ensuring that the desired product is isolated in high quality.

References

1. Chemical Development, Bristol-Myers Squibb Co., 1 Squibb Drive, New Brunswick, NJ 08903, tamas.benkovics@bms.com.
2. Comely, A. C.; Eelkema, R.; Minnaard, A. J.; Feringa, B. L. *J. Am. Chem. Soc.* **2003**, *125*, 8714–8715.
3. Achmatowitz, O. Grynkiewitz, G. *Carb. Res.* **1977**, *54*, 193–198.
4. Sugawara, K.; Imanishi, Y.; Hashiyama, T. *Tetrahedron: Asymmetry* **2000**, *11*, 4529–4535.

5. For selected examples, see (a) Kolb, H. C.; Hoffmann, H. M. R. *Tetrahedron: Asymm.* **1990**, *1*, 237–250; (b) Fürstner, A.; Feyen, F.; Prinz, H.; Waldmann, H. *Angew. Chem. Int. Ed.* **2003**, *42*, 5361–5364; (c) Sugawara, K. Imanishi, Y.; Hashiyama, T. *Heterocycles* **2007**, 597–607; (d) Jones, R. A.; Krische, M. J. *Org. Lett.* **2009**, *11*, 1849–1851.

6. Treatment of **3** and **4** with either acid or base conditions results in the formation of the oxopyrilium species which can then initiate polymerization of the substrate or 5+2 with a suitable dipolarophile. In addition, lactol **2** undergoes rapid polymerization in presence of base.

7. van den Heuvel, M.; Cuiper, A. D.; van der Deen, H.; Kellogg, R. M.; Ferigna, B. L. *Tetrahedron Lett.* **1997**, 1655–1658.

8. Our attempts to upgrade the enantioselectivity of the acetate via crystallization were not successful.

9. Fu, G. C. *Acc. Chem. Res.* **2004**, *37*, 542–547.

Appendix
Chemical Abstracts Nomenclature (Registry Number)

Furfurol; (98-00-0)
meta-Chloroperbenzoic acid; (937-14-4)
Benzoic anhydride; (93-97-0)
4-Dimethylamino pyridine; (1122-58-3)
Lipases; (9001-62-1)

Tamas Benkovics, a native of Hungary, obtained his B.S. in chemistry from Colorado State University in 2003. After spending two years with the process development group of Amgen in Thousand Oaks, CA, he joined the research group of Professor Tehshik P. Yoon at the University of Wisconsin–Madison. During graduate school, his research was focused on oxaziridine-mediated functionalizations of hydrocarbons. After receiving his Ph.D. in 2010, he joined the process research and development group of Bristol-Myers Squibb.

Adrian Ortiz received his B.S. in chemistry from the University of Arizona in 2004 where he performed research under the guidance of Professor Dominic V. McGrath. He then joined the group of Professor K.C. Nicolaou at the Scripps Research Institute in San Diego, CA where he studied the total synthesis of natural products. Upon completion of his Ph.D. in 2009, he joined the process research and development group at Bristol-Myers Squibb.

Zhiwei Guo obtained his B.S. in chemistry from Beijing University and his Ph.D. in medicinal Chemistry from West China University of Medical Sciences. Following his graduate studies, he has conducted post-doctoral studies at the University of Wisconsin–Madison under the guidance of Professor Charles J. Sih, and at Bristol-Myers Squibb with Ramesh N. Patel. Zhiwei spent the majority of his industrial career at Bristol-Myers Squibb, where he made key contributions using his skills in biocatalysis and knowledge of organic chemistry until his retirement in 2013.

Dr. Animesh Goswami earned his Ph.D. from the University of North Bengal in India and has conducted post-doctoral studies at Arizona State University and University of Iowa in USA. Dr. Goswami has 26 years of industrial experience in the application of biocatalysis for the synthesis of organic compounds. He has joined Bristol-Myers Squibb in 1998, and is currently a research fellow leading the Biocatalysis group of Chemical Development department in Research and Development. Before Bristol-Myers Squibb, he worked in Rhone-Poulenc in USA for 11 years. He is co-author of 45 publications and 15 patents and patent applications.

Dr. Prashant P. Deshpande received his Ph.D. from the State University of New York in 1991 under the supervision of Prof. Olivier R. Martin. After completion of postdoctoral fellowship at the University of Tennessee with Prof. D. C. Baker and at Memorial Sloan Kettering Cancer Center with Prof. S. J. Danishefsky, Prashant joined Bristol Myers Squibb Co. in 1997. Since then he is working as the multi-disciplinary pharmaceutical development & chemical development team leader with 16 years of industrial experience in advancing development of drug candidates from early phase to commercialization. He is co-author of 18 publications and 9 patents and patent applications.

Dr. Magnus Eriksson was born in Stockholm, Sweden. He received his undergraduate degree in Chemical Engineering and his Ph.D. in Organic Chemistry from Chalmers University of Technology in Gothenburg in 1995 under the guidance of Professor Martin Nilsson working on copper-promoted 1,4-additions to carbonyl compounds. After post-doctoral work at Boehringer Ingelheim Pharmaceuticals and at MIT with Professor Stephen Buchwald, he joined Boehringer Ingelheim Pharmaceuticals in 2000 where he is currently a Principal Scientist. His research interests include Process Research, catalytic transformations and synthetic methodology.

Mr. Suresh Kapadia was born in Mumbai, India. He received his B.Sc. in 1972 and M.Sc. in 1974 in organic chemistry from University of Mumbai. He worked as a chemist at New England Nuclear Corporation in Boston from 1977 to 1979. He joined Boehringer Ingelheim Pharmaceuticals in 1979 as Scientist II in medicinal chemistry department. He is currently a senior scientist in chemical development. His research interests include Process Research, Process Development and synthetic methodology.

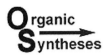

Iron-Catalyzed Selective Conjugate Addition of Aryl Grignard Reagents to 2,4-Alkadienoates: *tert*-Butyl (*Z*)-5-Phenyl-3-hexenoate

Takeshi Hata, Hideyuki Goto, Tomofumi Yokomizo, and Hirokazu Urabe[1]*

Department of Biomolecular Engineering, Graduate School of Bioscience and Biotechnology, Tokyo Institute of Technology, 4259-B-59 Nagatsuta-cho, Midori-ku, Yokohama, Kanagawa 226-8501, Japan

Checked by Liangbing Fu and Huw M. L. Davies

Procedure

A. *tert-Butyl (2E,4E)-2,4-hexadienoate* (**1**). An oven-dried, 500-mL, single-necked, round-bottomed flask equipped with an egg-shaped magnetic stirring bar (3 cm x 1.5 cm) is charged with sorbic acid (18.5 g, 165 mmol, 1.00 equiv) (Note 1). After being fitted with a rubber septum and evacuated by a needle connected to the manifold inserted through the septum, the flask is flushed with argon. The flask is then charged with dry dichloromethane (150 mL) (Note 2) by syringe through the septum, oxalyl chloride (28.3 mL, 330 mmol, 2.00 equiv) (Note 3) is added dropwise by syringe pump through the septum over 10 min at room temperature. The septum is removed and three drops of DMF (Note 4) are added using a pipette. A drying tube filled with calcium chloride is attached. After the evolution of gas ceases (about 30 min) and the resulting solution is stirred at room temperature for an additional 2 h, the solvent and excess oxalyl chloride are removed first on a rotary evaporator (in a water bath maintained at 40 °C) (Note 5) and finally with a vacuum pump (3.6 mmHg at room temperature for 5 min) to afford the desired sorbic acid chloride (Note 6) as a crude oil, which is directly used in the next step.

A separate, oven-dried, 1-L, three-necked, round-bottomed flask equipped with an octagonal magnetic stirring bar (4 cm x 1 cm) is fitted with an internal thermometer and two rubber septa. After the flask is evacuated by a needle connected to the manifold inserted through the side septum, the flask is flushed with argon. This operation is repeated twice. A solution of *tert*-butyl alcohol (12.2 g, 165 mmol, 1.00 equiv) (Note 7) in dry tetrahydrofuran (150 mL) (Note 8) is added through the septum via syringe. The septum in the middle neck is quickly removed and exchanged for a 200-mL pressure-equalizing addition funnel, the top of which is capped with a septum, and the flask is then cooled in an ice bath. *n*-Butyllithium (114 mL, 1.46 M in hexane, 165 mmol, 1.00 equiv) (Note 9) is transferred to the pressure-equalizing funnel by syringe and then is added to the reaction flask dropwise in *ca.* 5 min at the same temperature. After the ice-bath is removed and the mixture is stirred at room temperature for 1 h, it is cooled again in an ice bath and the crude sorbic acid chloride prepared above in dry tetrahydrofuran (100 mL) (Note 8) is added through the septum in one portion via syringe. After the mixture is stirred at room temperature for 1 h, the reaction is terminated by the addition of H_2O (300 mL) slowly in one portion while stirring at room temperature. The contents of the flask are transferred to a 1 L seperatory funnel. The organic layer is separated and the aqueous layer is extracted with ethyl acetate (once with 150 mL and twice with 100 mL each). The combined organic layers are washed with brine (100 mL), dried over Na_2SO_4 (40 g), filtered, and concentrated by rotary evaporation to give a crude oil, which is chromatographed on silica gel (Note 10) to afford the isomerically pure *tert*-butyl (2E,4E)-2,4-hexadienoate (**1**) (26.4–26.7 g, 95–96% overall yield from sorbic acid) as a colorless oil (Notes 11 and 12).

B. *tert-Butyl (Z)-5-phenyl-3-hexenoate (2)*. An oven-dried 500-mL, three-necked, round-bottomed flask is equipped with an egg-shaped magnetic stirring bar (3 cm x 1.5 cm) and a 200 mL pressure-equalizing addition funnel, the top of which is fitted with a rubber septum. The other necks of the flask are fitted with an internal thermometer and a rubber septum. The flask is charged with iron(II) chloride (1.01 g, 8.0 mmol, 0.1 equiv) (Note 13). A needle, which is connected to a manifold, is inserted through the septum, the flask is evacuated, and then argon is flushed through the flask. This evacuation and flush process is repeated twice. After *tert*-butyl (2*E*,4*E*)-2,4-hexadienoate (**1**) (13.5 g, 80.0 mmol) in dry tetrahydrofuran (80 mL) (Note 8) is added through the septum all at once via syringe at room temperature, the resulting suspension is stirred and cooled in a bath maintained at *ca.* –45 °C (Note 14). Phenylmagnesium bromide (144 mL, 1.0 M solution in THF, 144 mmol, 1.80 equiv) (Note 15) is added dropwise to the cold suspension through the pressure-equalizing addition funnel over 90 min. A cooling bath is used to maintain the mixture between –45 and –30 °C (Note 14) for 4 h, after which time the reaction is terminated at the same temperature by the addition of 1 M HCl solution (150 mL) by means of an addition funnel over *ca.* 10 min. The organic layer is separated and the aqueous layer is extracted with ethyl acetate (3 x 100 mL). The combined organic layers are washed successively with aqueous saturated NaHCO$_3$ solution (150 mL) and brine (100 mL), dried over Na$_2$SO$_4$ (60 g), filtered, and concentrated by rotary evaporation under reduced pressure at 40 °C to give a crude oil. Analysis by ^{1}H NMR does not reveal the presence of regio- or alkene stereoisomers. The crude product is chromatographed on silica gel (Note 16) to afford *tert*-butyl (Z)-5-phenyl-3-hexenoate (**2**) (16.5–16.6 g, 84%) as a colorless oil (Notes 17 and 18).

Notes

1. Sorbic acid was purchased from Sigma-Aldrich Co. (USA), and was observed to be isomerically pure and used as received.
2. Anhydrous dichloromethane was purchased from Sigma-Aldrich Co. (USA) and used as received.
3. Oxalyl chloride was purchased from Sigma-Aldrich Co. and used without purification.
4. Anhydrous DMF was purchased from Sigma-Aldrich Co. (USA) and used as received.
5. The stirring bar must be removed from the flask with a teflon-coated magnet before this operation.

Org. Synth. **2014**, *91*, 307-321   **309**   DOI: 10.15227/orgsyn.091.0307

6. As the boiling point of sorbic acid chloride is 74 °C/14 mmHg (Ongoka, P.; Mauze, B.; Miginic, L. *J. Organomet. Chem.* **1987**, *322*, 131–139), exhaustive evaporation may result in the decrease of product quantity.

7. *tert*-Butyl alcohol was purchased from Sigma-Aldrich Co. (USA) and used without purification.

8. Anhydrous tetrahydrofuran was purchased from Sigma-Aldrich Co. (USA) and used as received.

9. A 1.6 M solution of *n*-BuLi in hexane was purchased from Sigma-Aldrich Co. (USA) and the concentration was determined by titration to be 1.46 M prior to use according to a reported method: Ireland, R. E.; Meissner, R. S. *J. Org. Chem.* **1991**, *56*, 4566–4568.

10. The crude oil is charged onto a column of 5.5-cm diameter, which is packed with 100 g of silica gel (P60 from SiliCycle Inc., 230-400 mesh (40-63 μm)). Fractions were collected with Fisherbrand disposable culture tubes (Borosilicate Glass 16 x 50 mm). Approximately 1300 mL of the eluent (hexane/ethyl acetate: initially 100:0 (100 mL) and then 99:1 (1200 mL)) is required.

11. The product (**1**) has the following physicochemical properties: $R_f = 0.70$ (hexane/ethyl acetate = 90:10, TLC: Silica gel 60 F254 obtained from Merck and visualized with 254 nm UV lamp); ^{1}H NMR (400 MHz, CDCl$_3$) δ: 1.47 (s, 9 H), 1.83 (d, $J = 6.2$ Hz, 3 H), 5.69 (d, $J = 15.2$ Hz, 1 H), 6.07 (dq, $J = 15.2, 6.2$ Hz, 1 H), 6.16 (dd, $J = 15.2, 10.4$ Hz, 1 H), 7.14 (dd, $J = 15.2, 10.4$ Hz, 1 H). The *E,E*-diene stereochemistry is confirmed by ^{1}H NMR coupling constants. ^{13}C NMR (100 MHz, CDCl$_3$) δ: 18.6, 28.2, 80.0, 120.9, 129.8, 138.5, 143.9, 166.7; IR (neat) 2962, 2932, 1708 (C=O), 1646, 1617, 1455, 1367, 1330, 1280, 1249, 1170, 1135, 999 cm^{-1}; HRMS (FTMS + p NSI) [M + H] calcd for C$_{10}$H$_{17}$O$_2$: 169.1223. Found: 169.1221. Anal. Calcd for C$_{10}$H$_{16}$O$_2$: C, 71.39; H, 9.59. Found: C, 71.05; H, 9.55.

12. The checkers found the product to be >95% pure by GC (HP 5890 SERIES II, column: HP-5; 5% phenyl methyl siloxane; 30 m x 320 μm x 0.25 μm); Injection temperature: 150 °C; Oven program: Starting temperature, 50 °C for 5 min; heating to 250 °C by a rate of 10 °C per min; 250 °C for 10 min. Retention time of the product was 11.42 min and gave a single peak.

13. Iron(II) chloride (anhydrous, beads, ~10 mesh, 99.99% purity) was purchased from Sigma-Aldrich Co. (USA) and used as received.

14. The temperature of the acetone bath is maintained by a cryogenic reactor.

15. A 1.0 M solution of PhMgBr in THF was purchased from Sigma-Aldrich Co. (USA) and used as received.

16. Column chromatograph is carried out on a column of 5.5-cm diameter and packed with 225 g of silica gel (P60 from SiliCycle Inc., 230-400 mesh (40-63 μm)). Fractions were collected with Fisher-brand disposable culture tubes (Borosilicate Glass 16 x 50 mm). Approximately 2.0 L of the eluent (hexane/ethyl acetate: initially 100:0 (500 mL), 99:1 (500 mL), 98:2 (500 mL), and finally 96:4 (500 mL)) is used.

17. The product (2) has the following physicochemical properties: R_f = 0.62 (hexane/ethyl acetate = 90:10, TLC: Silica gel 60 F254 obtained from Merck and visualized with 254 nm UV lamp); ^{1}H NMR (400 MHz, CDCl$_3$) δ: 1.33 (d, J = 6.8 Hz, 3 H), 1.42 (s, 9 H), 3.00 (ddd, J = 16.8, 7.2, 1.6 Hz, 1 H), 3.10 (ddd, J = 16.8, 7.2, 1.6 Hz, 1 H), 3.71 (dq, J = 9.2, 6.8 Hz, 1 H), 5.55 (dt, J = 10.8, 7.2 Hz, 1 H), 5.68 (b dd, J = 10.8, 9.2 Hz, 1 H), 7.15-7.30 (m, 5 H); The Z-stereochemistry is confirmed by the ^{1}H NMR coupling constants and listed above (Z:E = >99:<1) and also by the ^{1}H NMR NOESY experiments showing the correlation between the peaks at δ: 5.55 and 5.68 ppm. ^{13}C NMR (100 MHz, CDCl$_3$) δ: 22.0, 28.1, 34.5, 37.5, 80.6, 120.5, 126.0, 126.9, 128.5, 137.7, 145.8, 171.0. IR (neat) 2974, 2930, 1729 (C=O), 1601 (C=C), 1452, 1392, 1367, 1328, 1256, 1144, 950, 843, 698 cm^{-1}; HRMS (FTMS + p NSI) [M + Na] calcd for C$_{16}$H$_{22}$O$_2$Na: 269.1512. Found: 269.1507. Anal. Calcd for C$_{16}$H$_{22}$O$_2$: C, 78.01; H, 9.00. Found: C, 78.09; H, 9.09.

18. The checkers found the product to be >95% pure by GC (HP 5890 SERIES II, column: HP-5; 5% phenyl methyl siloxane; 30 m x 320 μm x 0.25 μm); Injection temperature: 150 °C; Oven program: Starting temperature, 50 °C for 5 min; heating to 250 °C by a rate of 10 °C per min; 250 °C for 10 min. Retention time of the product was 17.58 min and gave a single peak.

Working with Hazardous Chemicals

The procedures in *Organic Syntheses* are intended for use only by persons with proper training in experimental organic chemistry. All hazardous materials should be handled using the standard procedures for work with chemicals described in references such as "Prudent Practices in the Laboratory" (The National Academies Press, Washington, D.C., 2011; the full text can be accessed free of charge at http://www.nap.edu/catalog.php?record_id=12654). All chemical waste should be disposed of in accordance with local regulations. For general guidelines for the management of chemical waste, see Chapter 8 of Prudent Practices.

In some articles in *Organic Syntheses*, chemical-specific hazards are highlighted in red "Caution Notes" within a procedure. It is important to recognize that the absence of a caution note does not imply that no significant hazards are associated with the chemicals involved in that procedure. Prior to performing a reaction, a thorough risk assessment should be carried out that includes a review of the potential hazards associated with each chemical and experimental operation on the scale that is planned for the procedure. Guidelines for carrying out a risk assessment and for analyzing the hazards associated with chemicals can be found in Chapter 4 of Prudent Practices.

The procedures described in *Organic Syntheses* are provided as published and are conducted at one's own risk. *Organic Syntheses, Inc.*, its Editors, and its Board of Directors do not warrant or guarantee the safety of individuals using these procedures and hereby disclaim any liability for any injuries or damages claimed to have resulted from or related in any way to the procedures herein.

Discussion

The transition metal-catalyzed conjugate addition of organometallic reagents to electron-deficient olefins is one of the most versatile methods for selective C-C bond formation. Nonetheless, the selective addition to $\alpha,\beta,\gamma,\delta$-unsaturated carbonyl compounds has not been amply solved, because they have multiple reaction sites (e.g., 1,2-, 1,4-, and 1,6-addition) and there are additional issues on the regio- and stereoselectivities of the remaining olefinic bond as shown in Scheme 1.[2]

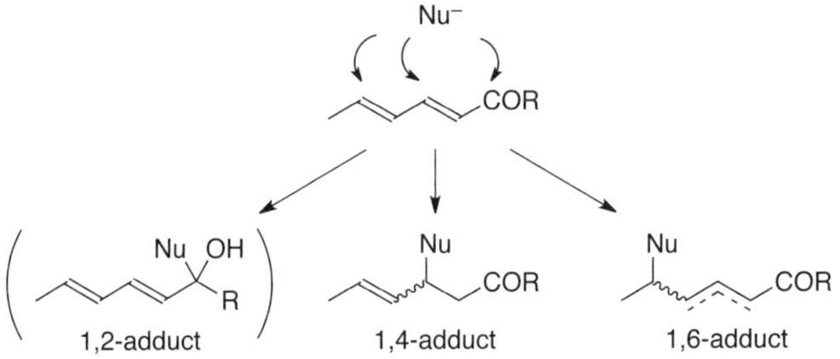

Scheme 1. Conjugate addition to α,β,γ,δ-unsaturated carbonyl compounds

Although copper salts have been mainly used as the catalysts in these reactions,[3,4] we reported a notable role of an iron catalyst in the 1,6-selective addition of aryl Grignard reagents to 2,4-dienoates or dienamides **3**, giving stereo-defined *cis*-4-aryl-2-alkenoates or -amides **6** (Scheme 2).[5] We proposed that the reaction involves the intermediary formation of the *s-cis*-diene-iron complex **4**, which effects the aryl transfer from iron to the

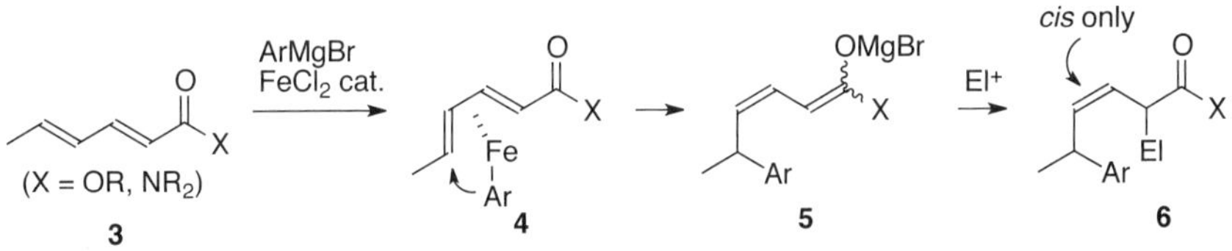

Scheme 2. Iron-catalyzed 1,6-addition of aryl Grignard reagents to α,β,γ,δ-unsaturated carbonyl compounds

terminal position of the dienoate to give the observed *cis*-product after hydrolysis (El⁺ = H⁺). The magnesium dienolate **5**, most likely generated *in situ*, could be also used for the reactions with other electrophiles (El⁺). Table 1 summarizes the products prepared from various 2,4-dienoates, aryl Grignard reagents, and electrophiles.[5a] In all cases, the corresponding 1,6-addition products are obtained virtually as a single isomer. The *cis* stereoselectivity regarding the ester group and the incoming aryl group is always exclusively high.

Table 1. Preparation of *cis*-5-aryl-3-alkenoates via iron-catalyzed selective 1,6-addition

entry	substrate	Ar	El+	El	product	isolated yield (%)
1		Ph	H+	H		78
2		Ph	D+	D		>98% d[a]
3		Ph	MeI	Me		70[b]
4		2-MeC$_6$H$_4$-	H+	H		65
5		Ph	H+	H		84
6		Ph	H+	H		84
7		Ph	H+	H		86

[a]Diastereoselectivity was 58:42. [b]Diastereoselectivity was 61:39.

A recent report claimed that the iron-catalyzed reactions might be actually catalyzed by copper impurities involved in the iron reagents.[6] We have confirmed that the present 1,6-addition was actually catalyzed by iron on the following basis: (i) Several representative copper catalysts did not furnish the desired products at our hands;[5a] (ii) Other groups reported that copper-catalyzed 1,6-additions gave the products exclusively with *trans*-olefin;[3a,7] (iii) The use of 99.998%-pure iron chloride, rather than the routinely used iron chloride (99.9% purity), did not alter the above outcome, giving virtually the same yield and regio- and stereoselectivities of the 1,6-adduct.[8]

In addition to esters, 2,4-dienamides are also good substrates for this iron-catalyzed 1,6-addition, the results of which are shown in Table 2.[5a,b]

Table 2. Iron-catalyzed 1,6-addition to 2,4-dienamides

ArMgBr (1.8~2.5 equiv)
FeCl$_2$ (10 mol%)

El+ (2.5 equiv)

THF, −45 ~ −35 °C

(1 equiv)

entry	substrate	Ar	El+	El	product	isolated yield (%)	d.s.[a]
1		Ph	H+	H		72	—
2		4-(MeO)C$_6$H$_4$-	H+	H		73	—
3		Ph	H+	H		79	—
4		Ph	H+	H		86	—
5		Ph	H+	H		73	95:5
6		Ph	MeI	Me		67	95:5
7		Ph	Br		58	95:5
8		Ph	Br		55	94:6
9		Ph	C$_6$H$_{13}$I	C$_6$H$_{13}$		69	95:5
10		4-(MeO)C$_6$H$_4$-	C$_6$H$_{13}$I	C$_6$H$_{13}$		63	96:4
11		3-(MeO)C$_6$H$_4$-	C$_6$H$_{13}$I	C$_6$H$_{13}$		70	94:6
12		Ph	C$_6$H$_{13}$I	C$_6$H$_{13}$		71	96:4
13		Ph	MeI	Me		68	97:3

[a]For entries 6-13, d.s. refers to the ratio of two major diastereoisomers. Two other minor isomers, which were formed in trace amounts and could not be isolated nor characterized, are omitted.

When the reaction was started with optically active amides derived from (2S,5S)-2,5-diphenylpyrrolidine, both the iron-catalyzed 1,6-addition and the subsequent alkylation of the intermediate enolate took place in a highly diastereoselective manner to afford the corresponding optically active three component-coupling products (Table 2, entries 6-13).[5b] A proposed stereochemical course giving the observed products **8** from **7** is shown in Scheme 3, where the *s-cis*-diene-iron intermediate appears to play an important role to control the stereochemistry of the first aryl addition at the remote position from the chiral auxiliary.

Scheme 3. Proposed stereochemical course of the reaction

The nature of iron to effect the selective 1,6-addition is also found in the conjugate addition to 2-en-4-ynoates and -amides.[5c] The results are summarized in Table 3, where methyl as well as aryl Grignard reagents took part in the reaction to give exclusively 3,4-dienoates or -amides after hydrolysis.

In summary, the new iron-catalyzed 1,6-conjugate addition of aryl Grignard reagents to 2,4-dienoates and -amides proceeded in a regio- and stereoselective manner to give virtually single *cis*-olefinic adducts. As iron is an inexpensive, non-toxic, and ubiquitous metal, this process fulfills the recent demand for an environmentally friendly process.[9]

Table 3. Iron-catalyzed 1,6-addition to 2-en-4-ynoates and -amides

entry	substrate	R²	product	isolated yield (%)
1	H₁₃C₆—≡≡—CO₂Bu-t	Me	H₁₃C₆, Me, CO₂Bu-t	84
		Ar	H₁₃C₆, Ar, CO₂Bu-t	
2		Ar = Ph		60
3		4-MeC₆H₄-		48
4		2-MeC₆H₄-		80
5		2-(MeO)C₆H₄-		74
6	(t-Bu)—≡≡—CO₂Bu-t	Me	(t-Bu), Me, CO₂Bu-t	68
7	H₁₃C₆—≡≡—CO₂Et	Me	H₁₃C₆, Me, CO₂Et	72
8		Ph	H₁₃C₆, Ph, CO₂Et	66
	H₁₃C₆—≡≡—CON(R)oxazolidinone	Me	H₁₃C₆, Me, CON(R)oxazolidinone	
9	R = H		R = H	75
10	Ph		Ph	74[a]
11[b]	Bn		Bn	85[a]

[a]Diastereoselectivity was 67:33. Absolute stereochemistry of the allene part has not been determined. [b]This reaction was performed at −78 °C for 3 h.

References

1. Department of Biomolecular Engineering, Graduate School of Bioscience and Biotechnology, Tokyo Institute of Technology, 4259-B-59 Nagatsuta-cho, Midori-ku, Yokohama, Kanagawa 226-8501, Japan. E-mail: hurabe@bio.titech.ac.jp

2. For reviews on conjugate additions to electron-deficient dienes, see: (a) Csákÿ, A. G.; De la Herrán, G.; Murcia, M. C. *Chem. Soc. Rev.* **2010**, *39*, 4080–4102. (b) Silva, E. M. P.; Silva, A. M. S. *Synthesis* **2012**, *44*, 3109–3128.

3. For recent examples of the copper-catalyzed conjugate addition to α,β,γ,δ-unsaturated carbonyl compounds, see: (a) den Hartog, T.; Harutyunyan, S. R.; Font, D.; Minnaard, A. J.; Feringa, B. L. *Angew. Chem., Int. Ed.* **2008**, *47*, 398–401. (b) Hénon, H.; Mauduit, M.; Alexakis, A. *Angew. Chem., Int. Ed.* **2008**, *47*, 9122–9124. (c) Lee, K.; Hoveyda, A. H. *J. Am. Chem. Soc.* **2010**, *132*, 2898–2900. (d) Wencel-Delord, J.; Alexakis, A.; Crévisy, C.; Mauduit, M. *Org. Lett.* **2010**, *12*, 4335–4337. (e) Tissot, M.; Hernández, A. P.; Müller, D.; Mauduit, M.; Alexakis, A. *Org. Lett.* **2011**, *13*, 1524–1527.

4. For recent examples of the selective conjugate addition to α,β,γ,δ-unsaturated carbonyl compounds with catalysts other than copper, see: Under Rh or Ir catalysis: (a) De la Herrán, G.; Csákÿ, A. G. *Synlett* **2009**, 585–588. (b) Nishimura, T.; Yasuhara, Y.; Sawano, T.; Hayashi, T. *J. Am. Chem. Soc.* **2010**, *132*, 7872–7873. With organocatalyst: (c) Oliva, C. G.; Silva, A. M. S.; Paz, F. A. A.; Cavaleiro, J. A. S. *Synlett* **2010**, 1123–1127. Without catalyst: (d) Ocejo, M.; Carrillo, L.; Badía, D.; Vicario, J. L.; Fernández, N.; Reyes, E. *J. Org. Chem.* **2009**, *74*, 4404–4407.

5. (a) Fukuhara, K.; Urabe, H. *Tetrahedron Lett.* **2005**, *46*, 603–606. (b) Okada, S.; Arayama, K.; Murayama, R.; Ishizuka, T.; Hara, K.; Hirone, N.; Hata, T.; Urabe, H. *Angew. Chem., Int. Ed.* **2008**, *47*, 6860–6864. (c) Hata, T.; Iwata, S.; Seto, S.; Urabe, H. *Adv. Synth. Catal.* **2012**, *354*, 1885–1889.

6. Buchwald, S. L.; Bolm, C. *Angew. Chem. Int. Ed.* **2009**, *48*, 5586–5587.

7. (a) Ganem, B. *Tetrahedron Lett.* **1974**, 4467–4470. (b) Corey, E. J.; Kim, C. U.; Chen, R. H. K.; Takeda, M. *J. Am. Chem. Soc.* **1972**, *94*, 4395–4396. (c) Corey, E. J.; Chen, R. H. K. *Tetrahedron Lett.* **1973**, 1611–1614. (d) Corey, E. J.; Boaz, N. W. *Tetrahedron Lett.* **1985**, *26*, 6019–6022.

8. Iron(II) chloride (99.998% pure beads (~10 mesh) purchased from Sigma-Aldrich Co., USA.) afforded the product of the same quality in 82% yield.

9. For reviews on iron-catalyzed and mediated organic reactions, see: (a) Plietker, B., Ed. *Iron Catalysis in Organic Chemistry*; Wiley-VCH: Weinheim, 2008. (b) Correa, A.; Mancheño, O. G.; Bolm, C. *Chem. Soc. Rev.* **2008**, *37*, 1108–1117. (c) Bauer, E. B. *Curr. Org. Chem.* **2008**, *12*, 1341–1369. (d) Enthaler, S.; Junge, K.; Beller, M. *Angew. Chem., Int. Ed.* **2008**, *47*, 3317–3321. (e) Bolm, C.; Legros, J.; Le Paih, J.; Zani, L. *Chem. Rev.* **2004**, *104*, 6217–6254. For enyne cyclization, see: (f) Michelet, V.; Toullec, P. Y.; Genêt, J.-P. *Angew. Chem., Int. Ed.* **2008**, *47*, 4268–4315. For coupling reactions, see: (g) Plietker, B. *Synlett* **2010**, 2049–2058. (h) Plietker, B.; Dieskau, A. *Eur. J. Org. Chem.* **2009**, 775–787. (i) Czaplik, W. M.; Mayer, M.; Cvengroš, J.; Jacobi von Wangelin, A. *ChemSusChem* **2009**, *2*, 396–417. (j) Sherry, B. D.; Fürstner, A. *Acc. Chem. Res.* **2008**, *41*, 1500–1511. (k) Fürstner, A.; Martin, R. *Chem. Lett.* **2005**, *34*, 624–629. For iron carbonyl complexes, see: (l) Semmelhack, M. F. In *Organometallics in Organic Synthesis. A Manual*, 2nd ed.; Schlosser, M., Ed.; John Wiley & Sons: Chichester, 2002; pp 1006–1121. (m) Wilkinson, G., Stone, F. G. A., Abel, E. W., Eds. *Comprehensive Organometallic Chemistry*; Pergamon Press: Oxford, 1982; Vol. 4, pp 243–649.

Appendix
Chemical Abstracts Nomenclature (Registry Number)

Sorbic acid; (110-44-1)
Oxalyl chloride; (79-37-8)
Sorbic acid chloride; (2614-88-2)
tert-Butyl alcohol; (75-65-0)
n-Butyllithium; (109-72-8)
tert-Butyl (2E,4E)-2,4-hexadienoate; (81838-85-9)
Iron(II) chloride: Ferrous chloride; (7758-94-3)
Phenylmagnesium bromide; (100-58-3)

Hirokazu Urabe was born in 1958 in Kanagawa, Japan. He received his Ph.D. under the supervision of Prof. Isao Kuwajima at Tokyo Institute of Technology in 1985. After he became Assistant Professor at the same institute in 1986, he was a postdoctoral fellow in Prof. B. M. Trost's group at Stanford University during 1988-1990. After he returned to the Department of Biomolecular Engineering, Tokyo Institute of Technology, he was promoted to Associate Professor (2000) and to Professor (2004). His research focuses on the development of new synthetic methods and their application to the synthesis of medicinally important organic compounds and naturally occurring products.

Takeshi Hata was born in 1972 in Kanagawa, Japan. He received his Ph.D. in 2000 from Kyoto University under the supervision of Prof. Tamejiro Hiyama. After he worked as a medicinal chemist at Mitsubishi Pharma Corporation from 2000 to 2005, he joined the faculty at Tokyo Institute of Technology in 2005 as Assistant Professor and was promoted to Associate Professor in 2013. He is interested in development of new methodologies for synthetic reactions and synthesis of biologically important natural products.

Hideyuki Goto was born in Aichi, Japan in 1989. He received his B.S. from Tokyo Institute of Technology in 2011, under the guidance of Prof. Hirokazu Urabe. He is a graduate student in Prof. Hirokazu Urabe's group at the same institute and his research interests lie in natural product synthesis.

Tomofumi Yokomizo was born in 1987 in Tokyo, Japan. After he received his B.S. in 2010 and M.S. in 2012 from Tokyo Institute of Technology under the guidance of Prof. Hirokazu Urabe, he is a researcher at ADEKA Corporation, Japan.

Liangbing Fu was born in Hubei, China. He earned his B.Sc. in Pharmacy from Tongji Medical College, Huazhong University of Science and Technology in 2008. He moved to Guangzhou Institutes of Biomedicine and Health, Chinese Academy of Sciences and obtained his M.Sc. in Medicinal Chemistry under the guidance of Professor Ke Ding in 2011. In the same year he started his Ph.D. studies at Emory University under the mentorship of Professor Huw M. L. Davies. His current research mainly focuses on the development of novel donor-acceptor carbenoid transformations.

3-Hydroxymethyl-3-phenylcyclopropene

Ramajeyam Selvaraj, Srinivasa R. Chintala, Michael T. Taylor and Joseph M. Fox*[1]

Department of Chemistry and Biochemistry, University of Delaware, Newark DE 19716

Checked by Jonathan Ruchti and Erick M. Carreira

Procedure

A. *Methyl phenyl diazoacetate (1).* A 1-L, one-necked round-bottomed flask is equipped with a Teflon-coated magnetic stir bar (4 cm long, 1.6 cm

Published on the Web 10/13/2014
© 2014 Organic Syntheses, Inc.

diameter) and a septum. The apparatus is evacuated (0.5 mmHg) and flame-dried, after which the flask is filled with nitrogen and allowed to cool to room temperature. The septum is removed and the flask is charged with acetonitrile (200 mL) (Note 1). 4-Acetamidobenzenesulfonyl azide (38.9 g, 162 mmol, 1.23 equiv) (Note 2) is added as a solid in one portion. The septum is replaced, a needle connected to a nitrogen source is inserted through the septum, and methyl phenylacetate (18.6 mL, 19.9 g, 132 mmol, 1.00 equiv) (Note 3) is added through the septum via syringe over 1 min. The reaction mixture is cooled in an ice bath and then 1,8-diazabicyclo[5.4.0]undec-7-ene (23.8 mL, 24.0 g, 158 mmol, 1.20 equiv) (Note 4) is added over the course of 15 min (Note 5) via syringe. The ice bath is then removed, the exterior of the reaction flask is protected from light by aluminum foil, and the mixture is stirred for 16 h at room temperature (24 °C). After 16 h, the solution is concentrated on the rotary evaporator (30 °C, 20 mmHg) (Note 6). The residue is partitioned between brine (100 mL) and diethyl ether (200 mL) (Note 1) and is transferred to a 1 L separatory funnel. The organic layer is washed with additional brine (100 mL). The brine layers are combined and back-extracted with diethyl ether (3 x 60 mL). The organic extracts are combined, dried over anhydrous MgSO₄ (15 g), filtered through a fritted glass funnel, and concentrated on a rotary evaporator (30 °C, 110 mmHg). The residual oil is then purified by silica gel column chromatography (Note 7) to yield 17.2 g (74%) of **1** (Notes 8 and 9) as a red, clear liquid (Note 10).

B. *Methyl 1-phenyl-2-trimethylsilylcycloprop-2-ene carboxylate (2).* A 250-mL, one-necked round-bottomed flask is equipped with a Teflon-coated magnetic stir bar (2.5 cm long, 1.2 cm diameter) and a septum, through which is inserted a needle connected to a manifold. The apparatus is evacuated (0.5 mmHg) and flame-dried. The flask is filled with nitrogen and cooled to room temperature. The septum is removed and the flask is charged with rhodium(II) acetate dimer (245 mg, 0.555 mmol, 5.00 × 10⁻³ equiv) (Note 11). The septum is replaced, and the apparatus is evacuated and refilled with nitrogen three times. Trimethylsilylacetylene (68.8 mL, 48.2 g, 490 mmol, 4.42 equiv) (Note 12) is added through the septum via syringe. Two 20 mL syringes are used to add a mixture of **1** (19.5 g, 111.0 mmol, 1.00 equiv) dissolved in trimethylsilylacetylene (19.0 mL, 13.3 g, 135 mmol, 1.22 equiv) at room temperature over 20 h using a syringe pump. The reaction mixture is then stirred for an additional 5 h. Vacuum distillation at room temperature (24 °C) (Note 13) is used to recover excess trimethylsilylacetylene (39 g). The residue is purified by column chromatography (Note 14), and the fractions containing the product are combined and concentrated on the rotary evaporator (40 °C, 20 mmHg) and then dried at 0.5 mmHg for 2 h to yield 18.7 g (69%) of **2** (Notes 15 and 16) as a yellow, clear liquid.

C. *3-Hydroxymethyl-3-phenyl-2-trimethylsilylcyclopropene (3)*. A 3-L three-necked round-bottomed flask is equipped with a Teflon-coated magnetic stir bar (5 cm long, 2 cm diameter). The right neck is connected to a gas inlet adapter and the other two necks are sealed with septa. The apparatus is evacuated (0.5 mmHg) and then flame-dried. The flask is filled with nitrogen and cooled to room temperature. A temperature probe is pierced through the left septum. A solution of **2** (18.7 g, 76.0 mmol, 1.00 equiv) in dry THF (350 mL) (Note 1) is added via cannula into the reaction flask. The reaction flask is then cooled with a dry ice/acetone bath to an internal temperature of –78 °C. To the stirring solution, a solution of DIBAL-H in THF (1.0 M, 269 mL, 269 mmol, 3.54 equiv) (Note 17) is added slowly via cannula (using nitrogen pressure) over a period of 4 h at –78 °C (Note 18). The reaction mixture is stirred for an additional 2 h at –78 °C. A septum is removed, and a 10% aqueous potassium carbonate solution (18 mL) is added over the course of 30 min to the flask using a Pasteur pipette (Note 19). The septum is replaced, and the dry ice bath is removed. When the internal temperature reaches –55 °C, a saturated aqueous potassium sodium tartrate solution (100 mL) and diethyl ether (200 mL) are added. The stirred suspension is allowed to warm to 24 °C over 11 h, then saturated aqueous potassium sodium tartrate solution (200 mL) and distilled water (100 mL) are added. The mixture is transferred to a 2 L separatory funnel and extracted with diethyl ether (2 × 100 mL). The organic layers are combined, washed with brine (400 mL), dried over anhydrous MgSO$_4$ (16 g), filtered through a fritted glass funnel, and then concentrated using rotary evaporator (40 °C, 20 mmHg). The residue is purified by column chromatography (Note 20), and fractions containing the product are combined and concentrated on the rotary evaporator (40 °C, 20 mmHg) and then dried at 0.5 mmHg for 2 h to yield 14.4 g (87%) of **3** (Notes 21 and 22) as a pale yellow solid.

D. *3-Hydroxymethyl-3-phenylcyclopropene (4)*. A 1-L, one-necked round-bottomed flask is equipped with a Teflon-coated magnetic stir bar (4 cm long, 1.6 cm diameter) and a septum, through which is inserted a needle connected to a manifold. The apparatus is evacuated (0.5 mmHg) and flame-dried, after which the flask is filled with nitrogen and cooled to room temperature. The septum is removed and the flask is charged with racemic

3 (14.4 g, 65.9. mmol, 1.00 equiv) and methanol (290 mL) (Note 1). The septum is replaced, a needle connected to a nitrogen source is inserted, and a temperature probe is inserted through the septum. The solution is cooled by an ice bath to an internal temperature of 5 °C, and the mixture is stirred. Potassium carbonate (92.2 g, 667 mmol, 10.1 equiv) (Note 23) is added at 4 °C. During the addition, the internal temperature increased to 7 °C. The ice bath is then exchanged for an oil bath, and the reaction mixture is heated to 36 °C and stirred for 3 h. The reaction mixture is filtered through a pad of Celite (20 g) on a fritted glass funnel (7 cm internal diameter). The Celite bed is washed with diethyl ether (4 × 50 mL) and distilled water (200 mL). The filtrate is concentrated on the rotary evaporator (40 °C, 20 mmHg), and then transferred to a 1 L separatory funnel and extracted with diethyl ether (200 mL). The aqueous layer is further extracted with diethyl ether (2 x 50 mL). The organic layers are combined, washed with brine (200 mL), dried over anhydrous MgSO$_4$ (17 g), filtered through a fritted glass funnel, and concentrated on the rotary evaporator (40 °C, 20 mmHg). The residue is purified by column chromatography (Note 24), and pure fractions are combined and concentrated on the rotary evaporator (30 °C, 20 mmHg) and then dried at 0.5 mmHg to yield 8.65 g (90%) of **4** (Notes 25 and 26) as an off-white solid.

Notes

1. THF and acetonitrile (all ACS, Reag. Ph Eur) were purchased from Merck and dried with a SP-1 solvent purification system from LC Technology Solutions. Methanol (ACS, Reag. Ph Eur) was obtained from Merck and used as received. The following solvents were used for flash column chromatography and during work-up (all laboratory grade, all used as received): diethyl ether (purchased from Univar), hexanes (purchased from Thommen-Furler), dichloromethane (purchased from EGT Chemie) and ethyl acetate (purchased from Thommen-Furler). The submitters used the following solvents: THF (99.9%) was obtained from Fisher Scientific and distilled from a blue solution of benzophenone ketyl. Acetonitrile (99.9%) was obtained from Fisher Scientific, and dried with columns packed with activated neutral alumina using the solvent purification system described by Bergman.[2] Methanol (99.9%) was purchased from Fisher Scientific and was used as

received for the preparation of **4** (reaction D). Diethyl ether (laboratory grade), hexanes (99.9%), dichloromethane (99.9%) and ethyl acetate (99.9%) were purchased from Fisher and used as received.

2. 4-Acetamidobenzenesulfonyl azide[3] (>98%) was purchased from TCI and used as received. The submitters purchased 4-acetamidobenzenesulfonyl azide (97%) from Sigma-Aldrich and used it as received.

3. Methyl phenylacetate (≥99%) was purchased from Sigma-Aldrich and used as received. The submitters purchased methyl phenylacetate (99%) from Sigma-Aldrich and used it as received.

4. 1,8-Diazabicyclo[5.4.0]undec-7-ene (DBU) (≥97%) was purchased from Fluorochem and used as received. The submitters purchased DBU (98%) from Acros Organics and used it as received.

5. DBU was added in 2 mL portions in intervals of one minute.

6. Care should be taken to avoid bumping during the rotary evaporation.

7. Flash column chromatography (9.0 cm diameter) was performed using 300 g of silica gel [Fluka, pore size 60 Å, 230 - 400 mesh particle size]. The top surface of the column was covered with sand (0.5 cm). The residue was loaded onto the column as a neat liquid. A small amount of 2% diethyl ether in hexanes (5 mL) was used assist the final transfer of the material to the column. The column was eluted with 2.2 L of 2% diethyl ether in hexanes, and 250 mL fractions were collected. Fractions #7–14 contained the product. TLC analysis during flash column chromatography was performed using 10% ethyl acetate in hexanes as the eluent. (Rf of **1** = 0.6, visualized with UV light; Rf_{SM} (methyl phenylacetate) = 0.5, visualized with UV light).

8. A yield of 75% was obtained when the reaction was performed on half-scale.

9. Physical properties of **1**[4] are as follows: R_f 0.6 (10% EtOAc in hexanes, UV detection). ^{1}H NMR (400 MHz, CDCl$_3$) δ: 3.86 (s, 3 H), 7.19 (t, J = 7.4 Hz, 1 H), 7.39 (m, 2 H), 7.49 (d, J = 7.4 Hz, 2 H). ^{13}C NMR (101 MHz, CDCl$_3$) δ: 52.0, 63.3, 124.0, 125.5, 125.8, 129.0, 165.6. IR (film) 3060, 3026, 2953, 2845, 2082, 1699, 1598, 1498, 1434, 1352, 1287, 1247, 1192, 1153, 1051, 1026, 909, 754, 692, 669, 632 cm^{-1}. HRMS (EI) m/z calcd for C$_9$H$_8$N$_2$O$_2$ [M]$^+$ 176.0580, found 176.0583. Anal. calcd. for C$_9$H$_8$N$_2$O$_2$: C, 61.36; H, 4.58; N, 15.90. Found: C, 61.28; H, 4.80; N, 16.06.

10. The submitters observed that purified **1** can also contain methyl phenylacetate (<5%) by ^{1}H NMR analysis; however, the presence of methyl phenylacetate was not observed by the checkers. The submitters

report that rigorous removal of all traces of methyl phenylacetate is not necessary. The cyclopropenation of trimethylsilylacetylene (Step B) has been performed by the submitters with methyl diazophenylacetate contaminated with as much as 10% methyl phenylacetate, without a significant decrease in yield.

11. Rhodium(II) acetate dimer (≥97%) was purchased from Fluorochem and used as received. The submitters purchased rhodium(II) acetate dimer (99%) from Colonial Metals and used it as received.

12. Trimethylsilylacetylene (≥97%) was purchased from Fluorochem and used as received. The submitters purchased trimethylsilylacetylene (98%) from GFS Chemicals and used it as received.

13. Distillation was performed under vacuum (0.5 mmHg) using a short path distillation unit. The receiver flask (100 mL) was immersed in acetone/dry ice bath (–78 °C).

14. Flash column chromatography (9.0 cm diameter) was performed using 176 g of silica gel [Fluka, pore size 60 Å, 230 - 400 mesh particle size]. The top surface of the column was covered with sand (0.5 cm). The residue was loaded onto the column as a neat liquid. Hexanes was used to assist the final transfer of the material to the column. The column was sequentially eluted with 10% dichloromethane in hexanes (1.1 L), 25% dichloromethane in hexanes (1 L), 33% dichloromethane in hexanes (1 L), 40% dichloromethane in hexanes (1 L) and 50% dichloromethane in hexanes (1.2 L). During chromatography, 200 mL fractions were collected. Fractions 3-26 contained the product and were combined. In fractions 3-8 a less polar, minor impurity was observed by TLC analysis. TLC analysis during flash column chromatography was performed using 30% ethyl acetate in hexanes as the eluent. ($Rf = 0.75$, visualized with ceric ammonium molybdate stain).

15. A yield of 73% was obtained when the reaction was performed on half-scale.

16. Physical properties of **2**[5,6] are as follows: R_f 0.75 (30% EtOAc in hexanes, visualized with ceric ammonium molybdate stain). ^{1}H NMR (400 MHz, CDCl$_3$) δ: 0.20 (s, 9 H), 3.67 (s, 3 H), 7.15–7.21 (m, 1 H), 7.25–7.29 (m, 4 H), 7.44 (s, 1 H); ^{13}C NMR (101 MHz, CDCl$_3$) δ: –1.3, 31.5, 52.0, 116.0, 120.0, 126.1, 128.0, 128.4, 142.6, 176.2; IR (film) 3119, 3059, 3025, 2957, 2902, 1945, 1731, 1704, 1602, 1495, 1446, 1434, 1289, 1251, 1220, 1042, 1024, 1012, 844, 791, 758, 698 cm^{-1}. HRMS (ESI) m/z calcd for C$_{14}$H$_{19}$O$_2$Si [M+H]$^+$ 247.11454 found 247.1144. Anal. calcd. For C$_{14}$H$_{18}$O$_2$Si: C, 68.25; H, 7.36. Found: C, 68.05; H, 7.22.

17. DIBAL-H (1M in THF) was purchased from Sigma-Aldrich and used as received.

18. With this slow rate of addition, the internal temperature was maintained below –70 °C (to avoid cyclopropene reduction) during the addition of DIBAL-H.

19. Care should be exercised during the addition, as the quenching of DIBAL-H is exothermic.

20. Flash column chromatography (9.0 cm diameter) was performed using 187 g of silica gel [Fluka, pore size 60 Å, 230 - 400 mesh particle size]. The top surface of the column was covered with sand (0.5 cm). The residue was loaded onto the column as a neat liquid. 5% Ethyl acetate in hexanes was used assist the final transfer of the material to the column. The column was sequentially eluted with 5% ethyl acetate in hexanes (1.7 L), 9% ethyl acetate in hexanes (1 L), 17% ethyl acetate in hexanes (600 mL). During chromatography, 200 mL fractions were collected. Fractions 4-14 contained the product. TLC analysis during flash column chromatography was performed using 30% ethyl acetate in hexanes as the eluent. (Rf = 0.54, visualized with ceric ammonium molybdate stain).

21. A reaction run on half-scale provided a product yield of 88%.

22. Physical properties of **3**[7] are as follows: mp 35 °C; R$_f$ 0.54 (30% EtOAc in Hexanes, with ceric ammonium molybdate stain was used to visualize TLC plates). ^{1}H NMR (400 MHz, CDCl$_3$) δ: 0.18 (s, 9 H), 1.18 (t, J = 5.7 Hz, 1 H), 3.99 (dd, J = 11.0, 5.3 Hz, 1 H), 4.16 (dd, J = 11.1, 5.7 Hz, 1 H), 7.12–7.20 (m, 3 H), 7.25–7.30 (m, 2 H), 7.70 (s, 1 H); ^{13}C NMR (100 MHz, CDCl$_3$) δ: –0.9, 30.2, 68.9, 123.6, 124.1, 125.3, 126.1, 128.1, 146.9; IR (film) 3368, 3084, 3058, 3023, 2957, 2898, 2871, 1692, 1600, 1578, 1495, 1408, 1250, 1065, 1012, 866, 842, 757, 698 cm^{-1}. HRMS (EI) m/z calcd for C$_{12}$H$_{15}$OSi [M-CH$_3$]$^+$ 203.0887, found 203.0884. Anal. calcd. For C$_{13}$H$_{18}$OSi: C, 71.50; H, 8.31. Found: C, 71.39; H, 8.41.

23. Anhydrous potassium carbonate (EP) was purchased from Brenntag Schweizerhall and used as received. The submitters purchased anhydrous potassium carbonate (99.9%) from Fisher Scientific and used it as received.

24. Flash column chromatography (6.0 cm diameter) was performed using 144 g of silica gel [Fluka, pore size 60 Å, 230 - 400 mesh particle size]. The top surface of the column was covered with sand (0.5 cm). The residue was loaded onto the column as a neat liquid. Diethyl ether was used assist the final transfer of the material to the column. The column

was sequentially eluted with 10% diethyl ether in hexanes (1 L), 14% diethyl ether in hexanes (1 L), 25% diethyl ether in hexanes (400 mL), and 50% diethyl ether in hexanes (1.2 L). During chromatography, 80 mL fractions were collected. Fractions 28-43 contained the product. TLC analysis during flash column chromatography was performed using 50% diethyl ether in hexanes as the eluent. (Rf = 0.32, visualized with ceric ammonium molybdate stain).

25. When the reaction was performed on approximately half-scale, the isolate yield of pure product was 80%.

26. Physical properties of **4**[5] are as follows: mp 58–59 °C; R_f 0.32 (50% diethyl ether in hexanes, visualized with ceric ammonium molybdate stain). ^{1}H NMR (400 MHz, CDCl$_3$) δ: 1.54 (m, 1 H), 4.08 (d, J = 5.9 Hz, 2 H), 7.19–7.7.27 (m, 3 H), 7.31–7.34 (m, 4 H); ^{13}C NMR (101 MHz, CDCl$_3$) δ: 29.2, 68.0, 113.2, 125.9, 126.4, 128.3, 145.8. IR (film) 3217, 3102, 3050, 2923, 2880, 1645, 1599, 1575, 1493, 1473, 1446, 1307, 1222, 1175, 1067, 1009, 985, 969, 898, 757, 698, 635, 565, 533, 471 cm^{-1}. HRMS (EI) m/z calcd for C$_{10}$H$_9$O [M-H]$^+$ 145.0648, found 145.0647. Anal. calcd. For C$_{10}$H$_{10}$O: C, 82.16; H, 6.89. Found: C, 82.20; H, 6.75.

Working with Hazardous Chemicals

The procedures in *Organic Syntheses* are intended for use only by persons with proper training in experimental organic chemistry. All hazardous materials should be handled using the standard procedures for work with chemicals described in references such as "Prudent Practices in the Laboratory" (The National Academies Press, Washington, D.C., 2011; the full text can be accessed free of charge at http://www.nap.edu/catalog.php?record_id=12654). All chemical waste should be disposed of in accordance with local regulations. For general guidelines for the management of chemical waste, see Chapter 8 of Prudent Practices.

In some articles in *Organic Syntheses*, chemical-specific hazards are highlighted in red "Caution Notes" within a procedure. It is important to recognize that the absence of a caution note does not imply that no significant hazards are associated with the chemicals involved in that procedure. Prior to performing a reaction, a thorough risk assessment should be carried out that includes a review of the potential hazards associated with each chemical and experimental operation on the scale that is planned for the procedure. Guidelines for carrying out a risk assessment and for analyzing the hazards associated with chemicals can be found in Chapter 4 of Prudent Practices.

The procedures described in *Organic Syntheses* are provided as published and are conducted at one's own risk. *Organic Syntheses, Inc.*, its Editors, and its Board of Directors do not warrant or guarantee the safety of individuals using these procedures and hereby disclaim any liability for any injuries or damages claimed to have resulted from or related in any way to the procedures herein.

Discussion

The Rh-catalyzed reactions of α-diazoesters with alkynes to give cyclopropene carboxylates was first described in 1978 (Eq 1),[8] and remains one of the most effective methods for the preparation of cyclopropenes.

$$\equiv\!-C_4H_9 \;+\; \underset{N_2}{\overset{H\diagdown\,CO_2Et}{\big|}} \quad \xrightarrow[84\%]{\underset{(1.25\ mol\ \%)}{Rh_2(OAc)_4}} \quad \overset{CO_2Et}{\underset{C_4H_9}{\triangle}} \qquad (1)$$

The range of diazoesters that participate in Rh-catalyzed cyclopropenation reactions is broad, and includes diazomalonates[6,9] and diazoacetates without α-substitution[8-11] or with α-silyl-,[12] α-sulfonyl,[13,14] α-alkenyl,[15,16] α-aryl,[9,17-21] α-trifluoromethyl,[22,23] and α-alkyl substitutents.[24-28] For α-alkyl-α-diazoesters with β-hydrogens, selectivity over β-hydride migration can be achieved by using bulky carboxylate ligands at low temperature.[24,27,28] Cyclic α-diazocarbonyl compounds are particularly resilient to β-hydride migration,[26,29] even for substratres with tertiary β-hydrogens (Eq 2).[26]

$$\xrightarrow[CH_2Cl_2\atop -50\ ^\circ C]{Rh_2(OPiv)_4\atop (0.5\ mol\ \%)} \qquad (2)$$

71%
>95:5 dr

Enantioselective Rh-catalyzed cyclopropenation was first described in 1992.[10] Advances in chiral catalyst development have expanded the scope of Rh-catalyzed cyclopropenation to include diazoacetates,[10,11,30-32] α-aryl-α-diazoesters,[21] α-alkenyl-α-diazoesters,[15,16] and α-alkyl-α-diazoesters.[24,27] Intramolecular, enantioselective cyclopropenations have also been reported.[33] Cyclopropene carboxylates and derivatives have been employed as starting materials for manifold reactions,[34,35] including carbometallation,[7,19,36-41] hydrometallation,[5,6,42-44] dimetallation,[44] Heck,[20] Sila Morita–Baylis–Hillman,[45] Pauson-Khand,[46] Diels-Alder,[11,47,48] dipolar cycloadditions[26,49] and ring expansions.[17,50-52] The preparation of 3-hydroxymethyl-3-phenylcyclopropene was initially described by Rubina, Rubin and Gevorgyan,[6,43] and a modification was reported by our group.[7] This compound and related, prochiral cyclopropenes have served as platforms for a number of enantioselective reactions, including enantioselective hydrostannation,[43] hydroacylation,[53] hydroboration,[5] carbomagnesation[7,54] and carbozincation[54,55] reactions.

Cyclopropenes produced through Rh-catalyzed reactions of diazoesters and alkynes have been applied to total synthesis[56] and to bioorthogonal chemistry.[57-59] Enantioselective cyclopropenation catalyzed by the unsymmetrical (R,R)-Rh$_2$(OAc)(DPTI)$_3$,[11,31] has served as the key step in the enantioselective total synthesis of (–)-pentalenene[56] (Scheme 1). Furthermore, cyclopropene carboxylates and their derivatives have recently emerged as important reaction partners in bioorthogonal reactions involving tetrazines[57-59] and in photoclick reactions involving tetrazoles.[60]

Scheme 1. Enantioselective Rh-catalyzed cyclopropenation as a key step in the total synthesis of (–)-pentalenene

References

1. Department of Chemistry and Biochemistry, University of Delaware, Newark DE 19716. jmfox@udel.edu. For financial support we thank NSF CHE 1300329. Data were obtained with instrumentation supported by NIH S10RR026962-01, NIH P20 RR017716, and NSF CRIF:MU grants: CHE 0840401 and CHE-1229234.
2. Alaimo, P. J.; Peters, D. W.; Arnold, J.; Bergman, R. G. *J. Chem. Educ.* **2001**, *78*, 64.
3. Davies, H. M. L.; Cantrell, W. R., Jr.; Romines, K. R.; Baum, J. *Org. Synth.* **1992**, *70*, 93.

4. Davies, H. M. L.; Hansen, T.; Churchill, M. R. *J. Am. Chem. Soc.* **2000**, *122*, 3063.

5. Rubina, M.; Rubin, M.; Gevorgyan, V. *J. Am. Chem. Soc.* **2003**, *125*, 7198.

6. Rubin, M.; Gevorgyan, V. *Synthesis* **2004**, *5*, 796.

7. Liu, X. Z.; Fox, J. M. *J. Am. Chem. Soc.* **2006**, *128*, 5600.

8. Petiniot, N.; Anciaux, A. J.; Noels, A. F.; Hubert, A. J.; Teyssié, P. *Tetrahedron Lett.* **1978**, *14*, 1239.

9. Müller, P.; Gränicher, C. *Helv. Chim. Acta* **1993**, *76*, 521.

10. Protopopova, M. N.; Doyle, M. P.; Muller, P.; Ene, D. *J. Am. Chem. Soc.* **1992**, *114*, 2755.

11. Lou, Y.; Horikawa, M.; Kloster, R. A.; Hawryluk, N. A.; Corey, E. J. *J. Am. Chem. Soc.* **2004**, *126*, 8916.

12. Arrowood, T. L.; Kass, S. R. *Tetrahedron* **1999**, *55*, 6739.

13. Liu, Y.; Ma, S. *Org. Lett.* **2012**, *14*, 720.

14. Wang, Y.; Fordyce, E. A. F.; Chen, F. Y.; Lam, H. W. *Angew. Chem., Int. Ed.* **2008**, *47*, 7350.

15. Briones, J. F.; Hansen, J.; Hardcastle, K. I.; Autschbach, J.; Davies, H. M. L. *J. Am. Chem. Soc.* **2010**, *132*, 17211.

16. Briones, J. F.; Davies, H. M. L. *Tetrahedron* **2011**, *67*, 4313.

17. Xie, X. C.; Li, Y.; Fox, J. M. *Org. Lett.* **2013**, *15*, 1500.

18. Xie, X. C.; Yang, Z.; Fox, J. M. *J. Org. Chem.* **2010**, *75*, 3847.

19. Yang, Z.; Xie, X. C.; Fox, J. M. *Angew. Chem., Int. Ed.* **2006**, *45*, 3960.

20. Chuprakov, S.; Rubin, M.; Gevorgyan, V. *J. Am. Chem. Soc.* **2005**, *127*, 3714.

21. Davies, H. M. L.; Lee, G. H. *Org. Lett.* **2004**, *6*, 1233.

22. Sang, R.; Yang, H.-B.; Shi, M. *Tetrahedron Lett.* **2013**, *54*, 3591.

23. Müller, P.; Grass, S.; Shahi, S. P.; Bernardinelli, G. *Tetrahedron* **2004**, *60*, 4755.

24. Goto, T.; Takeda, K.; Shimada, N.; Nambu, H.; Anada, M.; Shiro, M.; Ando, K.; Hashimoto, S. *Angew. Chem., Int. Ed.* **2011**, *50*, 6803.

25. Müller, P.; Gränicher, C. *Helv. Chim. Acta* **1995**, *78*, 129.

26. DeAngelis, A.; Dmitrenko, O.; Fox, J. M. *J. Am. Chem. Soc.* **2012**, *134*, 11035.

27. Boruta, D. T.; Dmitrenko, O.; Yap, G. P. A.; Fox, J. M. *Chem. Sci.* **2012**, *3*, 1589.

28. Panne, P.; Fox, J. M. *J. Am. Chem. Soc.* **2006**, *129*, 22.

29. Sattely, E. S.; Meek, S. J.; Malcolmson, S. J.; Schrock, R. R.; Hoveyda, A. H. *J. Am. Chem. Soc.* **2009**, *131*, 943.

30. Müller, P.; Imogai, H. *Tetrahedron: Asymmetry* **1998**, *9*, 4419.

31. Lou, Y.; Remarchuk, T. P.; Corey, E. J. *J. Am. Chem. Soc.* **2005**, *127*, 14223.
32. Weatherhead-Kloster, R. A.; Corey, E. J. *Org. Lett.* **2006**, *8*, 171.
33. Doyle, M. P.; Ene, D. G.; Peterson, C. S.; Lynch, V. *Angew. Chem., Int. Ed.* **1999**, *38*, 700.
34. Rubin, M.; Rubina, M.; Gevorgyan, V. *Synthesis* **2006**, 1221.
35. Rubin, M.; Rubina, M.; Gevorgyan, V. *Chem. Rev.* **2007**, *107*, 3117.
36. Yan, N.; Liu, X.; Fox, J. M. *J. Org. Chem.* **2007**, *73*, 563.
37. Simaan, S.; Marek, I. *Org. Lett.* **2007**, *9*, 2569.
38. Liao, L.-a.; Fox, J. M. *J. Am. Chem. Soc.* **2002**, *124*, 14322.
39. Delaye, P.-O.; Didier, D.; Marek, I. *Angew. Chem., Int. Ed.* **2013**, *52*, 5333.
40. Tarwade, V.; Liu, X.; Yan, N.; Fox, J. M. *J. Am. Chem. Soc.* **2009**, *131*, 5382.
41. Didier, D.; Delaye, P.-O.; Simaan, M.; Island, B.; Eppe, G.; Eijsberg, H.; Kleiner, A.; Knochel, P.; Marek, I. *Chem. Eur. J.* **2014**, *20*, 1038.
42. Rubina, M.; Rubin, M.; Gevorgyan, V. *J. Am. Chem. Soc.* **2002**, *124*, 11566.
43. Rubina, M.; Rubin, M.; Gevorgyan, V. *J. Am. Chem. Soc.* **2004**, *126*, 3688.
44. Trofimov, A.; Rubina, M.; Rubin, M.; Gevorgyan, V. *J. Org. Chem.* **2007**, *72*, 8910.
45. Chuprakov, S.; Malyshev, D. A.; Trofimov, A.; Gevorgyan, V. *J. Am. Chem. Soc.* **2007**, *129*, 14868.
46. Pallerla, M. K.; Fox, J. M. *Org. Lett.* **2005**, *7*, 3593.
47. Fisher, L. A.; Smith, N. J.; Fox, J. M. *J. Org. Chem.* **2013**, *78*, 3342.
48. Battiste, M. A. *Chem. Ind.* **1961**, 550.
49. Diev, V. V.; Kostikov, R. R.; Gleiter, R.; Molchanov, A. P. *J. Org. Chem.* **2006**, *71*, 4066.
50. Komendantov, M. I.; Domnin, I. N.; Bulucheva, E. V. *Tetrahedron* **1975**, *31*, 2495.
51. Chuprakov, S.; Gevorgyan, V. *Org. Lett.* **2007**, *9*, 4463.
52. Zhu, Z.-B.; Shi, M. *Org. Lett.* **2010**, *12*, 4462.
53. Phan, D. H. T.; Kou, K. G. M.; Dong, V. M. *J. Am. Chem. Soc.* **2010**, *132*, 16354.
54. Nakamura, M.; Hirai, A.; Nakamura, E. *J. Am. Chem. Soc.* **2000**, *122*, 978.
55. Krämer, K.; Leong, P.; Lautens, M. *Org. Lett.* **2011**, *13*, 819.
56. Pallerla, M. K.; Fox, J. M. *Org. Lett.* **2007**, *9*, 5625.
57. Yang, J.; Liang, Y.; Šečkutė, J.; Houk, K. N.; Devaraj, N. K. *Chem. Eur. J.* **2014**, *20*, 3365.
58. Kamber, D. N.; Nazarova, L. A.; Liang, Y.; Lopez, S. A.; Patterson, D. M.; Shih, H.-W.; Houk, K. N.; Prescher, J. A. *J. Am. Chem. Soc.* **2013**, *135*, 13680.

334

DOI: 10.15227/orgsyn.091.0322

59. Elliott, T. S.; Townsley, F. M.; Bianco, A.; Ernst, R. J.; Sachdeva, A.; Elsasser, S. J.; Davis, L.; Lang, K.; Pisa, R.; Greiss, S.; Lilley, K. S.; Chin, J. W. *Nat. Biotechnol.* **2014**, *32*, 465.
60. Yu, Z. P.; Pan, Y. C.; Wang, Z. Y.; Wang, J. Y.; Lin, Q. *Angew. Chem., Int. Ed.* **2012**, *51*, 10600.

Appendix
Chemical Abstracts Nomenclature (Registry Number)

4-Acetamidobenzenesulfonyl azide: Benzenesulfonyl azide, 4-(acetylamino)-; (2158-14-7)
Methyl phenylacetate: Benzeneacetic acid, methyl ester; (101-41-7)
1,8-Diazabicyclo[5.4.0]undec-7-ene: Pyrimido[1, 2-*a*]azepine, 2, 3, 4, 6, 7, 8, 9, 10-octahydro-; (6674-22-2)
Rhodium (II) acetate dimer: Rhodium, tetrakis[μ-(acetato-κ*O*:κ*O*')]di-, (*Rh-Rh*); (15956-28-2)
Trimethylsilylacetylene: Silane, ethynyltrimethyl-; (1066-54-2)
DIBAL-H: Aluminum, hydrobis(2-methylpropyl)-; (1191-15-7)
Potassium carbonate: Carbonic acid, potassium salt (1:2); (584-08-7)

Ramajeyam Selvaraj was born in 1982 in Tiruchendur, TamilNadu, India. He received his bachelor's degree from University of Madras in 2002 and his master's degree from Indian Institute of Technology Madras in 2004, where he carried out his research with Prof. Sethuraman Sankararaman. In 2012, he received his Ph.D. from the University of Delaware under the guidance of Professor Joseph Fox. In the fall of 2014, he moved to the Purdue University, where he is currently a postdoctoral researcher working in the laboratory of Professor Christopher Uyeda.

Srinivasa Rao Chintala was born in 1987 in Donkinivalasa, Andhra Pradesh, India. He received his bachelor's degree from Andhra University in 2007 and his master's degree from University of Hyderabad in 2010, where he carried out his research with Dr. Rangarajan Balamurugan and is currently pursuing his Ph.D. from the University of Delaware under the supervision of Prof. Joseph Fox.

Michael Taylor received his bachelor's degree in Biochemistry from Salisbury University, where he conducted undergraduate research with Professor Elizabeth Papish. In 2012, he received his Ph.D. from the University of Delaware under the guidance of Professor Joseph Fox. In the fall of 2012, he moved to the University of Cambridge, where he is currently a postdoctoral fellow working in the laboratory of Professor Matthew Gaunt.

Joseph Fox received his bachelor's degree from Princeton University, where he conducted undergraduate research with Maitland Jones Jr. He received the Ph.D. in 1998 from Columbia University under the direction of Thomas Katz. From 1998–2001, he was an NIH postdoctoral fellow at MIT with Stephen Buchwald. In 2001, Fox joined the faculty at UD in the Department of Chemistry and Biochemistry, where he was promoted to the rank of Professor in 2011. His research interests include synthesis with strained molecules, asymmetric catalysis, catalyst design, and bioorthogonal reaction development.

Jonathan Ruchti received his B. Sc. in chemistry from the University of Basel, where he worked under the supervision of Professor Andreas Pfaltz. He obtained a M. Sc. degree in chemistry from ETH Zürich where he is currently pursing doctoral studies in synthetic organic chemistry under the guidance of Professor Erick M. Carreira.

Enantioselective Synthesis of α– and β–Boc-protected 6-Hydroxy-pyranones: Carbohydrate Building Blocks

Sumit O. Bajaj, Jamison R. Farnsworth and George A. O'Doherty[*1]

Department of Chemistry and Chemical Biology, Northeastern University, 102 Hurtig Hall, 360 Huntington Ave, Boston, MA 02115

Checked by Heemal Dhanjee, Michael T. Tudesco, and John L. Wood

A.

Noyori (1*S*, 2*S*) catalyst (0.3 mol%)
HCO₂Na (3.0 M)
CTAB, rt

1 → **2**

B.

NBS, H₂O
NaOAc, 0 °C

2 → **3**

C.

(Boc)₂O
DMAP (10 mol%)
−78 °C to rt

BocO **α-L-Boc-pyranone** + BocO **β-L-Boc-pyranone**

3 → **4** + **5**

Procedure

A. *(S)-1-(Furan-2-yl)ethanol (2)*. To a 500 mL Erlenmeyer flask equipped with a 5 cm octagonal magnetic stir bar is added sodium formate (55.0 g,

0.801 mol) (Note 1) followed by 267 mL of deionized water to make a 3.0 M solution. The solution is stirred until all of the solid dissolves and simultaneously degassed by bubbling nitrogen gas through the solution via a needle for 5 min at room temperature. Liquid 2-acetyl furan (**1**) (16.4 mL, 18.1 g, 0.163 mol, 1 equiv) (Note 2), CH_2Cl_2 (2 mL) (Note 3), and cetyltrimethylammonium bromide (CTAB) (5.9 g, 10 mol%) (Note 4) are added, the flask loosely capped with a polyethylene cap, and the solution stirred for 20 min. To the resulting brown suspension, Noyori asymmetric transfer hydrogenation catalyst (R)-Ru(η^6-mesitylene)-(S,S)-TsDPEN (304 mg, 0.3 mol%) (Note 5) is added, and the Erlenmeyer flask again capped loosely with a polyethylene stopper. The resulting suspension is stirred vigorously at room temperature for 24 h (Note 6). The progress of the reaction is monitored by TLC using p-anisaldehyde as stain (Compound **2** has an R_f=0.26 in 20% EtOAc in hexanes. The starting 2-acetyl furan (R_f=0.37) is UV active, whereas the product alcohol is not) (Note 7). The darker brown reaction mixture is diluted with water (200 mL), the magnetic stir bar removed, and the solution is transferred to a 2-L separatory funnel. The mixture is extracted with Et_2O (4 x 500 mL) and the layers are allowed to separate over a period of 30 min for each extraction (Note 8). The combined organic layers are washed with saturated $NaHCO_3$ solution (150 mL) (Note 9), saturated NaCl solution (150 mL), dried over Na_2SO_4 (22.5 g) (Note 10), and passed through a small plug of silica gel (50 g SiO_2, vacuum filtration, 6.5 cm in diameter, 3.2 cm height, 0.5 cm sand). The Na_2SO_4 is washed with Et_2O (2 x 150 mL) and successively passed through the silica gel plug. The filtrate is concentrated under reduced pressure using a rotary evaporator (20 °C, 150 mmHg initial to 25 mmHg) yielding 16.7 g of the crude furan alcohol (**2**) as a dark grey oil (Note 11). A small amount of the representative sample is purified using silica gel flash chromatography eluting with 20-25% Et_2O/hexanes to give pure furan alcohol **2** as a colorless oil (Note 12).

B. *(2S, 6R)-6-Hydroxy-2-methyl-2H-pyran-3(6H)-one (3).* The crude furan alcohol (**2**) (16.57 g, 0.148 mol, 1.0 equiv) in a 500 mL 2-neck Erlenmeyer flask equipped with a thermometer and 5 cm magnetic stir bar (Note 13) is dissolved in 248 mL of THF/H_2O (3:1) (Note 14) and cooled to 1.2 °C using an ice-water bath. Solid $NaHCO_3$ (24.85 g, 0.296 mol, 2.0 equiv) (Note 9) and NaOAc•3H$_2$O (20.2 g, 0.148 mol, 1.0 equiv) (Note 15) are added successively over the course of 1 min each, during which time the temperature lowers to –1.2 °C. After stirring for 5 min, solid N-bromosuccinimide (27.97 g, 0.156 mol, 1.05 equiv) (Note 16) is added to the solution portion-wise (approx. 7 g every 5 min). The resulting orange/red color resulting from each addition of NBS quickly subsides to a green/brown suspension and the temperature is maintained between 1 °C and 10 °C. Upon the last addition of NBS, the orange/red color persists through the remainder of the reaction. The reaction mixture is stirred for an additional 2 h and the temperature maintained between 1–2 °C. Product formation is monitored by TLC (Note 7) using *p*-anisaldehyde as stain (Compound **3** possesses an R_f=0.24 in 40% EtOAc in hexanes, while the starting material's R_f=0.52). The resulting slurry is diluted with Et$_2$O (100 mL) and transferred to a 2-L separatory funnel containing H$_2$O (200 mL) and an additional Et$_2$O (100 mL) that were precooled to 4.5 °C in an ice/water bath. The organic layer is separated. The bottom aqueous layer, which still contains some solid material, is once again washed with Et$_2$O (3 x 400 mL) and the combined organics are washed with saturated aqueous NaCl (200 mL), after which the red/orange organic phase changes to a grey/faint red solution. The organic layer is then dried over sodium sulfate (30 g) (Note 10) with stirring for 30 min to give a clear grey organic phase. The sodium sulfate is filtered, washed with Et$_2$O (2 x 50 mL), and the combined filtrate concentrated under reduced pressure by rotary evaporator (20 °C, 150 mmHg initial to 25 mmHg final) yielding a crude mixture of alcohol diastereomers **3** as a grey/faint red oil (24.33 g) (Note 17), which is stored in a freezer at –19 °C. A small amount of the representative crude sample is purified by silica gel flash chromatography, eluting with 25-30% EtOAc/hexanes, to give a mixture of diastereomers **3** as a colorless oil (Note 18).

C. *tert-Butyl ((2S,6S)-6-methyl-5-oxo-5,6-dihydro-2H-pyran-2-yl) carbonate (4) and tert-butyl ((2R,6S)-6-methyl-5-oxo-5,6-dihydro-2H-pyran-2-yl) carbonate (5).* An oven-dried, 500 mL 3-necked, round-bottomed flask with 3.5 cm octagonal stirbar is equipped on the left neck with a nitrogen inlet adapter, the middle neck a rubber septum, and the right neck a rubber thermometer adapter with thermometer (Note 19). The crude mixture of diastereomeric allylic alcohols (**3**) (24.2 g, 0.148 mol, 1 equiv) (Note 20) is added and dissolved in anhydrous DCM (185 mL) (Note 3). The solution is cooled to –67.2 °C (using a dry ice–acetone bath) under nitrogen atmosphere, the center septum removed, and solid (Boc)$_2$O (74.25 g, 0.273 mol, 2.3 equiv) (Note 21) added via funnel over the course of one min so as to maintain stirring. A catalytic amount of DMAP (1.82 g, 14.8 mmol, 0.1 equiv) (Note 22) is then added. The resulting solution is allowed to stir for 6 h at –66.8 °C and then warmed to 2.2 °C in an ice-water bath and stirred for a 3 h period (Note 23). The ice-water bath is then removed and the reaction stirred for 1 h, during which time the reaction warms to 7.2 °C. The reaction is monitored by TLC using potassium permanganate as stain (Compounds **4** and **5** possess R_f=0.48 and 0.44, respectively, in 20% EtOAc in hexanes, while the starting material's R_f=0.11). The reaction mixture is then brought to 2.4 °C (ice-water bath), diluted with Et$_2$O (100 mL), and quenched by the addition of saturated NaHCO$_3$ (150 mL) (Note 3) over the course of 1 min. The mixture is extracted with Et$_2$O (4 x 500 mL) using 2-L separatory funnel. The organic layer is washed sequentially with 1N solution of NaHSO$_4$ (to remove excess DMAP) (Note 24), saturated NaHCO$_3$ (150 mL), and saturated aqueous NaCl (150 mL), after which the solution is dried over sodium sulfate (15.0 g) and filtered. The sodium sulfate is rinsed with Et$_2$O (2 x 50 mL) and the combined filtrate is transferred into a 3-L round-bottomed flask and concentrated under reduced pressure by rotary evaporator (20 °C, 150 mmHg initially to 25 mmHg), which yields a crude mixture of Boc-protected pyranones as a red-brown oil.

The crude product is purified by SiO$_2$ flash chromatography using a solvent gradient of Et$_2$O in hexane (0.5% to 10% Et$_2$O in hexane v/v). The diameter of the chromatography column is 9.0 cm, and the column is packed using silica gel (500 g) (Note 7) to a height of 14.5 cm and wetted with 0.5% Et$_2$O/hexanes. The oil is loaded to the top of the column and absorbed on to the silica. After placing a 0.5 cm layer of sand on top of silica, 500 mL of 0.5% Et$_2$O/hexanes is passed through the column. After the initial solvent is collected additional amounts of eluent are added to the top of the column with slight incremental increases in solvent polarity (e.g., 500 mL of 1% Et$_2$O/hexanes, followed by 500 mL of 2% Et$_2$O/hexane, until 500 mL of 10% Et$_2$O/hexanes has been passed through the column. An additional 2.5 L of 10% Et$_2$O/hexanes is then passed through the column). The first 4-L of solvent is collected in 500 mL fractions in Erlenmeyer flasks, after which the eluent is collected in 200 mL fractions. Fractions are monitored by TLC for product (using potassium permanganate as stain) (Note 13). The α-L-Boc-pyranone **4** elutes off of the column with the 8-10% Et$_2$O/hexanes wash. The fractions containing the pure α-L-Boc-pyranone **4** (least polar diastereomer, R_f (20% EtOAc/hexanes) = 0.59) are collected, concentrated on a rotary evaporator (20 °C, 150 mmHg initially to 25 mmHg), and placed under high vacuum (1.5 mmHg)), to yield 13.02 g (57.0 mmol) of a faint yellow solid. Fractions containing a mixture of diastereomers (obtained during the 10% Et$_2$O/hexanes solvent system) are also collected to yield 0.50 g of a colorless oil, which can be separated by further chromatography. The β-L-Boc-pyranone **5** elutes off of the column with the later 10% Et$_2$O/hexanes wash. Fractions containing the pure β-L-Boc-pyranone **5** (more polar diastereomer, R_f (20% EtOAc/hexanes) = 0.50) (Note 13) are collected, concentrated on a rotary evaporator (20 °C, 150 mmHg initially to 25 mmHg), and placed under high vacuum (1.5 mmHg) to yield 5.23 g (22.9 mmol) of a faint yellow solid (Note 25). The Boc-protected pyranones are stored in a freezer. The overall yield for the three steps is 51% (Note 26).

Notes

1. Sodium formate, BioUltra, ≥99.0%, was purchased from Sigma-Aldrich Co. and used as received.
2. 2-Acetyl furan, 99%, was purchased from Sigma-Aldrich Co. and used as received.
3. The checkers used ACS grade dichloromethane (CH$_2$Cl$_2$) purchased from Fisher Scientific and filtered through a column of activated alumina. The submitters used CH$_2$Cl$_2$ purchased from VWR International, LLC. and dried prior to use by percolation through anhydrous Al$_2$O$_3$.
4. Cetyltrimethylammonium bromide, BioXtra, ≥99.0%, was purchased from Sigma-Aldrich Co. and used as received.
5. Noyori asymmetric catalyst (R)-Ru(η^6-mesitylene)-(S,S)-TsDPEN (95%) was purchased from Sigma-Aldrich Co. and used as received.[2]
6. Sodium formate is used as a hydrogen donor source for this reaction along with cetyltrimethylammonium bromide, which functions as a phase transfer catalyst. Vigorous stirring is important for this step.
7. Silica gel SilicaFlash® F60 (40-63 μm/230-500 mesh) was purchased from Silicycle. Glass-backed extra hard layer TLC plates, 60 Å (250 μm thickness) were also purchased from Silicycle containing F-254 indicator.
8. In one instance the checkers noted an emulsion resulted without separation of the organic and aqueous phases. In this instance, an additional 200 mL deionized water was added, followed by an additional 200 mL Et$_2$O. The separatory funnel was shaken and the layers were allowed to separate over the course of 30 min.
9. Sodium bicarbonate ACS reagent, 99.7–100.3%, was purchased from EMD Chemicals Inc. and used as received.

10. Sodium sulfate, ACS grade, ≥99.0%, anhydrous, granular was purchased from EMD Chemicals Inc. and used as received.

11. A ^{1}H NMR of the crude reaction mixture sample was obtained to confirm product formation prior to use in the next step. When the reaction was performed using 8.2 mL of 2-acetyl furan, 8.5 g of crude alcohol **2** was obtained.

12. The purified product **2** showed following data: R_f (25% Et$_2$O/hexanes) = 0.26, $[\alpha]^{24}_D$ = –22 (c = 1.0, CH$_2$Cl$_2$) (Note 27); IR (neat) 3351, 2980, 2934, 1505, 1467, 1371, 1229, 1149, 1008, 927, 877, 809, 734; 598 cm^{-1}; ^{1}H NMR (400 MHz, CDCl$_3$) δ: 1.55 (d, J = 6.6 Hz, 3 H), 1.93 (d, J = 5.0 Hz, 1 H), 4.88 (dq, J = 1.2, 6.4 Hz, 1 H), 6.23 (ddd, J = 3.2, 0.8, 0.8 Hz, 1 H), 6.32 (dd, J = 3.2, 1.8 Hz, 1 H), 7.37 (dd, J = 1.8, 0.8 Hz, 1 H); ^{13}C NMR (100 MHz, CDCl$_3$) δ: 21.6, 63.9, 105.4, 110.4, 142.2, 157.9; HRMS (ESI) Calcd for (M+Na)$^+$: 135.0422, Found: 135.0417. The enantiomeric excess of the product was determined by Mosher ester analysis (>95% ee). Absolute stereochemistry was also determined by Mosher Ester analysis according to a protocol by Hoye et al.[3] To a flame-dried vial was added R-(+)-α-methoxy-α-(trifluoromethyl)phenylacetic acid, which was dissolved by the addition of 0.3 mL anhydrous DCM. Oxalyl chloride (43 μL, 0.48 mmol) was added, followed by DMF (2 μL, 0.026 mmol). This solution was stirred for 30 min, after which the stir bar was removed, rinsed with 0.1 mL DCM, and the vial concentrated under reduced pressure. The crude Mosher's acid chloride was then placed under high vacuum for 10 min. In a separate flame-dried 3 mL vial the crude product **2** (16.9 mg, 0.151 mmol) was dissolved in methylene chloride (0.3 mL). A 10 % DMAP pyridine solution (60 μL) was added with stirring, followed by the addition of crude Mosher's acid chloride that was dissolved in 0.2 mL DCM. The reaction was monitored by TLC. Upon completion, the crude reaction mixture was diluted with ether (1 mL), extracted with 1 N HCl (3 x 1 mL), washed with sat. NaHCO$_3$ (3 x 1 mL), dried over Na$_2$SO$_4$ and concentrated. The enantiomeric excess was determined by inspection of the crude ^{1}H NMR spectra by integration of peaks at 1.62 (d, J = 6.7 Hz, CH$_3$) for the (R,R)-isomer and 1.70 (d, J = 6.8 Hz, CH$_3$) for the (R,S)-isomer.[4] The procedure was repeated with S-(–)-α-methoxy-α-(trifluoromethyl)-phenylacetic acid.

13. The setup is illustrated in the accompanying photograph. The checkers found it most efficient to momentarily remove the thermometer during addition of any reagents.

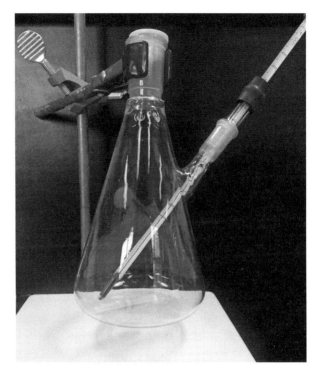

14. The checkers used non-stabilized tetrahydrofuran (THF) from Fisher Scientific and passed through a column of activated alumina. The submitters used THF purchased from VWR International, LLC. and dried by percolation through anhydrous Al_2O_3.

15. Sodium acetate trihydrate ReagentPlus®, ≥99.0% was purchased from Sigma-Aldrich Co. and used as received.

16. N-Bromosuccinimide ReagentPlus®, 99% was purchased from Sigma-Aldrich Co. and used as received.

17. A ^1H NMR of the crude reaction mixture sample was obtained to confirm product formation prior to use in the next step. When the reaction was performed at half scale, 14.0 g of crude material was obtained.

18. Product **3** possesses the following characterization data: R_f (40% EtOAc/hexanes) = 0.26; the mixture of α and β anomers displayed dextrorotatory chiroptic properties. The checkers observed a range of $[\alpha]^{24}_D$ = +76 to +103 (c = 1.0, CH_2Cl_2) (The submitters noted a 2.6:1 (α:β) selectivity with an $[\alpha]^{25}_D$ = + 44 (c = 1.0, CH_2Cl_2). The checkers observed

that leaving the sample in solution in an NMR tube caused the mixture to change from an initially 2.0:1 (α:β) mixture to a 3.3:1 (α:β) mixture. This variation in solution could account for the range of observed optical rotations.); IR (neat) 3377, 2988, 2940, 2875, 1690, 1447, 1373, 1234, 1159, 1016, 939 cm^{-1}; ^{1}H NMR (400 MHz, CDCl$_3$) major isomer (α) δ: 1.38 (d, J = 6.8 Hz, 3 H), 3.44 (bs, 1 H), 4.66 (q, J = 6.8 Hz, 1 H), 5.62 (m, 1 H), 6.09 (d, J = 10.2 Hz, 1 H), 6.89 (dd, J = 10.2, 3.4 Hz, 1 H); minor isomer (β) δ: 1.45 (d, J = 6.7 Hz, 3 H), 3.77 (bs, 1 H), 4.22 (dq, J = 6.7, 1.2 Hz, 1 H), 5.66 (bm, 1 H), 6.09 (dd, J = 10.3, 1.6 Hz, 1 H), 6.94 (dd, J = 10.3, 1.4 Hz, 1 H); ^{13}C NMR (100 MHz, CDCl$_3$) major isomer (α) δ: 15.7, 70.8, 88.0, 127.6, 144.7, 197.3; minor isomer (β)δ: 16.6, 75.6, 91.3, 128.9, 148.4, 197.3; HRMS (ESI) Calcd. for (C$_6$H$_8$O$_3$+Na)$^+$: 151.0371, Found: 151.0365

19. The setup is illustrated in the accompanying photograph.

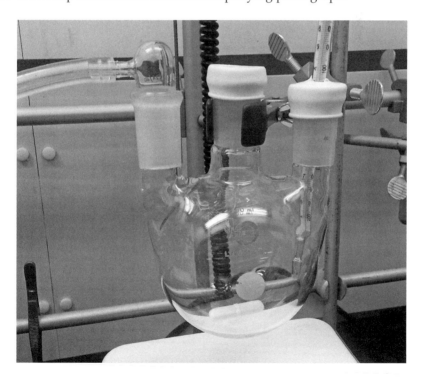

20. The number of equivalents and mmol in this step was calculated based on a theoretical quantitative yield in step B of the pyranone products **3**.

21. Di-*tert*-butyl dicarbonate ReagentPlus®, 99% was purchased from Sigma-Aldrich Co. and used as received. The number of equivalents was calculated based on a theoretical quantitative yield in step B.

22. 4-Dimethylaminopyridine ReagentPlus®, ≥99.0% was purchased from Sigma-Aldrich Co. and used as received.

23. The submitters note that in order to get good α-selectivity, the temperature should be carefully maintained between –78 °C to –30 °C. That is, the solution is stirred for 6 h in a dry ice-acetone bath and then warmed to – 30 °C over a 3 h period. Finally, the reaction mixture is brought to 0 °C (ice-water bath) diluted with Et₂O (100 mL) and quenched by the dropwise addition of saturated NaHCO₃ (150 mL).

24. Sodium bisulfate purum, anhydrous, ~95.0%, was purchased from Sigma-Aldrich Co. and used as received.

25. Two spots are observed on the TLC plate when using potassium permanganate as stain. The top spot is α-L-Boc-pyranone and the lower spot is β-L-Boc-protected pyranone. Pure fractions from the column are evaporated using rotary evaporator (25 °C, 150 mmHg initial to 25 mmHg final) and dried under high vacuum (1.5 mmHg) for 6 h. Both pure α-L-Boc-pyranone and β-L-Boc-pyranone are initially viscous liquids at room temperature, but they become solid after removing residual solvent under high vacuum. The products showed the following data: Compound **4:** α-L-anomer: R_f (20% EtOAc/hexanes) = 0.59; Diastereomeric Ratio: 99.6:0.4 (α:β); HRMS [α]^{24}$_D$ = +102 (*c* 1.0, CH₂Cl₂); IR (neat) 2983, 2940, 1747, 1700, 1371, 1274, 1255, 1154, 1105, 1090, 1056, 1028, 940, 859, 840 cm⁻¹; ¹H NMR (500 MHz, CDCl₃) δ: 1.41 (d, *J* = 6.7 Hz, 3 H), 1.52 (s, 9 H), 4.64 (q, *J* = 6.7 Hz, 1 H), 6.20 (d, *J* =10.2 Hz, 1 H), 6.33 (d, *J* = 3.6 Hz, 1 H), 6.87 (dd, *J* = 10.2, 3.7 Hz, 1 H); ¹³C NMR (125 MHz, CDCl₃) δ: 15.6, 28.0 (3C), 72.6, 84.0, 89.5, 128.8, 141.3, 152.2, 196.1; HRMS (+ESI) Calcd. for [C₁₁H₁₆O₅+Na⁺]: 251.0890, Found: 251.0892; Anal. Calcd. for C₁₁H₁₆O₅: C, 57.89; H, 7.07. Found: C, 58.08; H, 7.08. Compound **5:** β-L-anomer: R_f (20% EtOAc/hexanes) = 0.50; [α]^{24}$_D$ = –50 (*c* 0.3, CH₂Cl₂); IR (neat) 2994, 2940, 1749, 1700, 1370, 1273, 1250, 1157, 1032, 1007, 936, 852, 791 cm⁻¹. ¹H NMR (500 MHz, CDCl₃) δ: 1.50 (m, 12 H), 4.37 (q, *J* = 7.1 Hz, 1 H), 6.21 (d, *J* =10.4 Hz, 1 H), 6.37 (s, 1 H), 6.88 (dd, *J* = 10.3, 2.6 Hz, 1 H); ¹³C NMR (125 MHz, CDCl₃) δ: 19.0, 28.0 (3C), 76.1, 84.0, 90.2, 128.6, 143.1, 152.1, 196.3. Diastereomeric Ratio: 2.4:97.6 (α:β); HRMS (+ESI) Calcd. for [C₁₁H₁₆O₅+Na⁺]: 251.0890, Found: 251.0891. Anal. Calcd. for C₁₁H₁₆O₅: C, 57.89; H, 7.07. Found: C, 58.13; H, 7.03. Diastereomeric purity is

determined by GCMS analysis. GC conditions: (column: HP-5 5% Phenyl Methyl Siloxane, 30.0 m x 250 μm x 0.50 μm nominal), Rate: 25 °C, Oven set point: 40 °C, Hold Time: 2 min. AUX Heater on: Set point: 300 °C, Hold Time: 5 min., total run time: 17 min., retention time: α-L-Boc-pyranone 7.54 min, β-L-Boc-pyranone 7.65 min. The enantiomeric excess for the α-L-Boc-pyranone was determined to be 98% by chiral HPLC (conditions: Chiralpak AD column, eluent: hexane/i-PrOH = 93:7, flow rate: 1.0 mL/min). The peaks were visualized at 210 nm with retention times of 5.47 min (minor isomer) and 7.37 min (major isomer).

26. The submitters report the yields obtained over the three steps range from 55-60%. The checkers report that when the reaction was performed on half scale, a 56% yield was obtained over the three steps.

27. The S enantiomer appears numerous times in the literature. Among these, both the + and the – optical rotations are reported. The + rotation has been found to be in error.

Working with Hazardous Chemicals

The procedures in *Organic Syntheses* are intended for use only by persons with proper training in experimental organic chemistry. All hazardous materials should be handled using the standard procedures for work with chemicals described in references such as "Prudent Practices in the Laboratory" (The National Academies Press, Washington, D.C., 2011; the full text can be accessed free of charge at http://www.nap.edu/catalog.php?record_id=12654). All chemical waste should be disposed of in accordance with local regulations. For general guidelines for the management of chemical waste, see Chapter 8 of Prudent Practices.

In some articles in *Organic Syntheses*, chemical-specific hazards are highlighted in red "Caution Notes" within a procedure. It is important to recognize that the absence of a caution note does not imply that no significant hazards are associated with the chemicals involved in that procedure. Prior to performing a reaction, a thorough risk assessment should be carried out that includes a review of the potential hazards associated with each chemical and experimental operation on the scale that is planned for the procedure. Guidelines for carrying out a risk assessment and for analyzing the hazards associated with chemicals can be found in Chapter 4 of Prudent Practices.

The procedures described in *Organic Syntheses* are provided as published and are conducted at one's own risk. *Organic Syntheses, Inc.*, its Editors, and its Board of Directors do not warrant or guarantee the safety of individuals using these procedures and hereby disclaim any liability for any injuries or damages claimed to have resulted from or related in any way to the procedures herein.

Discussion

As part of a larger effort aimed at using the tools of asymmetric synthesis for medicinal chemistry, chemists have been seeking new methods for the de novo asymmetric synthesis of carbohydrates.[5] One broadly applicable asymmetric approach involves the use of α- and β-6-*t*-butoxycarboxy-2*H*-pyran-3(6*H*)-ones as carbohydrate precursors.[6] The route begins with the asymmetric synthesis of 6-*t*-butoxycarboxy-2*H*-pyran-3(6*H*)-ones from achiral acylfurans or vinylfurans.[7] The most practical approach begins with the acylation of furan at the 2-position and rely on the Noyori asymmetric hydrogen transfer reaction to install either the (*R*)- or (*S*)-furan alcohol stereochemistry (*i.e.*, D- or L-form respectively).[2] An NBS-promoted Achmatowicz oxidative rearrangement[8] of furan alcohols and subsequent *t*-butyl carbonate formation provides the 6-*t*-butoxycarboxy-2*H*-pyran-3(6*H*)-ones with variable C-6 substitution as a mixture of α- and β-diastereoisomers (Scheme 1).[9]

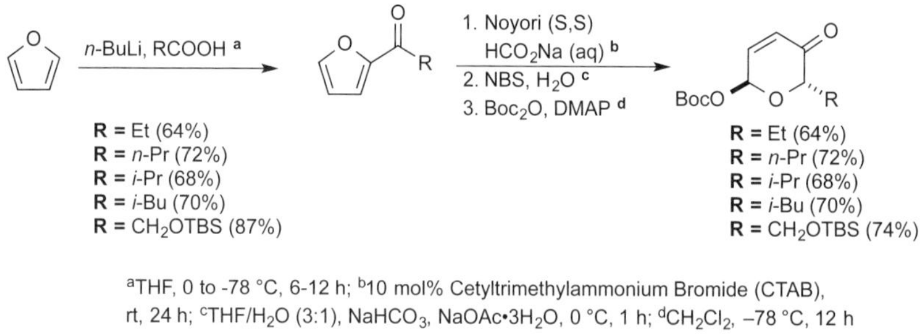

n-BuLi, RCOOH [a]	1. Noyori (S,S) HCO₂Na (aq) [b]
	2. NBS, H₂O [c]
	3. Boc₂O, DMAP [d]
R = Et (64%)	R = Et (64%)
R = *n*-Pr (72%)	R = *n*-Pr (72%)
R = *i*-Pr (68%)	R = *i*-Pr (68%)
R = *i*-Bu (70%)	R = *i*-Bu (70%)
R = CH₂OTBS (87%)	R = CH₂OTBS (74%)

[a]THF, 0 to -78 °C, 6-12 h; [b]10 mol% Cetyltrimethylammonium Bromide (CTAB), rt, 24 h; [c]THF/H₂O (3:1), NaHCO₃, NaOAc·3H₂O, 0 °C, 1 h; [d]CH₂Cl₂, −78 °C, 12 h

Scheme 1. Synthesis of variously substituted Boc-pyranones

The potential of the C-6 substituted α-6-*t*-butoxycarboxy-2*H*-pyran-3(6*H*)-ones as α-mannose precursors is demonstrated by the glycosylation/post-glycosylation sequence outlined in Scheme 2. The route begins with the Pd-π-allyl catalyzed glycosylation of benzyl alcohol to stereospecifically transfer (via a double inversion mechanism) the anomeric stereocenter to the α-pyranone product. Two highly stereoselective post-glycosylation reactions install the requisite *manno*-stereochemistry. Thus, a NaBH$_4$ promoted ketone reduction installs the C-4 alcohol and an OsO$_4$-catalyzed Upjohn dihydroxylation installs the C-2/3 diol stereochemistry.[10]

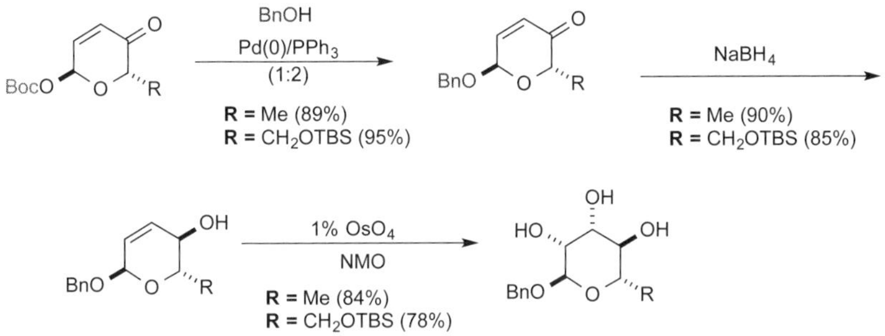

BnOH

Pd(0)/PPh$_3$
(1:2)

R = Me (89%)
R = CH$_2$OTBS (95%)

NaBH$_4$

R = Me (90%)
R = CH$_2$OTBS (85%)

1% OsO$_4$

NMO

R = Me (84%)
R = CH$_2$OTBS (78%)

Scheme 2. Boc-pyranones as α-mannose precursors

In addition to being mono-saccharide precursors, the Boc-pyranones are excellent building blocks for oligosacchirdes.[11] This is nicely exemplified in the syntheses of the oligo-rhamnose containing natural products, like the Cleistetrosides[12] and the Anthrax tetrasaccharide.[13] The structural range for which this methodology is applicable, is nicely demonstrated in the approach to the Anthrax tetrasaccharide, as it contains both α-L and β-D monosaccharides with the rare anthrose sugar.[14] The overall synthetic efficiency of this approach magnified when the glycosylation reaction was applied in an iterative bi-directional fashion. For example, an all *manno*-heptasaccharide with 25 stereocenters was produced in 12 steps from an achiral acylfuran.[15]

Figure 1. Oligosaccharides prepared via a Pd-catalyzed glycosylation

Finally, the ability to have myriad C-6 substitution along with the flexibility for variable substitution at the C-2,3,4 position makes these routes most readily adaptable to medicinal chemistry applications.[16] The applicability of these structurally and stereochemically divergent approaches is best suited for the synthesis of carbohydrate containing natural product libraries. Examples of this application can be seen in the syntheses of carbohydrate analogues of SL0101,[17] digitoxin[18] and methymycin[19] (Figure 2).

R = Me, Et, n-Pr, i-Bu
SL0101 analogues

R = Me, Et, n-Pr, i-Pr,i-Bu
Digitoxin Analogues

4-amino-manno-methymycin

Figure 2. Pd-catalyzed glycosylation

References

1. Department of Chemistry and Chemical Biology, Northeastern University, 102 Hurtig Hall, 360 Huntington Ave, Boston, MA-02115; *Email:* G.ODoherty@neu.edu. We are grateful to NIH (GM090259) and NSF (CHE-1213596) for financial support of this research.

2. (a) Ohkuma, T.; Koizumi, M.; Yoshida, M.; Noyori R. *Org. Lett.* **2000**, *2*, 1749. (b) Fujii, A.; Hashiguchi, S.; Uematsu, N.; Ikariya, T.; Noyori, R. *J. Am. Chem. Soc.* **1996**, *118*, 2521–2522. (c) Li, M.; O'Doherty, G. A. *Tetrahedron Lett.* **2004**, *45*, 6407–6411. (d) Li, M.; Scott, J. G.; O'Doherty, G. A. *Tetrahedron Lett.* **2004**, *45*, 1005–1009.

3. Hoye, T. R.; Jeffrey C. S.; Shao, F. *Nature Protocols*, **2007**, *2(10)*, 2451–2458.

4. Shi, X.; Leal, W. S.; Meinwald, J. *Bioorg. Med. Chem.* **1996**, *4(3)* 297–303.

5. Zhou, M. and O'Doherty, G. A. *Current Topics in Medicinal Chemistry*, **2008**, *8*, 114–125.

6. De Novo Synthesis in Carbohydrate Chemistry: From furans to Monosaccharides and oligosaccharides." in *ACS Symposium Series 990*,

Chemical Glycobiology, Xi Chen, Randall Halcomb and Peng George Wang Eds. *by Xiaomei Yu and George A. O'Doherty,* ACS, Washington, **2008**, pp. 3–22.

7. Harris, J. M.; Keranen, M. D.; O'Doherty, G. A. *J. Org. Chem.* **1999**, *64,* 2982–2983.

8. Achmatowicz, O.; Bielski, R. *Carbohydr. Res.* **1977**, *55,* 165–176.

9. Harris, J. M.; Keranen, M. D.; Nguyen, H.; Young, V. G.; O'Doherty, G. A. *Carbohydr. Res.* **2000**, *328*(1), 17–36.

10. Babu, R. S.; O'Doherty, G. A. *J. Am. Chem. Soc.* **2003**, *125,* 12406–12407.

11. (a) Babu, R. S.; Zhou, M.; O'Doherty, G. A. *J. Am. Chem. Soc.* **2004**, *126,* 3428–3429. (b) Babu, R. S.; O'Doherty, G. A. *J. Carb. Chem.* **2005**, *24,* 169–177.

12. Wu, B.; Li, M.; O'Doherty, G. A. *Org. Lett.* **2010**, *12(23),* 5466–5469.

13. (a) Guo, H.; O'Doherty, G. A. *Angew. Chem. Int. Ed.* **2007**, *46,* 5206–5208. (b) Wang, H.-Y. L.; Guo, H.; O'Doherty, G. A. *Tetrahedron* **2013**, *69,* 3432–3436.

14. Guo, H.; O'Doherty, G. A. *J. Org. Chem.* **2008**, *73,* 5211–5220.

15. Babu, R. S.; Chen, Q.; Kang, S-W.; Zhou, M.; O'Doherty, G. A. *J. Am. Chem. Soc.* **2012**, *134,* 11952–11955.

16. (a) Wang, H.-Y. L.; Xin, W.; Zhou, M.; Stueckle, T. A.; Rojanasakul, Y.; O'Doherty, G. A. *ACS Med. Chem. Lett.* **2011**, *2,* 73–78. (b) Hinds J. W.; McKenna, S. B.; Sharif, E. U.; Wang, H.-Y. L., Akhmedov, N. G.; O'Doherty, G. A. , *ChemMedChem.* **2013**, *8,* 63–69.

17. (a) Mrozowski, R. M.; Vemula, R.; Wu, B.; Zhang, Q.; Schroederd, B. R.; Hilinski, M. K.; Clarke, D. E.; Hecht, S. M.; O'Doherty G. A.; Lannigan, D. A, *ACS Med. Chem. Lett.* **2012**, *3,* 1086–1090. (b) Shan, M.; O'Doherty, G. A. *Org. Lett.* **2010**, *12,* 2986–2989. (c) Shan, M.; O'Doherty, G. A. *Org. Lett.* **2006**, *8,* 5149–5152.

18. (a) Elbaz, H.; Stueckle, T. A.; Wang, H.-Y. L.; O'Doherty, G. A.; Lowry, D. T.; Sargent, L. M.; Wang, L.; Dinu, C. Z.; Rojanasakul, Y. *Toxicol. Appl. Pharmacol.* **2012**, *258,* 51–60. (b) Wang, H.-Y. L.; Wu, B.; Zhang, Q.; Rojanasakul, Y.; O'Doherty, G. A. *ACS Med. Chem. Lett.* **2011**, *2,* 259–263. (c) Iyer, A.; Zhou, M.; Azad, N.; Elbaz, H.; Wang, L.; Rogalsky, D. K.; Rojanasakul, Y.; O'Doherty, G. A.; Langenhan, J. M. *ACS Med. Chem. Lett.* **2010**, *1,* 326–330. (d) Zhou, M.; O'Doherty, G. A. *J. Org. Chem.* **2007**, *72,* 2485–2493. (e) Zhou, M.; O'Doherty, G. A. *Org. Lett.* **2006**, *8,* 4339–4342. (f) Hinds, J. W.; McKenna, S. B.; Sharif, E. U.; Wang, H-Y. L.; Akhmedov, N. G.; O'Doherty, G. A. *ChemMedChem.* **2013**, *8,* 63–69.

19. Borisova, S. A.; Guppi, S. R.; Kim, H. J.; Wu, B.; Penn, J. H.; Liu, H-W.; O'Doherty, G. A. *Org. Lett.* **2010**, *12(22)*, 5150–5153.

Appendix
Chemical Abstracts Nomenclature (Registry Number)

(*S*)-1-(Furan-2-yl)ethanol (112653-32-4)
Sodium formate (141-53-7)
Cetyltrimethylammonium bromide (57-09-0)
2-Acetyl furan (1192-62-7)
(*R*)-Ru(η^6- mesitylene)-(*S,S*)-TsDPEN (174813-81-1)
N-Bromosuccinimide (128-08-5)
(2*S*, 6*R*)-6-Hydroxy-2-methyl-2H-pyran-3(6*H*)-one (138809-74-2)
(2*S*, 6*S*)-6-Hydroxy-2-methyl-2H-pyran-3(6*H*)-one (1385812-17-8)
tert-Butyl ((2*S*,6*S*)-6-methyl-5-oxo-5,6-dihydro-2*H*-pyran-2-yl) carbonate (865484-73-7)
tert-Butyl ((2*R*,6*S*)-6-methyl-5-oxo-5,6-dihydro-2H-pyran-2-yl) carbonate (916069-09-5)
N,N-Dimethylaminopyridine (1122-58-3)
(Boc)$_2$O: Di-*tert*-butyl dicarbonate; (24424-99-5)

Sumit O. Bajaj was born in Maharashtra, India. He received his Bachelor's degree (B.Sc.) in 2004 from Amravati University, Amravati and Master's degree (M.Sc. Organic chemistry) in 2006 from Sant. Gadge Baba Amravati University, Amravati, India. He began his doctoral research in the fall of 2010 at Northeastern University in Boston, MA. At Northeastern University, he is working on the synthesis of oligosaccharides natural products as anti-cancer agents under the guidance of Prof. George A. O'Doherty.

Jamison Farnsworth, originally from Northern Vermont, first attended Suffolk University, transferred after first year to Northeastern University. Majored in Chemistry while at Northeastern with a primary concentration in analytical and organic. In addition to research in the O'Doherty group, he participated in industrial CO-OPs, the first at Shire HGT and the second at Arsenal Medical. In 2012, he graduated from Northeastern with a BS in chemistry and now is currently employed in the biopharmaceutical industry as an analytical chemist.

George O'Doherty was born in Kilkenny Ireland in 1966 and received his undergraduate education from RPI with Professor Alan R. Cutler in 1987. After earning his Ph.D. with Professor Leo A. Paquette at OSU in 1993 he pursued postdoctoral studies with first Professor Barry M. Trost at Stanford and then Anthony G. M. Barrett. He began his independent career at Univ. of Minnesota in 1996 and in 2002, he moved to West Virginia University. He moved again in 2011, to Northeastern University where he has risen to the rank of Professor. His laboratory is interested in the use of asymmetric catalysis for the synthesis and medicinal chemistry study of biological important carbohydrate and natural products.

Michael T. Tudesco obtained his B.S. degree in Chemistry at the University of North Carolina at Chapel Hill in 2012, performing undergraduate research under the supervision of Professor Michel R. Gagné. After obtaining his Masters degree at Baylor University under the guidance of Professor John L. Wood, he took a position at Genentech in San Francisco.

Heemal Dhanjee obtained his BA in Mathematics and Molecular and Cell Biology at the University of California, Berkeley in 2007. He subsequently moved to California State University, Northridge and performed research in organic synthesis under the supervision of Professor Thomas Minehan. He is currently pursuing graduate research at Baylor University under the guidance of Professor John L. Wood.

AUTHOR INDEX VOLUME 91

CUMULATIVE SUBJECT INDEX FOR VOLUME 91

This index comprises subject matter for Volume **91**. For subjects in previous volumes, see the indices contained in the specific volume. Cumulative indices are also available in Collective Volumes I through XII or the single volume entitled *Organic Syntheses, Collective Volumes I-VIII, Cumulative Indices,* edited by J. P. Freeman.

Many compounds are named in the index in two forms. Most chemicals used in a procedure will appear in the index as written in the text, although the systematic name according to Chemical Abstracts nomenclature may also appear. A CAS Registry number is listed for all compounds for which the number has been assigned. Entries are listed for all starting materials, non-trivial reagents, products and important by-products. Products are indicated by the use of italics. General terms for classes of compounds and types of reactions are also included in the index.

ACETALS and HEMIACETALS **91**, 293, 338
4-Acetamidobenzenesulfonyl azide: Benzenesulfonyl azide, 4-(acetylamino)-;
    (2158-14-7) **91**, 322
Acetic anhydride; (108-24-7) **91**, 175
Acetic anhydride: Acetic acid, 1,1′-anhydride; (108-24-7) **91**, 39
2-Acetyl furan; (1192-62-7) **91**, 338
ACYLATION **91**, 83, 116, 175, 293, 338
ADDITION **91**, 72, 137, 307
ALCOHOLS **91**, 137
ALKYLATION **91**, 12, 260
ALLENES **91**, 12, 233
AMIDES AND LACTAMS **91**, 137, 201, 221
(S)-2-Amino-4-methyl-1,1-diphenylpentan-1-ol; (78603-97-1) 91, 137
4-Aminomorpholine: 4-Morpholinamine; (4319-49-7) **91**, 125
Aniline; (62-53-3) **91**, 211

Benzamide, *N*-(phenylmethyl)-; (1485-70-7) **91**, 201
Benzamide, N-hexyl-; (4773-75-5) 91, 201
Benzenemethanamine,
    N-(2,2,2-trifluoro-1-methoxyethyl)-N-[(trimethylsilyl)methyl];
    (1415606-26-6) 91, 162
Benzenemethanamine, N-[(trimethylsilyl)methyl]-; (53215-95-5) **91**, 162
Benzenemethanamine; (100-46-9) **91**, 162, 201

Benzenesulfonic acid, 4-methyl-; (104-15-4) **91**, 162
Benzoic acid; (65-85-0) **91**, 175
Benzoic anhydride; (93-97-0) **91**, 293
Benzoic acid, methyl ester; (93-58-3) **91**, 201
Benzo[h]quinoline; (230-37-3) **91**, 52
N-(Benzo[h]quinolin-10-yl)-4-dodecylbenzenesulfonamide; (1616119-88-0) **91**, *52*
Benzoyl peroxide; (94-36-0) **91**, 106
Benzyl chloride; (100-44-7) **91**, 116
N-Benzylcinchonidinium chloride: Cinchonanium,
    9-hydroxy-1-(phenylmethyl)-, chloride, (9S)-; (69221-14-3) **91**, 1
S-Benzyl isothiouronium chloride; (1334419-16-7) **91**, *116*
5-Benzyl-4-methyl-1,3-thiazol-2-amine hydrochloride; (95767-21-8) **91**, *185*
(±)-BINAP; 2,2′-Bis(diphenylphosphino)-1,1′-binaphthyl; (98327-87-8) *91*, *93*
1,1′-Bi-2-naphthol: [1,1′-Binaphthalene]-2,2′-diol; (602-09-5) **91**, 1
(*R*)-(+)-1,1′-Bi-2-naphthol: [1,1′-Binaphthalene]-2,2′-diol, (*R*)-; (18531-94-7)
    91, 1
1,1′-Bi-2-naphthol ditriflate: Methanesulfonic acid, trifluoro-,
    [1,1′-binaphthalene]-2,2′-diyl ester, (±)-; (128575-34-8) **91**, 1
(*R*)-(+)-2,2′-Bis(diphenylphosphino)-1,1′-binaphthyl[(*R*)-BINAP]: Phosphine,
    [1,1′-binaphthalene]-2,2′-diylbis[diphenyl-, (R)-; (76189-55-4) **91**, 1
[1,2-Bis(diphenylphosphino)ethane]nickel(II) chloride: Nickel,
    dichloro[1,2-ethanediylbis[diphenylphosphine]-P,P′]- (SP-4-2)-;
    (14647-23-5) **91**, 1
Bis(1,5-cyclooctadiene)rhodium(I) trifluoromethanesulfonate; (99326-34-8) *91*,
    93
1,3-Bis(2,4,6-trimethylphenyl)imidazolinium chloride; (173035-10-4) **91**, 83
Bis(triphenylphosphine)palladium(II) dichloride; (13965-03-2) **91**, 93, 283
Boron trifluoride diethyl ethereate; (109-63-7) **91**, 27
Bromobenzene; (108-86-1) **91**, 137
N-Bromosuccinimide; (128-08-5) **91**, 338
(±)-2,3-Butadien-1-ol, 4-cyclohexyl-; (153489-62-4) **91**, *233*
Butanaminium, *N*,*N*,*N*-tributyl-, fluoride, hydrate (1:1:3); (87749-50-6) **91**,
    233
3-Buten-1-yn-1-yl-Benzene; (13633-26-6) **91**, *93*
(S)-tert-Butyl (1-(benzylamino)-4-methyl-1-oxopentan-2-yl)carbamate:
    Carbamic acid,
    N-[(1S)-3-methyl-1-[[(phenylmethyl)amino]carbonyl]butyl]-,
    1,1-dimethylethyl ester; (101669-45-8) **91**, *201*
tert-Butyl (2E,4E)-2,4-hexadienoate; (81838-85-9) **91**, *307*
n-Butyllithium; (109-72-8) **91**, 307
tert-Butyl ((2R,6S)-6-methyl-5-oxo-5,6-dihydro-2H-pyran-2-yl) carbonate;
    (916069-09-5) **91**, *338*
tert-Butyl ((2S,6S)-6-methyl-5-oxo-5,6-dihydro-2H-pyran-2-yl) carbonate;
    (865484-73-7) **91**, *338*

tert-Butyl nitrite; (540-80-7) **91**, 106

tert-Butyl (Z)-5-phenyl-3-hexenoate; (1373407-82-9) **91**, 307

Cetyltrimethylammonium bromide; (57-09-0) **91**, 338

Chlorobis(ethylene)rhodium(I) dimer: Rhodium,
    di-µ-chlorotetrakis(η2-ethene)di-; (12081-16-2) **91**, 150

(Chloromethyl)trimethylsilane; (2344-80-1) **91**, 162

meta-Chloroperbenzoic acid; (937-14-4) **91**, 293

3-Chloro-4-phenylbutan-2-one; (20849-77-8) **91**, *185*

N-Chlorosuccinimide; (128-09-6) **91**, 116, 185

CONDENSATION **91**, 150, 162, 185, 201, 211

Copper(I) bromide; (7787-70-4) **91**, 233

Copper(II) chloride; (7447-39-4) **91**, 221

Copper(I) iodide; (7681-65-4) **91**, 27, 93, 283

Copper(II) acetate monohydrate; (6046-93-1) **91**, 211

Copper(II) acetate, anhydrous; (142-71-2) **91**, 211

COUPLING **91**, 83, 93, 106, 137, 273

CYCLIZATION **91**, 93, 185, 221, 283

CYCLOADDITION **91**, 27, 150, 162, 175, 322

Cyclohexanecarboxaldehyde; (2043-61-0) **91**, 233

(Diacetoxyiodo)benzene; (3240-34-4) **91**, 27

DABCO: 1,4-Diazabicyclo[2.2.2]octane; (280-57-9) **91**, 1, 125

DABSO, DABCO-*bis*(sulfur dioxide): 1,4-Diazoniabicyclo[2.2.2]octane,
    1,4-disulfino-, bis(inner salt); (119752-83-9) **91**, 125

1,8-Diazabicyclo[5.4.0]undec-7-ene: Pyrimido[1, 2-*a*]azepine, 2, 3, 4, 6, 7, 8, 9,
    10-octahydro-; (6674-22-2) **91**, 150, 322

DIAZOTIZATION **91**, 322

Di-*tert*-butyl dicarbonate; (24424-99-5) **91**, 137, 338

DIBAL-H: Aluminum, hydrobis(2-methylpropyl)-; (1191-15-7) **91**, 322

Dichloro(pentamethylcyclopentadienyl)iridium (III) dimer; (12354-84-6) **91**, 185

Dichloro($η^5$-pentamethylcyclopentadienyl)rhodium(III) dimer; (12354-85-7) **91**,
    52

Diethylzinc: FLAMMABLE LIQUID: Zinc, diethyl-; (557-20-0) **91**, 248

3,4-Dihydronaphthalen-2-yl pivalate: Propanoic acid, 2,2-dimethyl-,
    3,4-dihydro-2-naphthalenyl ester; (1192306-30-1) **91**, *83*

3,4-Dihydronaphthalen-2-yl trifluoromethanesulfonate: Methanesulfonic acid,
    1,1,1-trifluoro-, 3,4-dihydro-2-naphthalenyl ester; (143139-14-4) **91**, *39*

N, *N*-Dimethylaminopyridine; (1122-58-3) **91**, 175, 293, 338

(±)-Dimethyl 2-((-5-hydroxy-4-oxo-3,5-diphenylcyclopent-2-en-1-yl)methyl)
    malonate; (1401539-14-7) **91**, *93*

Dimethyl malonate; (108-59-8) *91*, *93*

1,3-Dimethyl-3-(p-tolyl)-1H-pyrrolo[3,2-c]pyridin-2(3H)-one; (1364652-24-3)
    ***91**, 221*

(E)-1,3-Diphenylhexa-3,5-diene-1,2-dione; (1401539-00-1) **91**, *93*

Iodine; (7553-56-2) **91**, 283

2-Iodoanisole; (529-28-2) **91**, 283

3-Iodoanisole: Benzene, 1-iodo-3-methoxy-; (766-85-8) **91**, 125

2-Iodobenzamide; (3930-83-4) **91**, 27

2-Iodylbenzoic acid; (64297-64-9) *91, 93*

Iodomethane; methyl iodide; (74-88-4) **91**, 260

3-Iodo-2-phenylbenzofuran; (246230-86-4) 91, 283

Iron(II) chloride: Ferrous chloride; (7758-94-3) **91**, 83, 307

(2S,3R)-3-Isopropyl-2,3-dihydronaphtho[1,2-b]furan-2,5-diyl diacetate.
   (1616409-78-9) 91, 175

L-Leucine, *N*-[(1,1-dimethylethoxy)carbonyl]-, methyl ester; (63096-02-6) **91**,
   201

L-Leucine methyl ester hydrochloride; (7517-19-3) **91**, 137

Lipases; (9001-62-1) **91**, 293

Lithium chloride; (7447-41-8) *91, 93*

Magnesium sulfate; (7487-88-9) **91**, 211

Magnesium turning; (7439-95-4) **91**, 137

METAL-CATALYZED REACTIONS **91**, 1, 27, 39, 52, 72, 83, 93, 125, 150,
   211, 260, 273, 283, 307

Methanesulfonic acid, 1,1,1-trifluoro-, trimethylsilyl ester; (27607-77-8) **91**, 162

3-Methoxy-N-morpholinobenzenesulfonamide: Benzenesulfonamide,
   3-methoxy-N-4-morpholinyl-; (1255365-27-5) 91, 125

3-Methoxyphenylmagneisum bromide: Magnesium, bromo(3-methoxyphenyl)-;
   (36282-40-3) **91**, 125

1-((3-Methoxyphenyl)sulfonyl)piperidine: Piperidine,
   1-[(3-methoxyphenyl)sulfonyl]-; (173681-65-7) 91, 125

(R)-5-(4-Methoxyphenyl)-2,3,8,8a-tetrahydroindolizin-7(1H)-one:
   7(1H)-Indolizinone, 2,3,8,8a-tetrahydro-5-(4-methoxyphenyl)-,(8aR)-;
   (913626-94-5) 91, 150

2-Methoxythiophene; (16839-97-7) **91**, 273

Methyl acetoacetate; (105-45-3) **91**, 211

4-(Methylamino)pyridine; (1121-58-0) **91**, 221

3-Methylbutanal; (590-86-3) **91**, 175

Methyl 4,4-dimethyl-3-oxopentanoate: Pentanoic acid, 4,4-dimethyl-3-oxo-,
   methyl ester; (55107-14-7) **91**, 248

Methyl 5,5-dimethyl-4-oxohexanoate: Hexanoic acid, 5,5-dimethyl-4-oxo-,
   methyl ester; (34553-32-7) 91, 248

Methylene iodide: Methane, diiodo-; (75-11-6) **91**, 248

2-Methyl-1*H*-indole; (95-20-5) **91**, 273

Methyl 3-methoxy-2-naphthoate; (1) (13041-60-6) 91, 260

Methyl 5-methyl-1-phenyl-3-(3-(trifluoromethyl)phenyl)-1H-pyrazole-4-
   carboxylate; (1259438-02-0) 91, 211

Methyl 2-naphthoate; (2) (2459-25-8) 91, 260

Thionyl chloride, (7719-09-7) **91**, 221

Thiourea; (62-56-6) **91**, 116, 185

p—Tolualdehyde; (104-87-0) **91**, 27

2-(*p*-Tolyl)propionic acid; (938-94-3) **91**, 221

Tri-*tert*-butylphosphonium tetrafluoroborate: Phosphine,
   tris(1,1-dimethylethyl)-, tetrafluoroborate(1-) (1:1); (131274-22-1) **91**, 125

2,4,6-Trichlorophenol: Phenol, 2,4,6-trichloro-; (88-06-2) **91**, 39

2,4,6-Trichlorophenyl 3,4-dihydronaphthalene-2-carboxylate:
   2-Naphthalenecarboxylic acid, 3,4-dihydro-, 2,4,6-trichlorophenyl ester;
   (1402012-58-1) **91***, 39*

2,4,6-Trichlorophenyl formate: Phenol, 2,4,6-trichloro-, 1-formate; (4525-65-9)
   91*, 39*

Tricyclohexylphosphine; (2622-14-2) **91**, 260

Triethylamine: Ethanamine, *N*,*N*-diethyl-; (121-44-8) **91**, 27, 93, 137, 150, 175,
   283

Triflic anhydride: Methanesulfonic acid, trifluoro-, anhydride (8,9); (358-23-6)
   91, 1

Trifluoroacetic acid: Acetic acid, 2,2,2-trifluoro-; (76-05-1) **91**, 60

3-(Trifluoromethyl)benzonitrile; (368-77-4) **91**, 211

Trimethylphosphine: Phosphine, trimethyl-; (594-09-2) **91**, 72

Trimethylsilylacetylene: Silane, ethynyltrimethyl-; (1066-54-2) **91**, 322

Triphenylphosphine; (603-35-0) **91**, 27

Vinyl bromide solution; (593-60-2) **91***, 93*

Xantphos: Phosphine,
   1,1'-(9,9-dimethyl-9*H*-xanthene-4,5-diyl)bis[1,1-diphenyl- ;
   (161265-03-8) **91**, 39

Yttrium (III) triflate; Yttrium(III) trifluoromethanesulfonate; (52093-30-8) **91***, 93*

Zinc bromide; (7699-45-8) **91**, 233

Zinc iodide; (10139-47-6) **91**, 233

Zinc perchlorate hexahydrate; (10025-64-6) **91**, 211